The Classification of Psychopathology

Neo-Kraepelinian and Quantitative Approaches

The Classification of Psychopathology

Neo-Kraepelinian and Quantitative Approaches

Roger K. Blashfield
University of Florida
Gainesville, Florida

Springer Science+Business Media, LLC

Library of Congress Cataloging in Publication Data

Blashfield, Roger K., 1944-
The classification of psychopathology.

Includes bibliographies and index.
1. Psychology, Pathological—Classification. I. Title. [DNLM: 1. Nervous system—Embryology—Congresses. 2. Neurons—Physiology—Congresses. 3. Nerve growth factors—congresses. 4. Molecular biology—congresses. WL 102.5 S9885c 1983]
RC455.2.C4B53 1984 616.89'0012 83-26871

ISBN 978-1-4612-9660-7 ISBN 978-1-4613-2665-6 (eBook)
DOI 10.1007/978-1-4613-2665-6

Cover illustration from "Do Psychiatric Patients Fit Their Diagnoses? Patterns of Symptomatology as Described with the Biplot" by J. S. Strauss, K. R. Gabriel, R. F. Kokes, B. A. Ritzler, A. VanOrd, and E. Tarana, *Journal of Nervous and Mental Disease*, 1979, *167*, 105–113. Courtesy of Williams and Wilkins Company.

Originally published by Plenum Press, New York in 1984
Softcover reprint of the hardcover 1st edition 1984

To my teachers

Moulton DeWalt
Jim Kelly
Rick Price
Alex Buchwald
John Castellan
Mark Aldenderfer
Les Morey

Prologue

> Because I am I, an odd piece of Egotism who could not make the riffle of living according to the precepts and standards society demands of itself, I find myself locked up with others of my kind in a "hospital" for the insane. There is nothing wrong with me—except I was born at least two thousand years too late. Ladies of Amazonian proportions and Berserker propensities have passed quite out of vogue and have no place in this too damned civilized world.
>
> Jefferson, 1948, p. 11

These words introduce a first-person account of a patient's experiences in a mental hospital. The book, *These Are My Sisters,* was written by Lara Jefferson while she was a patient on a unit that she called the "Hydro." She had developed the "delusion" that she was Shakespeare, and she used the process of writing as if she were Shakespeare as a form of self-therapy to help herself "learn to think differently." The publisher of this remarkable work comments that the manuscript was presented to him as an odd assortment of penciled writings on scraps of paper.

Lara Jefferson told relatively little about her past. She did say that she was the only surviving child of parents who were quite old when she was born (her father was 70; her mother, 46). She related that her mother had had a long history of psychiatric problems. Her mother was determined that Lara should become an evangelist or a musician dedicated to God in order to ensure her mother's salvation. Lara became neither. Lara also reported that she had been married. Physically, Lara described herself as a huge woman with a face like a fence.

What was wrong with Lara Jefferson? At the time, she was probably diagnosed as schizophrenic. She described herself as crazy, and certainly the odd logic in her thinking would imply the presence of a thought disorder. Nonetheless, understanding what Lara was saying often is quite easy, and her writings make it clear that she was an acute and sensitive observer of her environment:

> The nurse just now picked up one of the sheets I have written. She read it—looked at me oddly—and asked what in the hell I thought I was doing. And because she expected an answer in keeping with my strange occupation—I did not have the heart to disappoint her. So I gave her an answer that fitted. I told her that I was Shakespeare, the reincarnation of Shakespeare trying to sidestep a strait-jacket. (I'll admit that I feel queer enough to be the reincarnation of something but I doubt if Shakespeare would claim me.) But hurray! She came back down the aisle with a whole ream of paper and said to me: "Go to it, Shakespeare."
>
> Verily, verily, Shakespeare, I had no idea you could be called from your quiet English grave with so little effort. In my present predicament, I know of no one who could be quite such a fortunate choice for a delusion of grandeur. So welcome! I hope you will be as pleased with the arrangement as I am. Poor fellow, this is surely a come-down from your former position.
>
> Perhaps this is a penance—an expiation—an atonement you must make for filling so many pages of drama in your former existence with madmen. (Jefferson, 1948, pp. 25–26)

Lara herself was not certain what was wrong with her. The term she preferred to describe her problem was "Madness."

> This is the real thing, Mr. Shakespeare. Madness. Stark, senseless, maniacal Madness. It is like a mad bull in a pasture—much safer when you can survey it through a tall fence, from a safe distance. Unless you write faster; you will be tossed on the horns of the creature: to say nothing of being gored and trampled. You are seeing Madness here, Mr. Shakespeare; at first hand—and you have nothing but the stub of a chewed up pencil to protect yourself with. You will sure as hell have to write faster. I'm going—I feel that crazy "light" feeling in the temples and my eyes are not seeing rightly. Write, damn it. Write something. Anything, it does not matter. (p. 193)

This book concerns the classification of psychopathology. At the present, as was true 40 years ago when Lara Jefferson was in the mental hospital, we know relatively little about psychopathology. In Lara's words, psychopathology is an "unmarked sea" for which there exist no clear maps and charts.

Classification is a simple—but fundamental—step that is mandatory if we are to understand psychopathology. Classification is a basic human process that we all use to form abstract concepts in order to understand our environment. In the mental health field, we need a classification of psychopathology in order to help us understand what we clinicians deal with—people who have problems in living. The classification of psychopathology has been a neglected topic during much of this century, but the topic has experienced a resurgence of interest in the past two decades. This book focuses on two schools of thought in the United States that have been associated with this resurgence of interest; these schools of thought are called here the *neo-Kraepelinian* and the *quantitative.* The former term refers to a currently popular approach to classification within American psychiatry. The latter term refers to a group of researchers, mostly psychologists, who have used multivariate statistical techniques to create classificatory systems.

The reader will find that the topic of classification, as it is fundamental to the sciences, can be viewed from many different perspectives. History, statistics, sociology of science, professional politics, and scientific empiricism—these are all perspectives on classification that are represented in this book.

The discussions of classification in this book are directed at the associated theoretical issues. No attempt is made to propose a new classificatory system or to teach the reader about the clinical aspects of the mental health field. Nonetheless, the reader should not lose sight of the fact that the ultimate goal of this book is to improve our understanding of psychopathology. The patients with whom we deal as clinicians sail an "unmarked sea," and both they and we need understanding if there is to be hope.

Acknowledgments

A number of specific persons have contributed in various ways to this book. Most important have been the assistance of June Sprock and Lynn Robbins. June is a thoughtful, hard-working graduate student who has criticized all of the chapters, cross-checked many of the details, and helped me search various literatures. Lynn has criticized the details of my writing style, is responsible for most of the editing, and has insisted on making this book as good as possible. The ideas of June and Lynn are woven into mine across the entire book.

The publisher sent an initial version of the book to Nancy Andreasen, Craig Edelbrock, and Robert Kendell. Their critical feedback led to a much improved version of the book. Three other persons who read the book have had a major influence on its content: Denise Flick provided excellent criticisms of the arguments I had written originally in Chapters 1 and 3: Sue Williams was the first person to read the entire manuscript and to respond to its gestalt; and Walt Cunningham did an excellent job of criticizing the clarity and the technical accuracy of Chapters 6, 7, and 8. Other persons who have helped by reading various chapters are Ken Director, John Helzer, John Kuldau, Les Morey, Kathy Pinkston, Jane Samuelsen, Harvey Skinner, Yvonne Sparling, and Robert Spitzer.

A number of other people added to the book in various ways: Corwin Boake contributed many items for the cluster analysis bibliography; Joan Magyari, Susan Linn, and Gale Korn were the secretaries whose organizing skills and pleasantness kept me operational during this time-consuming effort; and Eliot Werner and Leonard Pace were the senior editors at Plenum who were always responsive to any inquiries I directed to them about the creative process of writing a book.

Finally, I want to thank my wife and children, as well as the patients and staff of the eighth floor, whose love has helped me grow throughout this time.

Roger K. Blashfield

Contents

Part I

Neo-Kraepelinian Approach to Classification

The intent of this book is to review, discuss, and critique the two major approaches to psychiatric classification that have dominated the American mental health profession during the last two decades: the neo-Kraepelinian perspective and the quantitative perspective. The first of these is the perspective that led to the creation of the third edition of the *Diagnostic and Statistical Manual of Mental Disorders* (DSM-III; American Psychiatric Association, 1980). The neo-Kraepelinians are a group of researchers and clinicians, mostly in psychiatry, who have advocated the use of explicit diagnostic criteria to overcome the problem of diagnostic unreliability and who have supported the generation of a new classification system that is much more consistent with and amenable to research than previous traditional psychiatric classifications.

In order to present the neo-Kraepelinian approach, this section begins with an analysis of Kraepelin and the early twentieth-century history of classification in the United States. The second chapter emphasizes the growth of the school of thought in American psychiatry that is being described by the term *neo-Kraepelinian*. The third chapter is the most polemic; it critically discusses the arguments among American mental-health professionals about the medical model. The research on diagnostic reliability is reviewed in Chapter 4. The final chapter of the section analyzes the ultimate product of the neo-Kraepelinians, the DSM-III.

Classification is both a simple and a complex process. The reader will discover that this book contains simpleminded ideas and complex arguments. The juxtaposition may, at points, be disconcerting. Karl Popper once asserted, "The business of science is not to make things clear, but clearer." I hope that this book will help the reader develop a clearer understanding of contemporary thinking about the classification of psychopathology.

Chapter 1

Kraepelin and His Influence on Modern Classification

Something has happened to me—I do not know what. All that was my former self has crumbled and fallen together and a creature has emerged of whom I know nothing. She is a stranger to me—and has an egotism that makes the egotism that I had look like skimmed milk; and she thinks thoughts that are—heresies. Her name is insanity. She is the daughter of madness—and according to the doctor, they each had their genesis in my own brain. I do not know—and I doubt if the doctors are as sure of what they think they know as they would like others to believe—or would like to believe themselves. I know nothing about such things, at least not in the way their knowledge runs.

They have a list of long Greek and Latin words and when they observe such and such symptoms in one of us, they paste the label for our phobia on us—and there is the end of the matter. I cannot see that they have accomplished so much in merely being able to remember all those long-handled names for our madness. We, who have learned what madness is by going through it—(and you cannot have closer knowing than that)—are separated from all others by a gulf so wide that it cannot be bridged. And there the matter lays—divided, split, and sundered.

Jefferson, 1948, pp. 14–15

Classification became an unpopular topic among American mental-health professionals during the mid-twentieth century. The best summary of the pejorative connotation of classification is contained in Szasz's (1961) statement, "Classifiers should be classified, not people."

There are three major reasons for the unpopularity of psychiatric classification in the United States. First, there have been a number of empirical studies demonstrating that traditional psychiatric classification systems had unacceptably low reliability values. These negative results were considered important because "there is no guarantee that a reliable system is valid, but assuredly an unreliable system must be invalid" (Spitzer & Fleiss, 1974, p. 341). These studies were reviewed by Kreitman (1961), Beck (1962), Zubin (1967), and Spitzer and Fleiss (1974).

Second, a number of criticisms were aimed at the traditional systems because of their association with a medical model perspective. The medical perspective in many ways conflicts with other perspectives that have gained influence in the mental health profession, such as the humanistic, the behavioral, and the psychoanalytic. These attacks on the medical model have come both from within psychiatry (from the antipsychiatrists such as Szasz and Laing) and from outside it (especially from clinical psychologists such as Schact & Nathan, 1977).

The third attack on traditional psychiatric classification grew from sociology and the development of labeling theory (Goffman, 1959; Scheff, 1966; Matza, 1968). This theory suggested that labels could become self-fulfilling prophecies that had strong effects on the persons being labeled. A dramatic example of these theoretical ideas is presented in Rosenhan's classic paper "Being Sane in Insane Places" (1973). The responses to this paper provided a valuable commentary on the degree of controversy surrounding the current use of classification in the mental health field (Crown, 1975; Million, 1975; Rosenhan, 1975; Spitzer, 1975; Weiner, 1975).

Despite these criticisms, the classification of psychopathology is a necessary and fundamental process. In the last decade, the importance of classification has been rediscovered, and the literature on classification showed a dramatic growth during the 1970s.

In order to understand why classification has experienced a resurgence of interest, it will be useful to enumerate five purposes that can be met by a good classification system. The first purpose of a classification system is providing the *nomenclature* necessary for communication among the people working in the field. For instance, scientists who are concerned with the study of corn must have a consistent set of classificatory terms to describe the different varieties of corn that exist. In the same way, mental health professionals need a system of basic terms to refer to the varieties of mental disorders with which they work.

A second purpose of a classification is furnishing a basis for *information retrieval.* There is a saying in botany that the name of a plant is the key to its literature. If a clinician is told that her new patient has Huntington's chorea, this diagnosis can be used to help the clinician to know where to look in a textbook for information about the patient's probable synptoms, the likely course of the disorder, and the neurophysiologic process involved in the degeneration of the central nervous system.

Third, a classification is useful for *description.* A classificatory system provides descriptive information about the primary entities being studied in a science. Knowing that a new patient has Huntington's chorea presents a fair amount of general information about the patient; the patient has a movement disorder, the patient will show a mild to moderate decrement of intellectual functioning, the patient is in his or her middle age, and the patient is likely to be depressed.

Another important purpose of a classification system in a science is providing a basis for making *predictions.* For instance, knowledge about the classification of chemical elements can be used to make predictions about the results of combining different inorganic compounds. In psychopathology, knowing that a patient has Huntington's chorea can lead to useful predictions about prognosis (but not to positive predictions about what treatment will arrest the disorder).

The fifth and final purpose of a classification is that it provides the basic concepts required for theory formulation within a science (*concept formation*). Linnaeus, in biology, developed an important classification of living organisms that provided Darwin with the outline for formulating the theory of biological evolution. Likewise, a good classification of mental disorders will form the basis of a theory about psychopathology. More details about these five uses of classification can be found in a paper by Blashfield and Draguns (1976b).

During the nineteenth century, there was a great deal of interest in formulating classifications of psychopathology. Not coincidentally, it was during this century that psychiatry became an organized profession and that insane asylums were developed as social institutions for the care of patients with mental disorders.

The classification of mental disorders in the early 1800s was crude by current standards. Early attempts at classification, such as those typically seen in the records of nineteenth-century asylums, were only broadly descriptive of the reasons that various patients were admitted. For instance, listed below is a classification of suicides as compiled from the records of the New York State Lunatic Asylum at Utica (Nichols, 1848):

Insanity
Melancholy
Domestic trouble
Delirium tremens
Pecuniary embarrassment
Disappointment in love
Intemperance
Seduction and desertion
Dissipation
Despair
Destitution
Loss of property
Discovery of his own fraud
Millerism
After discharge from work
After murder
After attempting to murder wife/intemperance
After receiving a disagreeable valentine
After enlisting in the army/had been a school teacher
Fear of want/being out of employment
Insanity caused by pecuniary embarrassment
Insanity at the same time drowned her child
Unknown

Clearly, some of these categories are quite recognizable to modern clinicians, even though different specific terms might be used. Other categories are highly idiosyncratic; they are hardly categories that would be seen as having general validity (e.g., after receiving a disagreeable valentine). In addition, the meanings of certain categories were not likely to be mutually exclusive (e.g., "insanity" and "insanity at the same time drowned her child").

Hurd (1881) published an article in the *Journal of Insanity* attacking the variation in terminology used across different insane asylums, and he argued for the adoption of a uniform system. In his article he gave the following as examples of variation in classificatory terms:

> The Hartford Retreat classifies the forms of insanity treated there, as follows: Mania acute, mania chronic, mania epileptic, mania suicidal, mania homicidal, mania periodical; melancholia acute, melancholia chronic, melancholia attonita; general paresis; methomania; dementia acute, dementia chronic, dementia senile; imbecility; moral insanity; not insane; unknown—eighteen separate forms.
>
> The Auburn Asylum for Insane Criminals gives the following: Mania acute, mania sub-acute, mania chronic, mania periodic, mania paroxysmal; melancholia; epilepsy, epilepsy with mania; paresis; not insane—ten forms.
>
> The Vermont Asylum gives the following forms: Mania, mania chronic; melancholia; dementia, dementia chronic; dipsomania—six forms only; no paresis, epilepsy, dementia after paralysis, etc. (p. 27)

Notice that these classifications seem more systematic than the classification of suicidal patients by Nichols. In particular, a set of diagnostic concepts appears rather consistently: mania, melancholy (now called *depression*), dementia, and paresis. Nonetheless, Hurd was correct in pointing out that the range and organization of classifications used by these mental institutions were different.

The interest in psychiatric classification during the 1800s culminated with the work of a German psychiatrist named Emil Kraepelin. Kraepelin published a series of popular textbooks on abnormal psychology, and his organization of mental disorders in these textbooks became the basis for modern classifications of mental disorders.

Despite the early momentum in the nineteenth century, classification has not been a popular topic during most of the twentieth century. There are many sources of the active lack of interest in classification. To some, classification has the political connotation of being associated with medicine and the medical-model approach to mental disorders; to others, it describes a "pigeon-holing" activity that dehumanizes the entities being classified—people; for still others, it represents a nineteenth-century approach to psychopathology that has shown little growth. As a result, classification has been largely ignored during most of the twentieth century. Recently, however, classification has attracted more attention, and a new literature on the topic is developing.

A major reason for the revived interest has been the realization that classification is a fundamental topic within a science. In addition, developmental psychologists have shown that it is a basic human conceptual activity. When children learn to think about their environment and to use language as a symbolic system to represent the external world, children classify the stimuli in their environment. With these classifications, they simplify the wide range of sensations that impinge on them, thus making it possible to focus their attention and to respond to the world. Classification is a developmentally simple, but necessary, step in the human analytical process.

In psychopathology, a classification is an organizing system that helps clinicians better understand the people who appeal to them for help. The current level of understanding about mental disorders is meager. Abnormalities in thinking, in affect, and in interpersonal relationships overwhelm the currently available knowledge from neurophysiology, psychology and sociology. Scientists, clinicians, and laypeople all understand very little about psychopathology.

This book is intended to improve the understanding about psychopathology by focusing entirely on classification. The book is divided into two sections: (1) a discussion of Kraepelin and the neo-Kraepelinian approach to classification; and (2) an overview of the empirical approaches to classification. It is hoped that as this book unfolds, the reader will better comprehend the fundamental issues associated with the classification of psychopathology.

EMIL KRAEPELIN

In reviewing the current literature in psychiatry and psychology, almost every discussion of classification makes a historical reference to Kraepelin. His work is often considered a summary of the many classificatory systems proposed during the nineteenth century. Most contemporary commentators agree that (1) Kraepelin was the originator of a classification system whose basic structure is still dominant in psychiatry; (2) Kraepelin adhered to a medical model; that is, he believed that mental disorders represent underlying disease states; and (3) Kraepelin's major organizing principle was that mental diseases could be isolated by grouping together those patients whose disorders had a similar course.

Emil Kraepelin was born in Neustrelitz, Germany, in 1856, the same year that another major figure in psychiatry, Sigmund Freud, was born. While in medical school, Kraepelin took classes from two professors who had a major influence on his later thinking. These professors were Bernhard von Gudden and Wilhelm Wundt.

Gudden was one of the great psychiatrists of Germany during the nineteenth century. He was the first German professor to receive an appointment in this discipline. Gudden received his clinical education from Christian Roller and was greatly influenced by Wilhelm Griesinger (who had claimed that all mental disorders were diseases of the brain). Gudden's early research was on eye movement and the crossing of the nerves at the optic chiasma. He also studied the connection between parasites and rabies.

Gudden established a successful insane asylum but soon left because of his dissatisfaction with administrative duties. He became a professor of psychiatry at the University of Zurich and editor of the leading psychiatric journal in Germany. Gudden was also the personal physician of King Ludwig II of Bavaria ("the mad king of Bavaria"). Gudden was best known for his study of the deterioration of the brain and for the formulation of a system that demonstrated the interdependence of nerve impulses and cell groups. Unlike Kraepelin, Gudden regarded the task of classifying the different forms of mental disorders as insoluble. Although he focused on neuroanatomy, he placed little importance on the search for identifiable changes in the brain tissue of the mentally ill. He also rejected intrapsychic theories of psychopathology such as Joseph Guislain's theory of inner struggle to compensate for psychic trauma. Unfortunately, Gudden's career was cut short in 1886 when he was killed by King Ludwig II, a manic who drowned both of them during a successful suicide attempt.

The other professor who influenced Kraepelin was Wilhelm Wundt, the father of experimental psychology. Wundt had been trained in medicine, physiology, and philosophy. During the mid-nineteenth century Wundt worked for Herman von Helmholtz and was placed in charge of a lab course on physiology. In 1874, he became a professor of philosophy at Leipzig. It was here that Wundt opened the world's first psychological laboratory. He also started the first psychological journal in 1881. Wundt's laboratory attracted students from all over the world. Instead of the functionalism that was popular in psychology during the nineteenth century, Wundt emphasized a structuralist view. He believed that psychology was a science and that its goal was the description of human consciousness. Wundt was concerned with observables, not the inferred mental states that other psychologists of the time attempted to study. Empiricism and observation were the cornerstones of his methodological approach to psychology.

Kraepelin was impressed by Wundt's methodology as opposed to the rather sterile approach taken by the neurologists of the time. While working in Wundt's laboratory, he intended to devote himself completely to experimental psychology. However, he eventually left, partially at the recommendation of Wundt, who, despairing about the future of psychology, urged his student to stay in medicine (Kahn, 1959).

After leaving Wundt's lab, Kraepelin worked in various German asylums. In 1886, he held a faculty position at the University of Dorpat along the Baltic Sea (Dorpat, now called Tortu, is located in the USSR, within Estonia). This university, not in the mainstream of German academic thought, was considered an academic haven for unconventional thinkers of the time (Harms, 1971). Finding in Dorpat the climate suitable for intellectual questing, Kraepelin organized a psychological laboratory using Wundt's methods for the study of patients with mental disorders. Kraepelin and his students performed reaction-time experiments, studies of simple mental abilities, and studies on mental and physical fatigue. He also published the initial volumes of a journal entitled *Psychologische Arbeiten* ("Psychological Studies"), which contained papers about experimental studies on abnormal patients. Additional volumes were completed throughout the 40 years of Kraepelin's academic career.

By 1891, Kraepelin had gained sufficient fame to be invited to a professorship at the University of Heidelberg. At this time, Germany was the world leader in medical research. The most important research on bacteriology, serology, pathology, neurology, and biochemistry was occurring in the German universities. Many of the best American physicians of the late nineteenth and early twentieth centuries were trained in Germany.

Kraepelin's initial influence in the psychiatric literature was through his psychological studies of mental disorders. One of the earliest references to Kraepelin in an English-language journal was made in the *Journal of Mental Science* (now called the *British Journal of Psychiatry*). A British psychiatrist, W. H. R. Rivers, who had studied with Kraepelin, presented a paper to the Royal Medico-Psychological Association of Great Britain providing a summary of the psychological methods that Kraepelin and his students at Heidelberg were using to study mental disorders (Rivers, 1895). He discussed a set of studies in which Kraepelin was using the techniques of Hermann Ebbinghaus, a famous psychologist, to study the mental abilities of insane patients. Interestingly, Rivers made no mention of classification or to Kraepelin's concern with such a topic.

During his years at Heidelberg, Kraepelin completed the majority of his influential works on psychiatric classification. Earlier, at Wundt's suggestion, Kraepelin had authored a brief textbook of slightly more than 300 pages called a *Compendium of Psychiatry* (1883). This book was simply a restatement of contemporary thinking in Germany about mental disorders. However, it attracted much interest, and further editions were published under the title of a *Short Textbook of Psychiatry* in 1887, 1889, and 1893 (Menninger, 1963). It was in the fourth edition (1893) that Kraepelin first used the term *dementia praecox,* which was grouped, along with *katatonia* and *dementia paranoides,* under the general category of "psychic degeneracy processes."

It is the sixth edition of his *Textbook of Psychiatry* for which Kraepelin has become known (Kraepelin, 1899, 1902). Two important and revolutionary concepts were emphasized in this edition: *manic-depressive insanity* (now called *manic-depressive disorder*) and *dementia praecox* (now *schizophrenia*). Kraepelin gained his widespread influence through the international controversy surrounding these two concepts. For instance, the first mention of Kraepelin's classificatory views in the *Journal of Mental Science* occurred in 1901, when a summary was given of a controversy at an Italian psychiatric meeting about the two concepts. By 1909, Kraepelin's views had become sufficiently well known in Great Britain so that four papers were published together in the *Journal of Mental Science*, which disputed his concepts (Drapes, 1909; Johnstone, 1909; Mickle, 1909; Nolan, 1909; see also the lengthy comments following these papers). Kraepelin's two concepts were revolutionary because (1) manic-depressive disorder was an amalgamation of two descriptive concepts (mania and melancholia) that had been separately recognized for over 2,000 years; and (2) dementia praecox represented a new descriptive concept that had many parallels to the well-recognized concept of dementia paralytica (paresis).

In 1904, Kraepelin moved to Munich to accept a chair at the local university and to head a new clinic and laboratory. Along with Kraepelin, important members of his research staff also moved (Alois Alzheimer, Franz Nissl, Felix Plaut—all of whom became important in academic psychiatry in their own right) (Harms, 1971). The seventh edition of Kraepelin's *Textbook* was published in 1904 (see Kraepelin, 1907). The eighth and final edition, four volumes totaling 3,068 pages, was published serially from 1909 to 1915. An English translation of a chapter about dementia praecox from Kraepelin's last edition was 328 pages in length, nearly the length of his first-edition "Compendium." During the 1920s, Kraepelin was working on a ninth edition, which did not reach completion before his death in 1927.

There are a number of important points to note about Kraepelin's approach to classification. The first is that Kraepelin, at least originally, did not intend to create new classifications. Kraepelin was an author of textbooks. And like all textbook authors, he organized his chapters around the major categories of psychopathology as they were generally accepted by contemporary authorities. What have become known as Kraepelin's classifications were literally nothing more than tables of contents to the various editions of his books. Thus, Kraepelinian classificatory systems were not structured with the intent of becoming official nomenclatures for use in institutions or outpatient clinics. Kraepelin organized his books in the way that he felt best helped him to present the information he wanted to communicate to students and to other professionals.

It is of interest to note that the reason for the initial interest in his textbooks was not his revolutionary ideas nor the diagnostic concepts he proposed. Instead, what initially stimulated interest in his textbooks was the clarity of his writing, particularly the clear and interesting clinical examples that are found throughout his books. For instance, the following quotation is excerpted from his discussion of the psychic symptoms of dementia praecox (Kraepelin, 1919/1971):

> And then there develops gradually or suddenly the symptom peculiarly characteristic of dementia praecox, namely, the *hearing of voices.* Sometimes it is only whispering, "as if it concerned me," as a patient says, a secret language, "taunting the captive"; sometimes the voices are loud or suppressed, as from a ventriloquist, or the call of a telephone, "children's voices"; a patient heard "gnats speak." Sometimes they shout as in a chorus or all confusedly; a patient spoke of "drumming in the ear"; another heard, "729,000 girls." Sometimes the voices appear to have a metallic sound, they are "resonant voices," "organ voices," or as of a tuning-fork. At other times they do not appear to the patients as sense perceptions at all; they are "voices of conscience" "voices which do not speak with words," voices of dead people, "false voices," "abortive voices." A patient said: "It appeared to me in spirit, as though they would find fault, without having heard it." There is an "inner feeling in the soul," an "inward voice in the thoughts"; "it is thought inwardly in me"; it "sounded as if thought"; "it was between hearing and forboding"—in such ways the patients express themselves about these sensory disturbances.
>
> The illusions not infrequently are connected with real noises. The clock speaks as if it were enchanted; the rushing of water is changed into words; each step under the patient speaks; a patient "heard the thoughts of others out of the soles of his boots." Here and there the voices have a rhythmical cadence, probably in connection with the carotid pulse. (pp. 7–8)

A third interesting feature to note about Kraepelin is that he was much more psychological in his approach to psychopathology than were his medically oriented contemporaries. He did believe, with these colleagues, that most mental disorders had an associated organic etiology. However, he advocated a careful behavioral analysis of patients in order to understand their clinical picture. He also was a proponent of using applied research to better understand abnormal behavior and its neurophysiologic bases. In both of these regards, Kraepelin was responding to the influence of his mentor, Wundt (Kahn, 1959).

Before Kraepelin, the most prominent classifier in Germany had been Wilhelm Greisinger (1817–1868), who asserted "all mental diseases are brain diseases" (Menninger, 1963). Followers of Greisinger attempted to support the validity of their classifications by showing that their categories were tied to neuropathologic findings on autopsy or were related to the genetic histories of the patients. Kraepelin, in contrast, believed that the most productive initial step in the study of psychopathology was making a careful behavioral description of patients in order to search for systematic patterns of similarities and differences. After a base of descriptive information was created, then the search for biological correlates could be undertaken.

Similarly, it is worth noting that Kraepelin was decidedly anti-Freudian. He believed that psychoanalysis was metapsychology at its worst and had no place in the scientific study of mental disorders. He was strongly opposed to the mentalistic and nonempirical orientation of Freud and Freud's followers (Kahn, 1959).

Regarding Kraepelinian classification, it was the sixth edition of his textbook (Kraepelin, 1899) that attracted the largest amount of international interest because of its promotion of the controversial concepts of dementia praecox and manic-depressive insanity. The 13 major categories in Kraepelin's text are listed below. Because of the importance of this edition, an entire listing of the 48 categories in this classification is shown in Appendix I.

Infection psychoses
Exhaustion psychoses
Intoxication psychoses
Thyroigenous psychoses
Dementia praecox
Dementia paralytica
Organic psychoses
Involutional psychoses
Manic-depressive psychosis
Paranoia
Psychogenic neuroses
Psychopathic conditions
Defective mental development

Note that most of the 13 major categories are easily recognizable in contemporary terminology, although some are not currently popular (e.g., psychopathic conditions and exhaustion psychoses). Also the prominence of certain disorders, such as dementia paralytica (paresis) and thyroigenous psychoses, reflects their being diagnostic categories that were the subject of active research in Kraepelin's time.

The major difference between Kraepelin's classification and currently accepted psychiatric classification primarily occurred at the subcategory level. The subcategories of "psychopathic conditions" (constitutional despondency, compulsive psychosis, impulsive psychosis, and contrary sexual instincts), for example, referred to a collection of concepts that are not obviously related to the contemporary subdivision of "personality trait disorders." However, many of the other subdivisions (e.g., of dementia praecox, manic-depressive insanity, alcoholism, and organic dementia) are generally recognizable (see Appendix I for details).

Kraepelin did not write directly about his views of classification nor about how he believed classificatory systems should be formed. Most of his ideas on this topic must be extrapolated from the principles inherent in his writings. First, as previously noted, Kraepelin believed that the best initial step in the study of psychopathology was to make careful observations of patients. From these observations, the symptoms of different disorders could be discerned. However, Kraepelin was aware that many symptoms can be misleading if overused in the diagnostic process: "Very similar symptoms can, at first glance, be apparent in the course of very different diseases" (Kraepelin, 1904). Thus, Kraepelin became concerned with differentiating those symptoms that, though striking, could come and go in different disorders from those symptoms that lead to accurate diagnoses of the individual disorders. He believed that the latter symptoms could be identified primarily by examining the course of the disorder and noting which symptoms were consistently present in the disorder.

Kraepelin also seemed to have implicitly assumed a Platonic perspective on diseases and classificatory concepts. Plato had advocated the notion of *ideal types* that exist beyond the world of material forms and natural objects. Any natural object is simply an approximation of a universal ideal type whose form is immutable. For instance, there is an ideal type called a *sphere.* A sphere does not exist in actuality. However, there are objects in the real world (e.g., baseballs) that represent approximations to the ideal type of sphere. Plato suggested that a major challenge to any science is to describe the ideal types that are basic to that science. In the area of medicine, diseases are the ideal types. Each disease, as it affects a given patient, is only an approximation to an essential disease process. It is the job of medical scientists, believed Kraepelin, to discover these essential diseases and their basic forms as expressed through patients' symptoms. Accomplishing this goal was Kraepelin's lifework.

> Such, in their essential features, are the points of view from which the clinical forms of insanity may be grouped today. Always starting from the relations of cause and effect, we may hope to discover clinical pictures such that the similarity or dissimilarity of the conditions under which they arise will find a clear expression in their phenomena, course, and termination. It is true that at present we are still met by obscurity and doubt at every step in forming a practical judgment on the material of clinical experience. We are still so far removed from a real knowledge of the causes, phenomena, course, and termination of the individual clinical forms that we cannot yet dream of a surely established edifice of knowledge at all. What we have formulated here is only a first sketch, which the advance of our science will

often have occasion to change and enlarge in its details, and perhaps even in its principal lines. (Kraepelin, *Lectures on Clinical Psychiatry,* 1904, p. 302)

ADOLF MEYER

During the early twentieth century, the United States, like most other countries, was coming under the influence of Kraepelinian concepts. One of the persons who brought Kraepelin's ideas to the United States was Adolf Meyer. Meyer was a Swiss physician who came to the United States at the end of the nineteenth century and who helped introduce German thinking to Americans about psychiatry and psychology, especially the ideas of Kraepelin and of Freud. Meyer became one of the truly great psychiatrists in the United States during the twentieth century. He was president of the American Psychiatric Association; he was an outstanding teacher who had very active students; and he developed his own holistic theory, which he called "psychobiology." Because of Meyer's approach to classification and his influence on American psychiatry, he is discussed here in some detail.

Adolf Meyer was born on September 13, 1866, in Niederweningen, Switzerland (near Zurich). Meyer passed his medical exams in 1890 in Switzerland; did two years of post-graduate work in Paris, London, Edinburgh, Berlin, and Zurich; and received his M.D. in 1892 from the University of Zurich. Meyer's primary professor at this university was August Forel, who was a student of Gudden and who replaced Gudden at the university of Zurich after he had been killed.

In 1892, Meyer came to the United States and began working as a pathologist in a hospital for the mentally ill. During the first 18 years of his career in the United States, Meyer worked at many universities, was an energetic teacher, and introduced many European ideas to his American colleagues and students. His geographic movement and active teaching help to account for his widespread influence during the formative years of American psychiatry. Meyer was a docent (lecturer) in neurology at the Illinois Eastern Hospital for the Insane from 1893 to 1895. From 1895 to 1902, he worked as a docent in psychiatry at Clark University in Worcester, Massachusetts, as well as being a pathologist and director of clinical and laboratory research at the Worcester Insane Hospital. In 1902, he was appointed director of the Pathological Institute of the New York State Hospitals, where he remained until 1910. Concurrently, Meyer was also a professor of psychiatry at

Cornell University Medical School. In 1909, Meyer was honored by Clark University with an honorary doctoral degree. Also honored at the same ceremony were Sigmund Freud and Carl Jung, who were making their celebrated visits to the United States.

Meyer was called to Baltimore in 1910 to chair the Department of Psychiatry at Johns Hopkins University, one of the most prestigious medical schools in the United States at that time. This medical school, founded in the 1890s, was designed to emulate the German approach to medical education. Meyer remained at Johns Hopkins until his retirement in 1941. Here, he had the responsibility of creating and directing the Henry Phipps psychiatric clinic. In 1911, Meyer published *Dementia Praecox,* one of the few of his books published before his death. In this book, Meyer discussed Kraepelin's celebrated diagnostic concept of dementia praecox, as well as the relevant ideas of psychoanalysts concerning this concept. Meyer was active and influential in the American Psychiatric Association during its development in the early twentieth century, and in 1927, he became president of this organization. Meyer died March 13, 1950. Two of his books, *Collected Papers* and *Psychobiology,* were published posthumously by his students.

Meyer was most famous for his holistic theory of psychiatry, which he called "psychobiology." Meyer conceived of the human being as a psychologically integrated biological unit. He emphasized the need to use knowledge from biology, physiology, psychology, and sociology to understand human behavior. He introduced the use of the comprehensive case history, which included data on hereditary, environmental, constitutional, social, economic, and psychological factors, to illuminate maladaptive behavior. He also acknowledged the influence of parental behavior on the development of personality.

Psychiatric disorders were conceptualized by Meyer as "reaction sets," that is, psychobiological reactions to multicausal life stress. Meyer rejected the idea that mental disorders were disease entities with established courses. He thought that illness represented a deviation in the process by which an individual attempted to adapt to a changing environment. Because mental disorders were seen as progressive habit formations, his treatment consisted of teaching new habits and eliminating faulty ones.

Meyer described "reactions sets" as groups of signs and symptoms that occur simultaneously and are characteristic of a disease. Further, he noted that these groups of symptoms occurred in typical patterns. After seven years of work at Worcester, Meyer developed a classification that described six typical patterns: *anergasia, dysergasia, thy-*

mergasia, parergasia, merergasia, and *oligerasia.* A superordinate division was between *holergasias,* or "sweeping" disorders of personality, and *merergasias,* in which the dysfunction is more limited.

Originally, Meyer was attracted to Kraepelin's famous work on classification and brought many of his ideas to the United States. In particular, one of Meyer's earliest publications in the United States was a review of the fifth edition of Kraepelin's textbook. Despite his early interest in Kraepelin and classification, Meyer later became extremely pessimistic about the utility of classification. Although he accepted the necessity of classification for statistical and administrative purposes, he noted the need for flexible classification systems and was discouraged about the hope of finding one comprehensive classification. He warned that the formulation of a clear-cut classification would be unlikely without research on an unwieldy number of groups, because the clinical experience of the time had shown that the complexity and the variability of human behavior overwhelmed any simple classificatory system. Later, Meyer suggested that people are so difficult to define that "we should give up the idea of classifying people like we do plants" (Lief, 1948, p. 105). He also stated in a lecture that "My chief effort was to eradicate the illusion that classification could precede accurate knowledge of facts. Psychiatry is craving for a premature classification. My first advice was to let classification rest or to give it a secondary place" (Lief, 1948, p. 101).

Meyer was particularly wary of diagnosis. In his posthumous book *Psychobiology* (1957), which contained a number of his lectures transcribed from students' notes, he was quoted as saying, "One word diagnoses in so plastic and complex field as human behavior are inadequate" (p. 126). He thought that, in real work with a patient, the diagnosis must be replaced by or supplemented with concrete specifications based on actual data. A diagnosis is a label, and no single label could possibly represent the complexity of a human being and his or her psychological environment. Meyer emphasized that it was the patient who was suffering rather than the disease. The real work is with the patient. He hoped to increase doctors' interests in their patients. He urged them to put the individual, not the label, in focus, and he stated that it does not matter what the patient is called as long as psychological events and mechanisms are used to create an understanding of the behavioral problem. In cases where the symptoms resemble a known nosological entity, diagnosis can be meaningful. However, in a large number of cases, diagnosis is not meaningful and may not be warranted. Watson (1951) reported that Meyer stated, "We understand the case, we don't need the diagnosis" (p. 22).

Meyer suggested that psychiatry put aside old, traditional classification and look at the facts and data as they group together. He recommended first determining all of the facts and dealing with the behaviors *per se,* rather than grouping facts according to meaningless names. Frequently, the facts are used to arrive at an arbitrary diagnosis, and then treatment planning is based on the label, rather than on the facts. Also, there is a "superstition" that once a problem is diagnosed, it can be solved. According to Meyer, it is the understanding of stresses, the "why" and the "how" of disease, not just its name, that is important.

Meyer was extremely influential in American psychiatry during the first half of this century, and he contributed to the decreased interest in classification by Americans during this time. Psychobiology replaced the Kraepelinian concern about classification with an emphasis on treatment and prevention. Why was Meyer so influential? First, his combining of biology and physiology with psychiatry, psychology, and sociology was appealing to many psychiatrists and mental health professionals. Other theories of psychopathology at that time were either purely biological or purely psychological. A second, related reason was his comprehensive evaluation of etiology, which was broader in scope than other theories. In addition, his emphasis on treatment and prevention was probably welcomed by American clinicians. Third, his theory was consistent with the dominant philosophies in America because Meyer was strongly influenced by pragmatism and the social activism of the mental hygiene movement. Meyer's attempts to increase the scope of psychiatry to include prevention, outreach, and follow-up can still be seen in the ideas behind the community-mental-health movement of the 1960s.

Perhaps another reason for Meyer's substantial influence was that his theory of psychobiology was introduced during a time when new behavioral theories were drastically changing views about psychology. Notice the many behavioral terms used by Meyer. He emphasized dealing with behavior, and he viewed psychopathology as maladaptive habits that had to be unlearned while new adaptive habits were learned. His reduced emphasis on diagnosis and classification and his increased emphasis on a careful evaluation of the patient and the environment were consistent with the principles advocated by behaviorists. However, Meyer had been affected by the psychoanalytic revolution, and unlike the behaviorists, Meyer did not reject concepts such as personality or the unconscious. His comprehensive evaluation of the individual included the examination of personality and of conscious/unconscious factors, as well as the analysis of behaviors. Therefore, his theory had more appeal than the "empty organism" approach of J. B. Watson, B. F. Skinner, and other behaviorists.

MODERN AMERICAN PSYCHIATRIC CLASSIFICATION

Despite the benign neglect with which American mental-health professionals have often viewed classification, various official classificatory systems have been proposed during the twentieth century. Consistently, these systems have reflected the importance accorded to Kraepelinian concepts. In 1917, the year that the United States entered World War I, a classification system was adopted that included the Kraepelinian concepts of dementia praecox, manic-depressive disorder, and paranoia (Menninger, 1963). This classification also made the fundamental Kraepelinian distinction between organic brain syndromes, functional psychoses, and neurotic/character disorders.

Between World Wars I and II, American interest in classification was not strong. Instead, more attention was focused on the expansion of psychoanalytic ideas, the development of personality theories, and the investigation of new forms of somatic and psychological therapies. Nevertheless, in 1932, the American Psychiatric Association did officially adopt a new system as part of the *Standard Classified Nomenclature of Diseases* (American Psychiatric Association, 1933). This classification contained 24 major categories, 19 of which had appeared in Kraepelin's sixth-edition classification (see Appendix I). This new classification, however, did not attract much attention and, according to Karl Menninger (1963), never gained popular acceptance in American Psychiatry.

World War II led to a renewed emphasis on classification. During the war, nearly 10% of all discharges before the armistices were for psychiatric reasons. At the termination of hostilities, there were four major, competing classification systems: (1) the "standard" system adopted by the American Psychiatric Association in 1932; (2) the U.S. Army classification; (3) the U.S. Navy classification; and (4) the Veterans Administration system (Raines, 1952).

In response to this chaotic and embarrassing state, the American Psychiatric Association decided to form a task force to create a system that would become standard in the United States. The result was the *Diagnostic and Statistical Manual of Mental Disorders* (first edition—DSM-I, 1952). By its contemporaries, this manual was known as the "gray book" because of the gray paper cover on the manual.

The DSM-I was important for a number of reasons. First, the major rationale behind the DSM-I was to create a classification that was a consensus of contemporary thinking. Care was taken to represent as nearly as possible the concepts that were popular in American psychiatry in the 1940s and 1950s. In other words, the DSM-I emphasized the communication purpose of classification. This purpose, which is the most basic reason for the existence of any classification system, underscores the use of a classification system as an accepted nomenclature that members of a profession can use to discuss their clinical and research knowledge. Consistent with this communication purpose, early editions of the DSM-I were revised through comments elicited in a questionnaire sent to 10% of the members of the American Psychiatric Association. The DSM-I was finally adopted by a vote of the membership of the American Psychiatric Association (Raines, 1952).

A second important aspect of the DSM-I was that its emphasis on communication led to a similar organizing movement in international psychiatric classification. A committee chaired by a British psychiatrist, E. Stengel, was formed to review the classificatory systems used by various countries and to make any necessary recommendations. What Stengel (1959) found was a hodgepodge of diagnostic systems across and sometimes within different countries. In 1951, the World Health Organization had proposed an international psychiatric classification as part of the *International Classification of Diseases* (sixth edition—ICD-6; World Health Organization, 1948). Before the sixth edition, the ICD had not contained a subdivision devoted to mental disorders because the previous editions were concerned only with causes of death. By 1959, only five member countries of the World Health Organization (Finland, Peru, New Zealand, Thailand, and the United Kingdom) had adopted the ICD-6 classification. Stengel despaired over the confused state of international classification, especially because the ICD-6 had not been generally accepted. The only positive note concerned the DSM-I, which Stengel considered an advance over the other national classifications.

As a result of Stengel's review, there was an international movement to create a consensual system that would be adopted by the World Health Organization countries. The final product was the mental disorder section of the *International Classification of Diseases* (eighth edition—ICD-8; World Health Organization, 1967). In the United States, psychiatrists worked energetically with the international movement to create this system (Kramer, 1968). The American version of the ICD-8 was the second edition of the *Diagnostic and Statistical Manual of Mental Disorders* (American Psychiatric Association, 1968), otherwise known as the DSM-II. Only small differences existed between the ICD-8 and the DSM-II. For example, the American DSM-II recognized the term *passive aggressive personality disorder*, whereas this term had not been accepted internationally.

The DSM-II represented the culmination of the movement to create a classification system that was a consensus of accepted concepts and that could be used for the purpose of communication. Stengel (1959) made clear why communication regarding classification is so important:

> A more serious obstacle to progress in psychiatry is difficulty of communication. Everybody who has followed the literature and listened to discussions concerning mental illness soon discovers that psychiatrists, even those apparently sharing the same basic orientation, often do not speak the same language. They either use different terms for the same concepts, or the same term for different concepts, usually without being aware of it. It is sometimes argued that this is inevitable in the present state of psychiatric knowledge, but it is doubtful whether this is a valid excuse. (p. 601)

RECENT CHANGES IN PSYCHIATRIC CLASSIFICATION

During the 1960s, there were three issues of relevance to psychiatric classification in addition to the emphasis on national and international communication. The three issues were related in that they were anticlassification in their expression and were associated with a general negative orientation to the diagnostic process. These three issues were interclinician reliability, the medical model, and the labeling perspective of diagnosis. All three were mentioned briefly in the beginning of this chapter and will be elaborated on in more detail later in the book.

In reaction to these attacks on classification, a "neo-Kraepelinian" movement developed that represented the reaffirmation of Kraepelin's interest in description and classification. This movement contains psychiatrists who believe in the importance of classification, who tend to support the "back to basics" approach to psychiatry, who oppose the psychoanalytic perspective, and who are very much interested in research.

A major event in this neo-Kraepelinian movement was the publication of research criteria for diagnosis by a group of St. Louis psychiatrists (Feighner, Robins, Guze, Woodruff, Winokur, & Munoz, 1972). In their paper, explicit criteria were proposed for 16 diagnostic categories. These criteria attracted a great deal of attention. Shortly after the publication of this paper, a group from New York and St. Louis jointly proposed the *Research Diagnostic Criteria* (Spitzer, Endicott, & Robins, 1975). These latter criteria used the same style as those of Feighner *et al.* in order to suggest precise definitions for a number of subgroups of schizophrenia and the affective disorders.

The major result of the neo-Kraepelinian movement has been the creation of the third edition of the *Diagnostic and Statistical Manual of Mental Disorders* (DSM-III; American Psychiatric Association, 1980). The original versions of the DSM-III were extensions of the *Research Diagnostic Criteria.* This new classification has undergone extensive testing and many revisions. Five major revisions of the DSM-III were promulgated before the final version was published by the American Psychiatric Association in 1980. These different versions were subjected to large-scale field trials and to criticisms by clinicians (Spitzer & Sheehy, 1976; Spitzer, Williams, & Skodol, 1979). In addition, modifications were made based on published research that had become available during its development.

The DSM-III differs markedly from its predecessors. Three main changes were involved: (1) the use of explicit diagnostic criteria; (2) a multiaxial diagnostic system; and (3) an attempt to create an internally consistent organization of clinical syndromes (Spitzer & Forman, 1979). The use of explicit diagnostic criteria was a response to the frequent criticisms of poor reliability aimed at the DSM-I and the DSM-II. The second innovation, a multiaxial system, was proposed in order to better represent the diversity of information available about patients. The final major change, the reorganization of the hierarchical system of relationships among categories, was made to enhance the internal consistency of the system.

SUGGESTED READINGS

Kendell, R. E. *The role of diagnosis in psychiatry.* Oxford, England: Blackwell Scientific Publications, 1975.

This book is the best discussion of psychiatric classification in the current literature. There are chapters on the alternate meanings of the concept *disease* as applied to mental disorders, the dimensional versus the categorical approach to classification, and cross-cultural differences in diagnostic perceptions. Kendell's chapter on "The Nature of Disease and Diagnosis" covers various ideas about the meaning of the disease concept. For instance, disease can be conceptualized as any process causing human suffering (i.e., disease). Also, disease has been defined as whatever a physician treats, but Kendell warns that this definition is probably worse than no definition at all. A third approach is to view a disease as a process that has led to "structural damage to the body." This definition raises difficulties because the human body in a "normal" condition can contain a variety of structural forms and because many commonly recognized diseases cannot be shown to be associated with structural damage. Other definitions that Kendell entertains are disease as a response to stress, disease as a statistical abnormality, and disease as a biological disturbance.

Kendell is a very careful thinker about the issues of classification. His numerous journal articles are also of interest, especially his thoughtful discussions on the classification of depression (Kendell, 1968, 1976). In many ways, the book you are reading is intended to supplement Kendell's book by emphasizing topics that were not the primary focus of his work.

Menninger, K. *The vital balance.* New York: Viking Press, 1963.

This book discusses Menninger's approach to the classification of mental disorders, as well as his rationale for a very simple classificatory system. Menninger ends by positing five orders of human dysfunction that make up his suggested classification: (1) "normal" adaptive repsonses to stress; (2) neurotic syndromes; (3) naked aggression; (4) psychotic states; and (5) malignant anxiety that can lead to death.

By far the most interesting section in this book, however, is the appendix, which summarizes over 300 different classificatory systems. The appendix is organized historically and starts with outlines of classificatory systems for mental disorders proposed by Hippocrates and Plato. A very large section of the appendix is devoted to nineteenth-century systems. Included are outlines of all eight editions of Kraepelin's classifications, with a commentary on the evolution of Kraepelin's ideas. It should be noted that the outlines of Kraepelin's classifications are not very accurate.

Feighner, J. P., Robins, E., Guze, S. B., Woodruff, R. A., Winokur, G., & Munoz, R. Diagnostic criteria for use in psychiatric research. *Archives of General Psychiatry,* 1972, *26,* 57–63.

This paper has attracted a great deal of attention in the psychiatric literature. During the first eight years after its publication, the article has gathered over 1,000 citations, which is about 75 times greater than the citation rate associated with most other articles in the same journal. The major innovation in the paper is the presentation of explicit rules for diagnosing 16 traditional categories: mania, schizophrenia, depression, alcoholism, anorexia nervosa, and so on. In addition, Feighner *et al.* argued that five types of information can be used to validate classificatory categories: (1) clinical presentation; (2) laboratory studies; (3) differentiation from other disorders; (4) follow-up studies; and (5) family studies.

Spitzer, R. L., & Wilson, P. T. Nosology and the official psychiatric nomenclature. In A. M. Freedman, H. I. Kaplan, & B. J. Sadock (Eds.), *Comprehensive textbook of psychiatry* (2nd ed.). Baltimore: Williams & Wilkins, 1975.

Spitzer was the chairman of the task force that created the DSM-III. This chapter is the clearest and the best written presentation of his orientation to classification. As such, the chapter represents an exposition of a school of thought, with which Spitzer is associated, called the *neo-Kraepelinian perspective.* The members of the neo-Kraepelinian school are researchers who adopt an empirical approach to the study of mental disorders, who emphasize the importance of classification, who eschew the psychoanalytic approach to psychiatry, and who generally believe in genetic and/or biochemical theories of the etiology of most mental disorders.

Chapter 2

The Neo-Kraepelinian Movement in American Psychiatry

Perhaps I believe that the world can get forward most by clearer and clearer definitions of fundamentals. Accordingly I propose to stick to the tasks of nomenclature and terminology, unpopular and ridicule-provoking though they may be.

Southard, 1919, as quoted by Menninger, 1963, p. 3

During the 1950s and 1960s, there was an increasing dissatisfaction with the topic of psychiatric classification. The reason for this dissatisfaction was the concurrence of three different attacks on classification. These were the ideas (1) that psychiatric diagnosis, as standardly practiced, was unreliable; (2) that classification was associated with the medical model; and (3) that diagnosis has the negative consequence of assigning labels, thereby potentially creating self-fulfilling prophecies. Each of these attacks is briefly discussed here in turn.

The concern about diagnostic reliability developed from a number of empirical studies, published during the 1950s and 1960s, that generally concluded that the diagnostic reliability of traditional psychiatric classificatory systems was unacceptably low. One of the earliest and most frequently cited studies addressing the topic of interclinician agreement was performed by Ash (1949). He found that three psychiatrists agreed on the same diagnosis for only 10 of 52 patients, a rate of agreement that was viewed as unsatisfactory. The predominance of negative conclusions from similar studies lent support to the claim that psychiatric diagnosis was unreliable. This assertion holds considerable importance in that adequate reliability is believed to be essential for any useful classificatory system: "There is no guarantee that a reliable system is valid, but assuredly an unreliable system must be invalid" (Spitzer & Fleiss, 1974, p. 341). The topic of diagnostic reliability is the central theme of Chapter 4, and further discussion of the issues surrounding this topic is deferred to that chapter.

A second reason for the criticism of traditional psychiatric classification is its association with the medical model. The medical model is a perspective that has implicitly permeated the mental health field. From the perspective of the medical model, all mental disorders are *diseases*. The persons afflicted with these diseases are called *patients*; they need treatment from *doctors*; *diagnosis* is an essential first step if one is to *prescribe* the best *therapy* and to predict the natural *course* of the patient's disorder. Severely disturbed patients need *medication* and perhaps *hospitalization*; their care should be paid for by *health insurance policies*. However, it is important to note that the medical model is not the only perspective that can be identified in the mental health field. Other major perspectives that have had strong influences on contemporary thinking about psychopathology are (1) the *humanistic* perspective, which stems from the work of Carl Rogers, Frederick Perls, and Charles Truax; (2) the *psychoanalytic* perspective, which grew from the innovative ideas of Sigmund Freud and his followers, such as Harry Stack Sullivan, Alfred Adler, and Carl Jung; and (3) the *behavioral* perspective, which was developed from the work on learning theory in psychology. A book by Price (1971) provides a good general introduction to these different perspectives. Professionals from the humanistic, psychoanalytic, and behavioral perspectives have all disparaged classification as a legitimate activity because of its association with the medical model. The next chapter of this book, Chapter 3, discusses the meaning of the medical model as it relates to psychiatric classification.

The third attack on traditional psychiatric classification grew from sociology and the development of labeling theory (Goffman, 1959; Scheff, 1966; Matza, 1968). This theory suggested that labels could become self-fulfilling prophecies that would strongly affect the persons being labeled. A dramatic example of some of the ideas in labeling theory are represented in a paper by Rosenhan (1973) entitled "Being Sane in Insane Places." Rosenhan reported a study in which eight normal persons sought admission to 12 different inpatient units. All eight persons correctly reported information about themselves except for two things. They lied about their names in order to avoid the stigma of having a mental hsopitalization record, and they reported having heard an auditory hallucination in which a voice said "thud," "empty," or "hollow." In all 12 instances the patients were admitted: 11 of the admissions were assigned the diagnosis of schizophrenia, and 1 was assigned the diagnosis of mania. All 12 discharges were followed with the diagnosis of schizophrenia in remission. The average length of stay in the institutions was about 20 days. Much of the paper discusses the depersonalization that these eight bogus patients noted during their stay on the units. From this research, Rosenhan concluded that modern mental health professionals cannot distinguish the sane from the insane. The Rosenhan article generated a great deal of controversy. Most of one issue of the *Journal of Abnormal Psychology* (October 1975) was devoted to papers that discussed the implications of the Rosenhan paper (see pp. 56–57).

These three attacks on the classification of psychopathology were quite influential and convinced many mental health professionals to belittle the use of diagnosis and classificatory systems. Subsequent to these attacks and directly in response to their influence, a reactionary movement developed in the 1970s that asserted the importance of classification for the understanding of psychopathology. This movement was associated with the back-to-basics trend in psychiatry during the 1970s, which emphasized the medical foundation of psychiatry. This movement was referred to as the *neo-Kraepelinian movement* (Klerman, 1978). The term *neo-Kraepelinian* was used because the movement promoted a resurgence of many of the ideas associated with Kraepelin and his view of mental disorders. As the quote from Southard at the beginning of this chapter suggests, the neo-Kraepelinians strongly believe in the importance of classification and have been determined to stimulate improvements in this area. Klerman (1978) proposed that the neo-Kraepelinians felt alienated by the dominance of the psychoanalytic perspective in psychiatry and were united through their mutual adherence to a biological orientation that emphasizes the relationship between psychiatry and traditional medical principles.

THE NEO-KRAEPELINIAN CREDO

In his analysis of this movement, Klerman (pp. 104–105) suggested that the ideas implicitly assumed by the neo-Kraepelinians could be summarized by nine propositions (see Table 1):

TABLE 1. Nine Propositions Associated with the Neo-Kraepelinian Movement

1. Psychiatry is a branch of medicine.
2. Psychiatry should utilize modern scientific methodologies and base its practice on scientific knowledge.
3. Psychiatry treats people who are sick and who require treatment for mental illness.
4. There is a boundary between the normal and the sick.
5. There are discrete mental illnesses. Mental illnesses are not myths. There is not one, but many mental illnesses. It is the task of scientific psychiatry, as of other medical specialties, to investigate the causes, diagnosis, and treatment of these mental illnesses.
6. The focus of psychiatric physicians should be particularly on the biological aspects of mental illness.
7. There should be an explicit and intentional concern with diagnosis and classification.
8. Diagnostic criteria should be codified, and a legitimate and valued area of research should be to validate such criteria by various techniques. Further, departments of psychiatry in medical schools should teach these criteria and not depreciate them, as has been the case for many years.
9. In research efforts directed at improving the reliability and validity of diagnosis and classification, statistical techniques should be utilized.

Note. From Klerman (1978).

1. "Psychiatry is a branch of medicine." Casually read, this proposition seems trivial. Of course, psychiatry is a branch of medicine. Such a claim appears to be truism. However, what is important in this assertion is the implication that psychiatry, as a discipline, has more in common with other areas of medicine (such as neurology and endocrinology) than it does with other disciplines among the mental health professions (such as clinical psychology, social work, and psychiatric nursing). In effect, the neo-Kraepelinian movement emphasizes the assertion that first and foremost, psychiatrists are physicians. Kraepelin probably would not have emphasized the association of psychiatry with medicine as strongly as the neo-Kraepelinians do. Kraepelin had been greatly influenced by Wundt and other experimental psychologists. Although Kraepelin clearly advocated neuroanatomical and neurophysiologic research in psychopathology, he also argued for a strong relationship between psychiatry and the science of psychology (Mayer-Gross, 1929).

2. "Psychiatry should utilize modern scientific methodologies and base its practice on scientific knowledge." The second proposition underscores the neo-Kraepelinian emphasis on science. Like Kraepelin, the members of this recent movement in psychiatry believe that a classificatory system is necessary for a scientific understanding of psy-

chopathology and that classificatory systems should be evaluated empirically (Blashfield & Draguns, 1976a; Skinner, 1981).

3. "Psychiatry treats people who are sick and who require treatment for mental illness." This proposition is an avowal of the medical-model perspective on psychopathology. This avowal has a number of scientific, philosophical, and political implications that will be discussed in some detail in the next chapter.

4. "There is a boundary between the normal and the sick." The assertion being made is that there is a clear, qualitative distinction between persons who have a mental disorder and persons who have not. This proposition contrasts with the view of Thomas Szasz (1961), for example, who believes that psychopathology can best be conceptualized as the study of "problems in living" that affect all people to a greater or lesser extent. The neo-Kraepelinians do not believe that recognized mental disorders such as schizophrenia, bipolar affective disorder, depression, and anorexia nervosa are simply extreme forms of behavioral and social problems that in milder ways are present in the lives of all of us. In effect, the neo-Kraepelinians argue that depression, when carefully defined as a clinical entity, is qualitatively different from the mild episodes of depression that everyone experiences at some point in his or her life.

5. "There are discrete mental illnesses. Mental illnesses are not myths. There is not one, but many mental illnesses. It is the task of scientific psychiatry, as of other medical specialities, to investigate the causes, diagnosis, and treatment of these mental illnesses." Psychiatry has continually been beset by identity problems (Schwartz, 1974; Clare, 1976). What has been unclear is whether psychiatry should emphasize being a profession or being a science, what theoretical perspectives are most germane to its subject matter, and what the appropriate relationship between psychiatry and other medical and behavioral disciplines should be. The neo-Kraepelinians strongly favor a particular identity for psychiatry, and they claim that this proposed identity should determine the priorities for future efforts in the field (cf. Lipowski, 1981).

6. "The focus of psychiatric physicians should be particularly on the biological aspects of mental illness." Consistent with Kraepelin's views, the neo-Kraepelinians are largely antipsychoanalytic in their approach to psychopathology. The psychoanalytic perspective focuses on the mental activities of patients and is concerned with an understanding of the dynamics associated with a case. The neo-Kraepelinians eschew the psychoanalytic perspective as a highly inferential approach that cannot be empirically studied and that requires a large

number of implicit, subjective judgments on the part of psychiatrists. In the interests of empirical scrutiny, the neo-Kraepelinians feel that the emphasis instead should be on the biological aspects of psychopathology.

7. "There should be an explicit and intentional concern with diagnosis and classification." This position contrasts with the general lack of interest in and negative orientation to diagnosis and classification that was standard among most mental health professionals during the 1960s (Clare, 1976). The positive assertion is being made that classification is important and deserves careful analysis.

8. "Diagnostic criteria should be codified, and a legitimate and valued area of research should be to validate such criteria by various techniques. Further, departments of psychiatry in medical schools should teach these criteria and not depreciate them, as has been the case for many years." The neo-Kraepelinians have seen themselves as an "embattled minority" for a number of years (Klerman, 1978). The unpopularity of classification has often led to negative attitudes' being triggered by their research on the definition of mental disorders. However, this movement believes that the definition of mental disorders through carefully codified diagnostic criteria is an important undertaking. As discussed later in this chapter, it is the neo-Kraepelinian movement that suggested the use of explicit diagnostic criteria (Feighner *et al.*, 1972) and that promoted the use of these diagnostic criteria in the DSM-III (see Chapter 5).

9. "In research efforts directed at improving the reliability and validity of diagnosis and classification, statistical techniques should be utilized." The last proposition implies that two important focuses for research on diagnostic categories are reliability and validity. The neo-Kraepelinians believe that the research attacking the reliability of the first two versions of the DSM offered important criticisms, and that any future changes in psychiatric classification should lead to distinct improvements in reliability. The neo-Kraepelinians also espouse the use of statistical procedures as part of the basic methodology of a recognized science. Thus, psychiatrists should be taught statistical concepts and their use in research. The neo-Kraepelinians have been active in the area of improving statistical methodology within psychiatric research. For instance, Spitzer and his colleagues have been largely responsible for suggesting and promoting the use of the statistic *kappa* as a measure of reliability in studies of diagnostic agreement (Cohen, 1960; Spitzer, Cohen, Fleiss, & Endicott, 1967; Fleiss, Cohen, & Everitt, 1969; Fleiss, Spitzer, Endicott, & Cohen, 1972). In addition, Spitzer *et al.* have frequently used factor analysis in their descriptive

research (Spitzer, Fleiss, Endicott, & Cohen, 1967; Spitzer, Endicott, Fleiss, & Cohen, 1970).

CENTRAL MEMBERS OF THE NEO-KRAEPELINIAN MOVEMENT

Another way to develop an understanding of the neo-Kraepelinian perspective is to learn about the professionals most closely associated with this perspective. The persons whom Klerman labeled as the central members of this perspective were Guze, Robins, Winokur, Klein, and Spitzer. The first two of these five men are Samuel Guze and Eli Robins, both of whom are at Washington University in St. Louis. Guze, the Vice Chancellor for Medical Affairs at Washington University, has been interested in a wide variety of research about psychopathology and has published almost 100 journal articles between 1965 and 1983. He has also published one book titled *Criminality and Psychiatric Disorders* (1976), and has been a coauthor of another, *Psychiatric Diagnosis* (Woodruff, Goodwin, & Guze, 1974). Because Guze has been a strong proponent of the medical-model approach to psychopathology, his ideas on this topic are discussed in detail in the next chapter.

The second central member of the neo-Kraepelinian perspective, Eli Robins, is also at Washington University, where he holds a chair as a distinguished professor in psychiatry. His major research interests have concerned classification, biochemistry, and manic-depressive disorders. The most cited journal article by Robins was one that he published jointly with Guze, entitled "Establishment of Diagnostic Validity in Psychiatry Illness: Its Application to Schizophrenia" (1970). Robins has also been the coauthor of one book, *Male and Female Homosexuality* (Saghir & Robins, 1973).

The third person mentioned as central to the neo-Kraepelinian perspective is George Winokur, who was a member of the faculty of Washington University during the 1960s. However, in the early 1970s, Winokur moved to the University of Iowa, where he became chairperson of the psychiatry department. Winokur has been a prolific researcher, publishing 167 journal articles in the years from 1965 to April 1983. Clearly his most influential work is the book *Manic Depressive Insanity* (Winokur, Clayton, & Reich, 1969). Winokur has continued to have as his major interest the area of affective disorders and has been involved in a number of descriptive, family history, and biochemical studies of the disorders of mood.

The last two men who are included by Klerman as members of the neo-Kraepelinian perspective are at the New York State Psychiatric

Institute. The first of these is Donald Klein, who is the director of research at the New York State Psychiatric Institute and is also a professor at Columbia University. Klein is well known in the field of psychiatry for his extensive research on psychopharmacology. He published nearly 170 journal articles from the mid-1960s to 1983. Klein's most cited publication is *Diagnosis and Drug Treatment of Psychiatric Disorders* (Klein & Davis, 1969). There is now a second edition of this book (Klein, Gittelman, Quitkin, & Rifkin, 1980).

Robert Spitzer is the last member of the neo-Kraepelinian perspective to be discussed (cf. Spitzer, 1982). Like Klein, he is currently at the New York State Psychiatric Institute. Because of Spitzer's importance to contemporary psychiatric classification, a relatively detailed description of his career is presented here. Spitzer was born in White Plains, New York, in 1932. He did his undergraduate work at Cornell University in psychology, graduating in 1953. He completed his M.D. at New York University in 1957 and did his psychiatry residency at the New York State Psychiatric Institute, where he has remained. Spitzer has also had training in psychoanalysis, as he attended the Columbia University Psychoanalytic Clinic for Training and Research. Spitzer's interest in research began early in his career. His first publication was a paper entitled "The Neck-Righting Reflex in Children," which appeared in the *Journal of Pediatrics* (Spitzer, Kramer, & Rabkin, 1958). After completing his residency, Spitzer gained a research fellowship in biometrics at the New York State Psychiatric Institute. Through this position, Spitzer worked with Joseph Zubin, a well-known clinical psychologist who had been publishing in psychiatry and psychology journals on the topics of schizophrenia, diagnosis, and epidemiology. During the 1960s, Spitzer collaborated with Jean Endicott, a psychologist, and Joseph Fleiss, a statistician, to create a number of standardized clinical interviews, such as the Mental Status Schedule (Spitzer, Fleiss, Endicott, & Cohen, 1967; Spitzer, Endicott, Fleiss, & Cohen, 1970). In addition, he was interested in computerized approaches to diagnosis, and he created the program DIAGNO-II, which could be used to assign a clinical diagnosis with complete Mental Status Schedule data (Spitzer & Endicott, 1968, 1969).

In the mid-1960s, Spitzer served as a consultant to the Task Force on Nomenclature and Statistics for the American Psychiatric Association, which created the DSM-II. He coauthored the major article in the *American Journal of Psychiatry* that introduced and defended the DSM-II (Spitzer & Wilson, 1968). Spitzer became a member of the Task Force on Nomenclature and Statistics in 1969.

A highly emotional and political issue appeared in the early 1970s within the American Psychiatric Association over the inclusion of the diagnosis of homosexuality within the DSM-II. Gay activist groups argued that American psychiatry was inappropriately labeling homosexuality as a disease. These gay groups demanded that psychiatry drop the category of homosexuality from the DSM-II, and they disrupted national meetings of the American Psychiatric Association in order to force an awareness of their concerns. A number of psychiatrists who believed in a psychoanalytic perspective on psychopathology were opposed to this change. Spitzer was a central figure in this debate, and he helped arrange the change in the DSM-II that satisfied the gay activists and that was eventually accepted by the majority of the American Psychiatric Association members (Spitzer, 1973).

Spitzer became chairperson of the classification task force in 1974. Having been impressed by the paper published by researchers at Washington University about explicit diagnostic criteria (Feighner *et al.*, 1972), Spitzer helped create the *Research Diagnostic Criteria* (RDC) (Spitzer *et al.*, 1975), the precursor of the DSM-III. The RDC emphasized diagnostic criteria for the schizophrenic and the affective disorders. Most of Spitzer's efforts during the late 1970s were devoted to the development and the final publication of the DSM-III. The history of the DSM-III is covered in Chapter 5.

Robins, Guze, Winokur, Klein, and Spitzer are not the only adherents of the neo-Kraepelinian perspective. In order to create an approximate list of such adherents, a search was made for all persons who were coauthors of any journal articles or books published by these five men since 1965. A total of 169 names was found. Of these persons, 67 had been coauthors of at least three publications with a central member of the neo-Kraepelinian movement. Figures 1, 2, and 3 are sociograms that display the coauthor relationships among these 67 persons for the five-year blocks of 1965–1969 (Figure 1); 1970–1974 (Figure 2); and 1975–1979 (Figure 3). A line is drawn between two persons if they were coauthors of any book or journal article during the five-year time period displayed.*

Figure 1 clearly shows the existence of three groups: the Hillside Hospital group of Klein and his colleagues; the "New York group" of Spitzer, Endicott, and Fleiss; and the "St. Louis group" of Guze, Robins, Winokur, and a few other researchers at Washington Univer-

*The locations shown in the three figures represent the geographic whereabouts of each author when coauthoring a paper with one of the five men central to the neo-Kraepelinian perspective. Thus, for example, Figure 2.3, which depicts publishing relationships from 1975 to 1979, has Munoz in San Diego. In fact, however, Munoz was located in Wisconsin from 1975 to 1977 and in San Diego from 1977 to 1979.

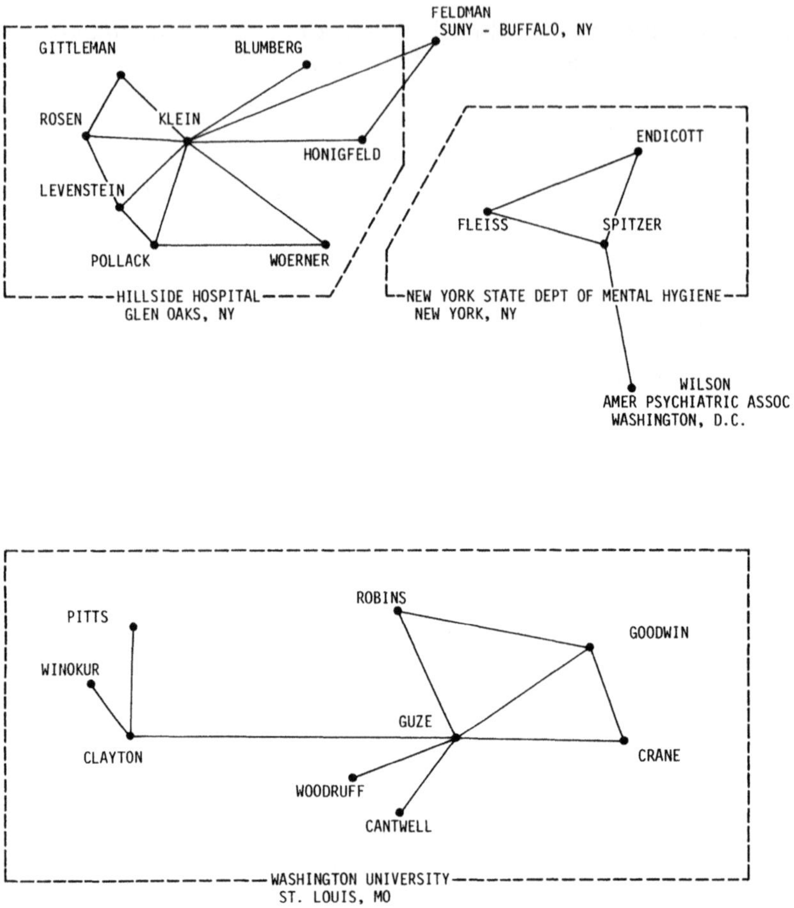

FIGURE 1. Coauthors with central Neo-Kraepelinians, 1965–1969.

sity in St. Louis. In the second five-year period, a number of new researchers were added to each subgroup. A major structural change was the development of an offshoot of the St. Louis group, which resulted from Winokur's move to the University of Iowa. In Figure 3, which depicts the publications from 1975 to 1979, all of the different subgroups are interconnected. In addition, the total number of authors increased as had the geographic diversity of the placement of the various coauthors. These figures should give the reader an indication of the growth of the neo-Kraepelinian movement and the geographic structure of this movement.

To summarize, the neo-Kraepelinians are a group of psychiatrists

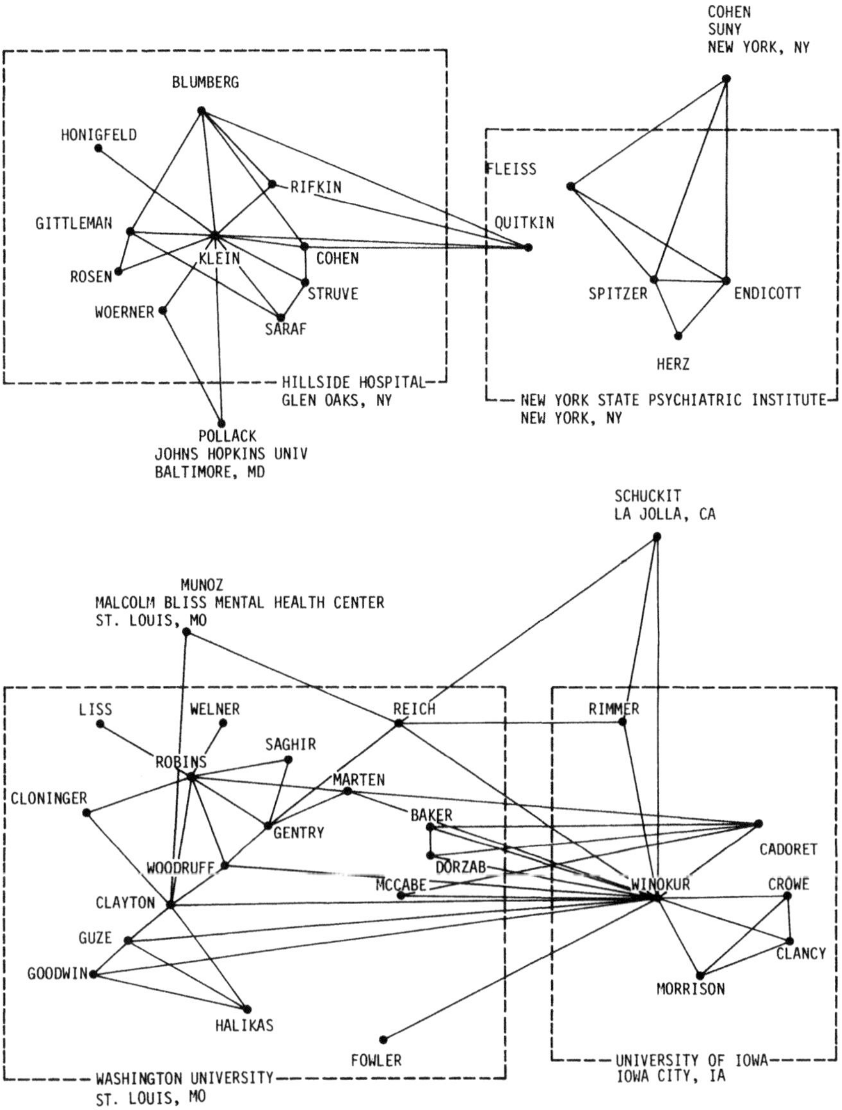

FIGURE 2. Coauthors with central Neo-Kraepelinians, 1970–1974.

who have been opposed to the anticlassification arguments of the 1950s and 1960s. The neo-Kraepelinians asserted the positive values of classification, advocated the medical model, and strongly favored a research approach to psychopathology based on genetics and biochemistry. These psychiatrists are located primarily in three places in

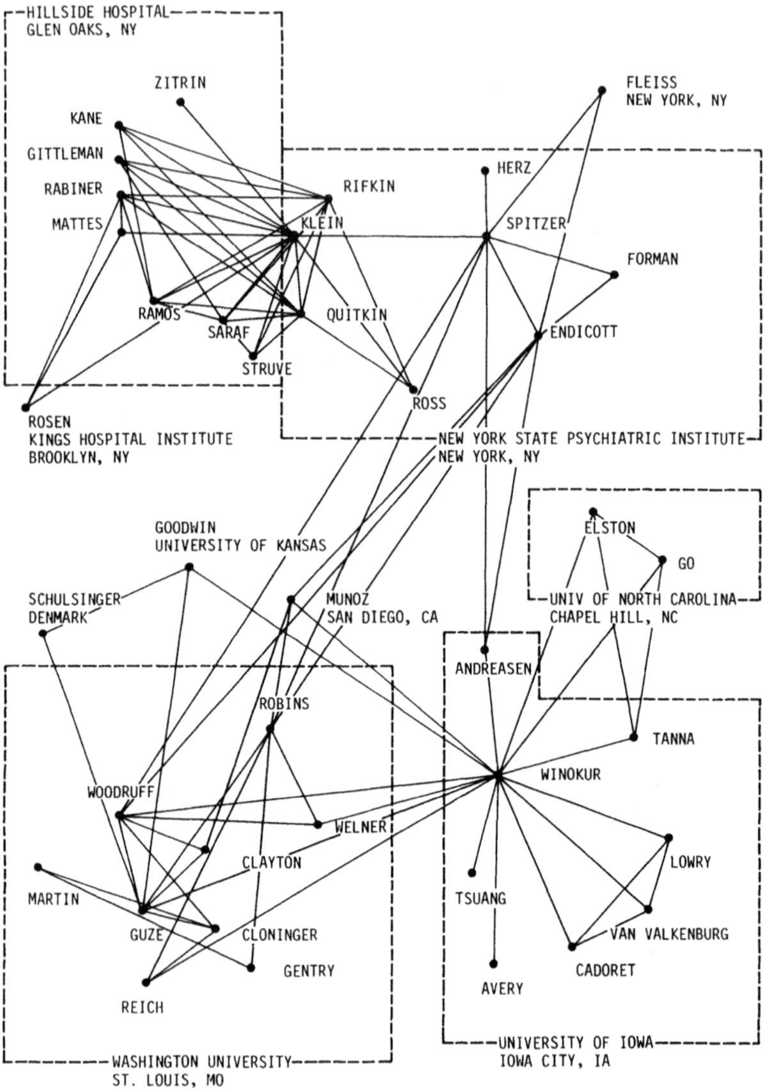

FIGURE 3. Coauthors with central Neo-Kraepelinians, 1975–1980.

the United States: New York City, St. Louis, and Iowa City. The central members of this perspective are Robins, Guze, Winokur, Klein, and Spitzer.

THE FEIGHNER PAPER

A further means of understanding the neo-Kraepelinian movement is to discuss one of the most influential articles published by adherents of this perspective. The publication was a journal article in the *Archives of General Psychiatry* titled "Diagnostic Criteria for Use in Psychiatric Research" (1972). The authors of the paper were John Feighner, Eli Robins, Samuel Guze, Robert Woodruff, George Winokur, and Rodrigo Munoz. At the time the article was written, all of the authors were at Washington University in St. Louis. The importance of the Feighner *et al.* paper stems from the following facts: it led to the formulation of the explicit diagnostic criteria in the DSM-III; it attracted an exceptionally large number of citations; and it demonstrated the growth of power of the neo-Kraepelinian movement.

One comment should be made before we proceed. The remainder of this chapter is devoted to a discussion of the Feighner paper. This extensive analysis of the Feighner paper has the danger of overemphasizing the importance of this paper and of suggesting that it was responsible for the impact of the neo-Kraepelinian perspective on classification. The Feighner paper was not that important, and certainly, the thinking of the neo-Kraepelinians has extended beyond the limits of the Feighner paper. The reason for the extensive analysis of this paper is that the analysis allows the introduction of principles from the sociology of science, principles that will be shown to be useful in understanding psychiatric classification.

The Feighner *et al.* paper is a reasonably brief journal article that contains primarily the explicit diagnostic criteria that define 16 diagnostic categories. An example of the criteria for one category is the definition of schizophrenia:

> For a diagnosis of schizophrenia, A through C are required.
>
> A. Both of the following are necessary: (1) A chronic illness with at least six months of symptoms prior to the index evaluation without return to the premorbid level of psychosocial adjustment. (2) Absence of a period of depressive or manic symptoms sufficient to qualify for affective disorder or probable affective disorder.
>
> B. The patient must have at least one of the following: (1) Delusions or hallucinations without significant perplexity or disorientation associated with them. (2) Verbal production that makes communication difficult be-

> cause of a lack of logical or understandable organization. (In the presence of muteness the diagnostic decision must be deferred.)
>
> (We recognize that many patients with schizophrenia have a characteristic blunted or inappropriate affect; however, when it occurs in mild form, interrater agreement is difficult to achieve. We believe that, on the basis of presently available information, blunted affect occurs rarely or not at all in the absence of B-1 or B-2.)
>
> C. At least three of the following manifestations must be present for a diagnosis of "definite" schizophrenia, and two for a diagnosis of "probable" schizophrenia. (1) Single. (2) Poor premorbid social adjustment or work history. (3) Family history of schizophrenia. (4) Absence of alcoholism or drug abuse within one year of onset of psychosis. (5) Onset of illness prior to age 40. (Feighner *et al.*, 1972, p. 59)

In addition to schizophrenia, the authors proposed diagnostic criteria for depression, mania, secondary affective disorders, anxiety neurosis, obsessive-compulsive neurosis, phobic neurosis, hysteria, antisocial personality disorder, alcholism, drug dependence (excluding alcholism), mental retardation, organic brain syndrome, homosexuality, transsexualism, and anorexia nervosa. These 16 categories represent diagnostic concepts that the St. Louis group believed had sufficient evidence to be considered valid. This observation is notable because the DSM-II contained about 180 categories, yet Feighner *et al.* implicitly suggested that less than 10% of the DSM-II categories were sufficiently valid for future research on psychopathology.

Besides presenting the criteria for the 16 diagnostic categories, Feighner *et al.* also proposed five "phases" that can be used to establish the diagnostic validity of specific mental disorders:

1. *Clinical description.* The first phase is a description of a collection of apparently co-occurring symptoms that define the clinical presentation of a mental disorder. Also of relevance are the demographic characteristics of the patients having this disorder.

2. *Laboratory studies.* During this phase, biochemical, neurological, psychological, and other tests are used to identify any additional correlates of the disorder.

3. *Delimitation from other disorders.* At this point, the general picture of the disorder should be emerging, and the next phase is to draw careful boundaries between this disorder and other disorders with which it could be confused.

4. *Follow-up studies.* These studies are used to look at the course of the disorder. Feighner *et al.* argued, as had Kraepelin, that if there are marked differences in outcome in patients having the same diagnosis, then it is likely that this diagnosis is obscuring the presence of more than one mental disorder.

5. *Family studies.* Another major step in demonstrating the validity of a disorder is to check the family histories of the patients with the disorder. These family histories should be homogeneous across the patients who actually have the same mental disorder.

The most remarkable feature of the Feighner paper was its impact on subsequent psychiatric literature. Sociologists have developed *citation analysis* to aid in their understanding of the dissemination of scientific knowledge (Menard, 1973). In citation analysis, sociologists look for the journal articles that cite a particular target publication. It is assumed that the impact, but not necessarily the importance, of a paper can be assessed through a study of its citations. If an article is frequently cited, then the article has probably had a wide readership and has affected the thinking of those who have read it.

By means of citation analysis, it is easy to show that most scientific journal articles have relatively little impact. Most published papers average only one or two citations per year in subsequent journal articles. For instance, the typical article published during 1973 in the *Archives of General Psychiatry* received 2.1 citations three years later.

In contrast, during the first ten years after its publication (1972–1982), the Feighner paper received about 1,650 citations. In order to give the reader some idea of how impressive this rate is, Figure 4 shows the number of citations per year between 1972 and 1981.* In addition, the number of citations per year of the Feighner paper grew rapidly and steadily. In 1973, there were 15 citations, and in 1980, there were over 200. Most papers have a peak in the number of citations four or five years after publication. After that time, the frequency of citations usually decreases. In 1981, nine years after publication, the frequency of citations had not yet reached a discernible final peak.

For the purpose of comparison, two other citation frequencies are shown. The first is a plot of the average number of citations per year of another article published in the *Archives of General Psychiatry.* This article was selected from the same volume as the paper by Feighner *et al.* (1972). The latter paper was listed under the same heading in the index of the journal (i.e., "Mental disorder, diagnosis"). The article was by Fleiss, Spitzer, Endicott, and Cohen and concerned a statistical measure of interclinician reliability when there are more than two diagnosticians. The Fleiss *et al.* paper had only had 15 citations in the eleven years after it was published, whereas during the same period the Feighner paper had nearly 90 times that number.

*The abscissa is defined as the number of years after publication. Thus *2* means "two years after publication."

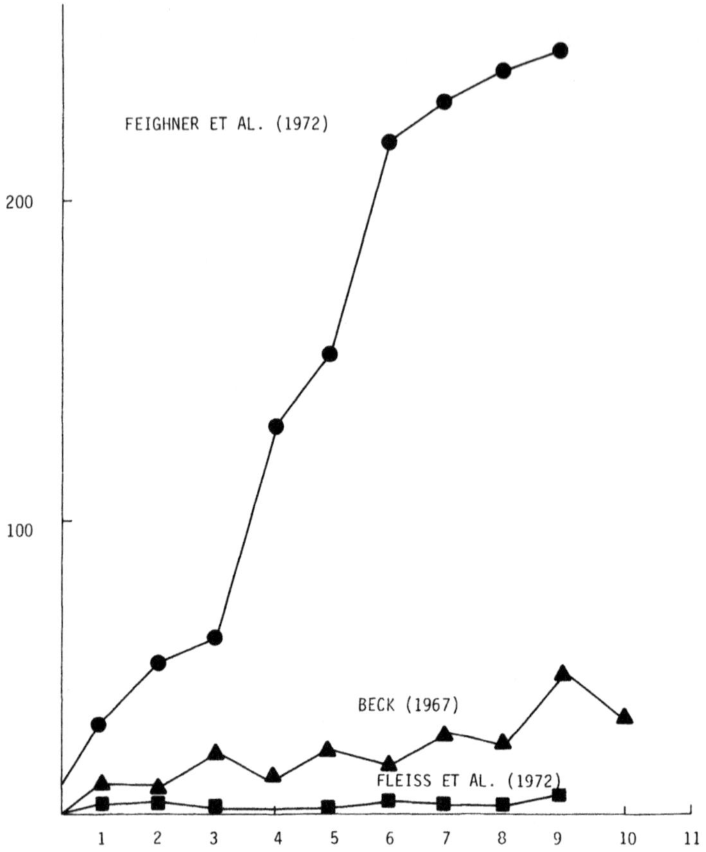

FIGURE 4. Citations of Feighner *et al.* (1972), Fleiss *et al.* (1972), and Beck (1967), as a function of years since publication.

For comparison purposes, the citation frequency of Aaron Beck's book *Depression* (1967) is also reported. The book by Beck was selected because it is considered a major contribution to knowledge about affective disorders and because it has become a standard reference in the study of psychopathology. Thus, Beck's book is important, and this importance should be reflected in a reasonably high rate of citations. As can be seen in Figure 4, the frequency of citations of Beck in the nine years after its publication was much higher than that of the average journal article (2.1 citations) in the *Archives*. Nonetheless, the Feighner paper still had over 10 times the total number of citations in its first seven years of the book by Beck.

Sociology of Science and the Feighner Paper

An obvious question follows from the citation data for the Feighner paper. Why has the Feighner paper had such a dramatic impact? Two answers are suggested.

One explanation concerns the sociological issues involved in the acceptance and the dispersion of new ideas within a scientific field. Sociological issues are generally ignored in commentaries on psychiatric classification. The reason is that classification is considered an integral part of the "science" of psychopathology. Thus, classification, like other aspects of science, is felt to be above such sociological influences as politics.

In contrast, modern historians of science, such as Hull, Tessner, and Diamond (1978), have suggested that there are two types of forces that can be used to account for the changes within any science. The first is the *cognitive* (internal, scientific) force, based on the quality and the amount of evidence supporting any innovation in a science before it is adopted. It is the cognitive aspect of science that researchers are trained to use in order to improve the state of knowledge within any area of a science. The cognitive aspects of science are (1) the accumulated scientific knowledge, such as that reported in scientific journals; (2) the standards of science by which new data and methods are evaluated; (3) the scientific theories that stimulate new discoveries and ideas; (4) the basic problems in need of solution through future research; and (5) the response by the scientific community to new contributions (Cole & Zuckerman, 1975).

The second force involved in scientific change is the *noncognitive* (external, extrascientific). Noncognitive issues include religious beliefs, metaphysical conceptions, socioeconomic issues, and internal politics, all of which Hull *et al.* asserted can be involved in the changes within a science. An example of a noncognitive factor influencing science is a debate between Newton and Leibniz over which had precedence in the discovery of calculus (Broad, 1981). Both had published works in the late seventeenth century explaining the principles of calculus. The book by Leibniz was actually published earlier than the publication by Newton. Newton, when starting his studies to develop calculus, had sent a letter to Europe explaining some of his ideas. Newton claimed that Leibniz had plagiarized his ideas, even though no contemporary evidence exists that shows that Leibniz had ever seen Newton's letter. In 1710, at Newton's request, the Royal Society of London decided to form an impartial panel to adjudicate any differences between these two prominent scientists. Despite the supposed impartial nature of the panel, Newton picked its membership in order to make his victory relatively certain. After this "unbiased" panel decided in Newton's favor, Newton wrote the final report by the panel and published it with only the names of the panel members as authors. As an aside, it is interesting to note that, in fact, probably neither man was the earliest person to publish a treatise on calculus. A Japanese mathematician named Seki Takakazu actually published an earlier work. Newton and Leibniz were not aware of the work by Takakazu (Kondo, 1981). This personal battle between two important scientists of the seventeenth and eighteenth centuries did not directly change calculus. However, it clearly did injure the prestige of Leibniz among his contemporaries, and it also had the effect of absorbing a great deal of the time of these two geniuses.

Differentiating the importance of the cognitive and the noncognitive forces associated with any particular scientific change is not an easy task. Generally, the extrascientific aspects of change have been ignored. However, a new speciality has developed called the *sociology of science*. This new discipline assumes that scientists are human beings, and that the interpersonal behavior of scientists is governed by the same factors that govern interpersonal behavior within other groups of human beings. A number of principles have been suggested by sociologists of science, only a few of which are briefly summarized here.

The first principle is that the literature of science grows exponentially. In particular, the number of scientific journal articles doubles every 15 years. This means that as much was published in the literature of science between 1965 and 1980 as was published in all the time before 1965—a rather staggering fact when one considers how much one will need to read in the next 15 years simply to keep abreast in one's chosen speciality. Not only the literature of science grows exponentially; so does the number of scientists. Price (1963) noted that "80 to 90 percent of all the scientists that have ever lived are alive now" (p. 1). By the mid-twentieth century, there were approximately 1 million persons in the United States who would define themselves as scientists. In 1900, this number was 100,000; in 1850, 10,000; and in 1800, only 1,000. This exponential growth of science has had numerous effects on our society and on the conduct of science itself. Currently, however, there is an indication that this exponential growth has reached its natural limits, and that future growth will occur at a less furious rate (Menard, 1973; Price, 1981).

The second principle of the sociology of science is what Merton

(1968) called the "Matthew effect." This effect is named for a verse in the Gospel according to Saint Matthew that says, "For unto every one that hath shall be given, and he shall have abundance; but from him that hath not shall be taken away even that which he hath" (Matthew 25:29). Because of the Matthew effect, most persons who publish journal articles receive very few citations to their articles (they hath not). On the other hand, there is a smaller number of persons who also publish articles in scientific journals and whose articles frequently attract a large number of citations (they hath abundance). It is the latter group of authors who are relatively likely to be well known, to be considered authorities in their fields, to be on journals' editorial boards, to receive grants, and so on. It is the latter persons who are relatively "rich" in terms of the medium of science; and as the verse of Matthew implies, "the rich get richer while the poor get poorer." A specific statement of the Matthew effect was made by Price (1965), who suggested that 50% of all citations are made of articles published by only 10% of the authors within a given field; on the other hand, about half of all authors in a literature receive only approximately 10% of the citations. The Matthew effect has been well documented and directly concerns any discussion of how advantage accumulates for certain researchers within a field.

The third principle in the sociology of science is what has been called the existence of *invisible colleges*. The concept of the *invisible college* was first used in the seventeenth century to refer to the collection of scientists who eventually formed the Royal Society of London (Cole & Zuckerman, 1975). The noun *college* was used to describe such a group of scientists because, like the faculty of a college, the group represented a collection of intellectuals who had a sense of allegiance to each other and frequently interacted both professionally and socially. The adjective *invisible* was used because the membership of the group was not obvious to persons who had little knowledge of seventeenth-century science. Price (1963) reintroduced the term *invisible college* in order to describe the existence of such groups in the modern sciences. These groups are the collections of scientists who often attend the same conferences, who publish in the same journals, who invite each other to give presentations at their home institutions, and who share preprints of their research endeavors. Price also argued that it is through the political power of such "colleges" that many of the important changes in a science are made.

The neo-Kraepelinian movement represents an invisible college within the field of psychiatry. A demonstrated earlier, this college apparently had nuclei in St. Louis and New York, which spread out to

TABLE 2. Persons Who Cited Feighner *et al.* (1972) More Than Once between 1972 and 1974

Briscoe, C. A.	Murphy, G. E.
Dunner, D. L.	Schildkraut, J. J.
Feighner, J. P.	Schuckit, M. A.
Goodwin, D. W.	Spitzer, R. L.
Halmi, K. A.	Tauson, V. B.
Kashiwag, T.	Welner, A.
Liss, J. L.	Winokur, G.
Maas, J. W.	Woodruff, R. A.
Morrison, J. R.	

Iowa and Southern California. Of the six coauthors who wrote the Feighner paper, three have been identified as central members of the neo-Kraepelinian perspective: Robins, Guze, and Winokur.

The neo-Kraepelinian college was largely responsible for the initial promotion of the Feighner paper. Table 2 contains a list of 17 authors who were the senior authors of two or more journal articles published between 1972 and 1974 that cited the Feighner paper. When the names in this table are compared with the names in Figures 1, 2, and 3, a correspondence of 8 of the 17 names is noted (Goodwin, Morrison, Murphy, Schuckit, Spitzer, Welner, Winokur, and Woodruff). Among the remaining nine names, six had been residents at or were located at Washington University, the University of Iowa, or New York State Psychiatric Institute (Dunner, Feighner, Halmi, Liss, Schildkraut, and Tauson). In short, all but three of the persons who began a multiple referencing of the Feighner paper soon after its publication can be directly connected with the invisible college described by the name *neo-Kraepelinian.*

Another interesting feature about the early citations of the Feighner paper is that most were made in articles published in the *Archives of General Psychiatry* (23% of the articles published in the years 1972–1974). This journal has continued to be the most frequent source of citations of the Feighner paper over the 1970s. There are 12 researchers on the editorial board of the *Archives.* As of 1974, none of these board members was from the neo-Kraepelinian movement. However, in 1975, two neo-Kraepelinians, Klein and Winokur, were added to the editorial board. In addition, one other member of the editorial board could be considered at least sympathetic with this invisible college (Gerald Klerman, the person who first gave this movement its name).

After its initial promotion by adherents of the neo-Kraepelinian perspective, the number of citations of the Feighner paper showed rapid growth during the mid to late 1970s (see Figure 4); the Matthew effect was very likely a contributing reason. The Feighner paper had three prolific coauthors whose publications had already been demonstrated as "having abundance." In addition, the relatively large number of early citations of this paper increased the probability that the article would attract the attention of other researchers, as would the fact that two adherents of the neo-Kraepelinian perspective became members of the editorial board of an influential psychiatric research journal. The initial popularity of the Feighner paper certainly helped add to its increasingly impressive popularity during the late 1970s.

One final note should be made about the Feighner paper. If the rate of growth of citations from Figure 4 is calculated, one will note that the number of citations was doubling about *once a year* for the first four years after its publication. By the end of the 1970s, this rate of growth had slowed to doubling approximately *once every seven years.* In other words, the citations of the Feighner paper followed the exponential growth pattern of all scientific literature as noted by sociologists of science (Price, 1963). Additionally, the rate of this exponential growth was much faster than would be expected from normative studies of the literature growth, which show a doubling rate of about once per 15 years. This rapid growth is particularly notable because it has been well recognized that the classification of mental disorders has been a topic of relatively little interest during most of the twentieth century (White, 1973); that is, the literature on psychiatric classification had shown very little growth during the 1900s. Therefore, the rapid growth of citations of the Feighner paper is in striking contrast to the prior lack of interest in classification, and it underscores the impact of the neo-Kraepelinian perspective in helping to stimulate a resurgence of interest in classification.

Diagnostic Criteria and the Feighner Paper

The second major explanation for the popularity of the Feighner paper is the fact that it used relatively explicit diagnostic criteria. The diagnosis of various mental disorders has always been a troublesome issue in the mental health field. Most of the time, clinicians must depend on the self-report of patients. This is a problem because patients may find it difficult to accurately characterize the issues that have led them to see a clinician, and they may consciously or unconsciously provide deceptive information to the clinician. Psychiatry, unlike most medical specialities, must make most diagnoses without the assistance of definitive laboratory studies. The Feighner paper held an attraction in that it proposed relatively explicit diagnostic criteria. The definitions of categories in the glossaries of the DSM-I, the DSM-II, the ICD-8, and the ICD-9 were vaguely worded prose statements. Consider, for example, the definition of schizophrenia in the DSM-II:

> This large category includes a group of disorders manifested by characteristic disturbances of thinking, mood and behavior. Disturbances in thinking are marked by alterations of concept formation which may lead to misinterpretation of reality and sometimes to delusions and hallucinations, which frequently appear psychologically self-protective. Corollary mood changes include ambivalent, constricted and inappropriate emotional responsiveness and loss of empathy with others. Behavior may be withdrawn, regressive and bizarre. The schizophrenias, in which the mental status is attributable primarily to a *thought* disorder, are to be distinguished from the *major affective disorders* which are dominated by a mood disorder. The *Paranoid states* are distinguished from schizophrenia by the narrowness of the distortions of reality and by the absence of other psychotic symptoms. (American Psychiatric Association, 1968, p. 33)

Notice that the general symptoms that the DSM-II uses to define schizophrenia (i.e., "disturbance of thinking, mood and behavior") are so broad that all mental disorders could be subsumed under the heading of schizophrenia. The DSM-II definition contains 14 symptoms in its definition of schizophrenia:

A = disturbance of thinking
B = disturbance of mood
C = disturbance of behavior

D = alteration of concept formation
E = misinterpretation of reality
F = delusions
G = hallucinations
H = ambivalent emotional responsiveness
I = constricted emotional responsiveness
J = inappropriate emotional responsiveness
K = loss of empathy
L = withdrawn behavior
M = regressive behavior
N = bizarre behavior

Symptoms D through G imply A; symptoms H through K imply B; and symptoms L through N imply C. A problem with the above DSM-

II definition is that it fails to specify how many of which symptoms are sufficient for a diagnosis of schizophrenia. What if the patient exhibits all 14 symptoms? Clearly, in this case, the patient would meet the DSM-II definition. But requiring that all 14 of these symptoms be present is too restrictive. Few patients, including those that clinicians would agree are schizophrenic, have all 14 symptoms. If requiring 14 is too restrictive, then how many are necessary? Would 1 symptom be sufficient? Probably not. Most clinicians would not be willing to diagnose a patient who is exhibiting "withdrawn behavior" as schizophrenic simply on the basis of this one symptom. Depressed patients, patients with an organic brain syndrome, normal persons under psychological stress—all tend to withdraw. Diagnosing any person exhibiting behavioral withdrawal as schizophrenic would lead to an overly inclusive definition of the category. If requiring all 14 symptoms leads to an overly restrictive definition and requiring only 1 symptom yields too broad a definition, exactly how many symptoms are necessary: 3? 6? 11? The DSM-II, with its prose definition of diagnostic categories, does not provide adequate answers to these questions. Thus, the failure of the DSM-II to provide the clear conditions under which a diagnosis should be made is a fundamental fault that obscures its clarity when applied in daily clinical practice.

Compare the DSM-II definition with the definition of schizophrenia as proposed in the Feighner paper (see pp. 37–38). Feighner *et al.* listed nine characteristics associated with schizophrenia. Their decision rules clearly specify which of these characteristics are necessary and how many of the characteristics are sufficient for the diagnosis of schizophrenia. In short, the Feighner criteria appear to have more clearly specified meanings than do the prose definitions of the DSM-II.

The relative attractiveness of the Feighner criteria is supported by analyzing 50 randomly selected articles which cited Feighner *et al.*; 38 of these 50 articles used the Feighner criteria to assign diagnoses to patients in the research sample. For example, White, McLeod, and Davidson (1976) used the Feighner criteria for schizophrenia in order to define their research sample in a study of monoamine-oxidase activity in schizophrenics. A related use of the Feighner criteria was made by Kelly, Koch, and Buegel (1976), who diagnosed mania *à la* Feighner *et al.* in an adolescent patient and subsequently began lithium therapy. A third study, by Liss, Welner, and Robins (1973), used the Feighner criteria to assign diagnoses in a study comparing the psychiatric symptoms of white and black inpatients. From the 50 articles selected, it would appear that the Feighner criteria have been most commonly used to define depression, mania, schizophrenia, and anorexia nervosa.

An important effect of the Feighner paper is that it was responsible for stimulating the use of explicit criteria in the DSM-III. Shortly after the Feighner paper was published, one of its coauthors, Robins, collaborated with Spitzer and Endicott to create a new classification of schizophrenic and affective disorders. This classification was titled the *Research Diagnostic Criteria* (RDC) (Spitzer *et al.*, 1975). The RDC definitions of classificatory categories were modeled after the Feighner criteria.

Along with the RDC, a structured interview was created to elicit the symptoms necessary for arriving at a diagnosis with the RDC. This interview was the Schedule for Affective Disorders and Schizophrenia (SADS). Both the RDC and the SADS became heavily used in research concerning schizophrenia and depression that was sponsored by the National Institute of Mental Health.

The RDC was also important because it became a precursor of the DSM-III. The first version of the DSM-III incorporated the RDC definitions of many categories. Thus, the DSM-III used the Feighner-type definition for all classificatory categories. The use of explicit criteria has been noted as one of the major changes in the DSM-III when it has been compared with the DSM-II (Spitzer & Sheehy, 1976; Spitzer, Williams, & Skodol, 1979).

To summarize, there are two major reasons that the Feighner paper became so popular. The first is that the Feighner paper was written by a collection of authors who were prominent in the neo-Kraepelinian movement and that many of the initial citations of the paper came from persons sympathetic with this movement. The second reason is that the paper's diagnostic criteria were perceived as providing distinctly more explicit definitions of classificatory categories than the prose definitions in the DSM and the ICD.

Diagnostic Criteria and Reliability

The intent of the Feighner paper was to overcome a major problem in psychiatry associated with the use of vague and nonspecific definitions. The problem of vague definitions has been recognized as a serious issue for some time. For instance, in 1952, the same year that the DSM-I was published, Reid and Finesinger published a paper in the *American Journal of Psychiatry* titled "The Role of Definitions in

Psychiatry." Reid and Finesinger started their paper with this paragraph:

> "If you would talk with me," said Voltaire, "you must define your terms." This advice, while often heard, is seldom heeded. But perhaps we psychiatrists, particularly, should take it to heart and mend our verbal ways. For the commonest criticism leveled against us by medical students, by professional colleagues, and by the general public, is that it is impossible to understand our technical verbiage. We are likely to ascribe these difficulties in communication to ignorant prejudice and to unconscious resistance. No doubt these factors are important; but psychiatrists, like other mortals, are given to projection as a form of self-defense and, if we look at the facts squarely, some of the responsibility for this failure to communicate more effectively must be laid to our own use or misuse of language. (p. 413)

If the Feighner criteria have improved psychiatric classification by increasing the specificity of category definition, then the reliability of a classification using the 16 Feighner categories should be better than that noted with DSM-I or DSM-II definitions. Good reliability of diagnostic categories is essential if a classification system is to be used as a basis for communication. A classification system whose usage varies more from user to user than from category to category is of little value as a basis for communication among clinicians (Blashfield & Draguns, 1976a).

The most detailed study of the reliability of the Feighner criteria was performed by members of the St. Louis group (Helzer, Robins, Taibleson, Woodruff, Reich, & Wish, 1977; Helzer, Clayton, Pambakian, Reich, Woodruff, & Reveley, 1977). These investigators administered a structured interview that contained 150 items designed to elicit information for rating the symptoms listed in the Feighner criteria. One hundred and one patients randomly selected from consecutive admissions to a private, short-term inpatient unit at Renard Hospital in St. Louis were interviewed twice. One clinician interviewed a patient on the day after admission; a second clinician performed the same structured interview on the next day. Three clinicians (Clayton, Helzer, and Pambakian) were systematically rotated into all possible pairings. Two other clinicians (Reich and Woodruff) checked the interviews for completeness.

The results of the Helzer *et al.* study are shown in Table 3. Four columns are presented in this table: percentage of overall agreement, percentage of specific agreement, kappa values for that study, and average kappa values computed from earlier reliability studies. The percentage of overall agreement ("overall agreement %") refers to the total percentage of cases in which the two clinicians agreed on the

TABLE 3. Reliability of the Feighner Criteria

Diagnosis	Overall agreement %	Specific agreement %	Kappa	Kappa from earlier studies
Depression	86	84	.66	.33–.41
Mania	96	73	.82	.33–.41
Anxiety neurosis	89	73	.76	.45
Schizophrenia	96	43	.58	.57
Antisocial personality	95	72	.81	.53
Alcoholism	88	77	.74	.71
Drug dependence	96	76	.84	
Hysteria	88	68	.72	
Obsessional illness	97	67	.78	
Homosexuality	99	75	.85	
Organic brain syndrome	92	20	.29	.77
Undiagnosed psychiatric illness	76	20	.19	
Average concordance	92	62	.66	

Note. These results are from Tables 2 and 3 of Helzer, Clayton, Pambakian, Reich, Woodruff, and Reveley (1977). The values of "Kappa from earlier studies" are the summary values from a review of reliability studies by Spitzer and Fleiss (1974).

diagnosis. For instance, there were 29 cases of the total 101 patients that both clinicians agreed were alcoholic. In addition, there were 60 cases that both agreed were not alcoholic. Thus, the percentage of overall agreement was 100 × (29 + 60)/101 = 88% for alcoholism.

Another statistic reported was the percentage of specific agreement. Notice that most of the cases in which agreement occurred concerning alcoholism were cases that neither clinician considered alcoholic. In other words, most agreement concerned negative cases. The second statistic, "specific agreement" refers to the rate of agreement for cases in which both clinicians positively diagnosed the disorder in question. In the Helzer study, there were 12 cases that one clinician, but not the other, considered alcoholic. Thus, the percentage of specific agreement was computed as 100 × 29/(29 + 12) = 71%. (Remember that there were 29 cases that both clinicians had diagnosed as alcoholic.)

The third statistic reported is kappa. This statistic was suggested by Cohen (1960) as a measure of classification reliability. Essentially, it is a correlation-like measure that has a value of $\kappa = 0.0$ when clinicians show random agreement and $\kappa = 1.0$ when clinicians agree perfectly. The exact meaning of kappa and its strengths and weaknesses are discussed in Chapter 4. The value of kappa as an estimate of the reliability of the diagnostic criteria for alcoholism was $\kappa = .74$.

The fourth column shows the average kappa value as compiled in a review by Spitzer and Fleiss (1974). These authors reviewed nine previous reliability studies that used the DSM-I or the DSM-II to define diagnostic categories. The average reliability for the diagnosis of alcoholism in these studies was $\kappa = .71$ (range: $\kappa = .56$ to $\kappa = .87$). Thus, the Helzer study suggested that the use of the Feighner criteria did not lead to any major improvement in the diagnosis of alcoholism, which already had reasonably high reliability. Notice, however, that the use of the Feighner criteria seems to have led to a distinct improvement in the diagnosis of depression (.41 in earlier studies, compared with .55 found by Helzer) and of mania (.33 in earlier studies, had improved to .88). The affective disorders had been the major source of clinician disagreement in earlier reliability studies (Sandifer, Pettus, & Quade, 1964).

In fact, the Helzer *et al.* study found equal or higher reliability values than those usually observed in earlier studies for all categories except one. The only exception was the major category of organic brain syndrome (OBS). Spitzer and Fleiss (1974) found OBS to be the category with the highest average reliability values in previous reliability studies ($\kappa = .77$, with range of $\kappa = .62$ to $\kappa = .87$). However, in the Helzer study, the kappa value for OBS was only $\kappa = .29$. Helzer *et al.* argued that a reason for the low reliability associated with the Feighner definition of OBS could be that key terms such as *impairment of orientation, impairment of memory,* and *deterioration of mental functioning* were not defined.

Important to note here is that the use of diagnostic criteria does not guarantee high reliability. The danger of using such criteria is to assume, without empirical data, that the use of these criteria must necessarily improve the definition of a diagnostic category. Compare, for instance, the definition of organic brain syndrome as offered by Feighner *et al.* with the definition from the DSM-II.

Feighner *et al.* (1972) Criteria for Organic Brain Syndrome

This diagnosis is made when either criterion A or criterion B is present.

A. Two of the following manifestations must be present. (In the presence of muteness the diagnosis must be deferred.) (1) Impairment of orientation. (2) Impairment of memory. (3) Deterioration of other intellectual functions.

B. This diagnosis is also made if the patient has at least one manifestation (A) in addition to a known probable cause for organic brain syndrome. (p. 61)

DSM-II Definition of Organic Brain Syndromes

> These disorders are manifested by the following symptoms: (a) Impairment of orientation; (b) Impairment of memory; (c) Impairment of all intellectual functions such as comprehension, calculation, knowledge, learning, etc.; (d) Impairment of judgment; and (e) Lability and shallowness of affect.
>
> The organic brain syndrome is a basic mental condition characteristically resulting from diffuse impairment of brain tissue function from whatever cause. Most of the basic symptoms are generally present to some degree regardless of whether the syndrome is mild, moderate or severe.
>
> The syndrome may be the only disturbance present. It can be associated with psychotic symptoms and behavioral disturbances. The severity of the associated symptoms is affected by and related to not only the precipitating organic disorder but also the patients' inherent personality patterns, present emotional conflicts, his environmental situation, and interpersonal relations. (American Psychiatric Association, 1968, p. 22)

No increase in clarity is obvious when contrasting the Feighner definition with the DSM-II definition for organic brain syndrome. If anything, the DSM-II is more specific. The reason for making the comparison is to emphasize the point that the use of diagnostic criteria *à la* Feighner *et al.* does not guarantee high specificity and reliability.

Diagnostic Criteria and Coverage

Another issue concerning the use of diagnostic criteria can also be noted from an analysis of the Helzer *et al.* study: the coverage of classificatory categories that are defined by means of specific criteria. *Coverage* has been defined as the percentage of patients in a general psychiatric sample to which a given diagnosis has been applied (Blashfield, 1973). The coverage of the 16 categories defined by the Feighner criteria are shown in Table 4. These values were extrapolated from tables presented by Helzer *et al.* (i.e., Helzer and his associates did not address the issue of coverage in their article, and no data were presented relevant to coverage).

Depression was the category in the Helzer *et al.* study that had by far the greatest coverage. Both clinicians used this diagnosis on 74 of the 101 patients; depression was mentioned as a diagnosis by at least one clinician for 88 of the 101 patients. Only 13 patients were not given a diagnosis of depression by at least one interviewer.

This strikingly high rate of coverage has two possible explanations. The first possibility is that depression was in fact a very common syndrome on the inpatient unit at Renard Hospital (St. Louis is a depressing city?). The other explanation is that the definition of depression offered by Feighner *et al.*, although reliable, is too broad and

TABLE 4. Coverage of Feighner Criteria

Diagnosis	Percentage of patients to which this diagnosis was assigned by both clinicians
Depression	74
Mania	11
Anxiety neurosis	30
Schizophrenia	3
Antisocial personality	3
Alcoholism	29
Drug dependence	13
Hysteria	8
Obsessional illness	6
Homosexuality	3
Organic brain syndrome	2
Undiagnosed psychiatric illness	6

"covered" a number of patients who would not normally be diagnosed as having this syndrome. Helzer *et al.* did not directly comment on the high rate of coverage for depression. However, they did say that multiple diagnoses were common in their study. In fact, each patient received a mean of 2.5 diagnoses.

Elsewhere, however, members of the St. Louis group have commented on the wide coverage associated with their definition of depression. They have noted that depression is the major reason that people seek psychiatric assistance in their clinical settings. Here, the St. Louis group appears to have been confusing the use of the term *depression* to refer to a symptom and its use when referring to a diagnostic category. Certainly, their claim that depression is a major reason for self-referral is correct if depression refers to a symptom. Depression is a subjectively unpleasant state, and the symptom of depression is common in almost all mental disorders. However, the fact that depression as a symptom is omnipresent among psychiatic patients does not mean that depression as a mental disorder is widespread. This comment on the confusion between the possible meanings of the term *depression* is made here simply to demonstrate the pervasive nature of the nomological problems associated with words, concepts, and classificatory categories in discussions of psychopathology. The St. Louis group contains persons who think relatively precisely about classification, yet even they can apparently fall prey to these nomological confusions. In fact, it is the presence of a depressive syndrome that generally stimulates alcoholics and neurotics to refer themselves for psychiatric or psychological treatment (Woodruff, Clayton, & Guze, 1975). In their article "Is Everyone Depressed?", Woodruff *et al.* studied 900 relatives of 500 patients in their clinics. Of these 900, 55% were found to have no mental disorders. Only 10% of the relatives met the Feighner criteria for depression. Woodruff *et al.* concluded that "everybody is *not* depressed" when their criteria are used.

Concerns about the coverage of the Feighner criteria are not limited just to the broad coverage of depression. The diagnosis of schizophrenia was assigned by both interviewers to only 3 of the 101 admissions. This was a low rate of coverage for an inpatient unit, as inpatient units usually contain a relatively large number of schizophrenics. Again Helzer *et al.* did not comment on this result.

The observation of relatively low coverage associated with the Feighner criteria for schizophrenia was not restricted to the Helzer study. For instance, Morrison, Clancy, Crowe, and Winokur (1972) used the Feighner criteria retrospectively to assign diagnoses to patients admitted to the Iowa State Psychopathic Hospital from 1935 to 1944. This study focused on patients who were diagnosed as having affective or schizophrenic disorders. Morrison *et al.* found that the Feighner criteria for schizophrenia could be applied to only 37% of the patients who had a chart diagnosis of schizophrenia. The reasons for most exclusions were (1) the episodic course of the patient's disorder (17% of the exclusions); (2) the patient's symptomatology's lasting less than the required six months (53%); and (3) the possibility of a concurrent affective disorder (14%).

Recently, two other studies have commented on the poor coverage of the Feighner definition of schizophrenia. Brockington, Kendell, and Leff (1978) compared various definitions of schizophrenia on two samples gathered in Great Britain. The samples contained those patients for whom the authors believed that functional psychosis was a diagnostic possibility. When the Feighner criteria were used, only 6% of the sample could be diagnosed as schizophrenic. This was relatively low coverage; that is, 23% of the sample had a hospital diagnosis of schizophrenia. Additionally, Brockington *et al.* found that the Feighner definition did not demonstrate high concordance with any other definition of schizophrenia. The kappa values used to compute the agreement between patients diagnosed as schizophrenic by the Feighner *et al.* definition and diagnosed by other definitions (e.g., the RDC, Schneider, and Wing) ranged from $\kappa = .08$ to $\kappa = .44$, with an average of $\kappa = .24$. The interrater reliability of the Feighner definition was only $\kappa = .44$. In contrast, most other definitions had reliability values exceeding $\kappa = .60$. Like Morrison *et al.*, Brockington *et al.* concluded that the Feighner definition was overly restrictive, particularly in requiring a six-month duration of symptoms.

In a related study, Overall and Hollister (1979) compared six definitions of schizophrenia using diagnostic criteria. The sample consisted of 166 patients involved in a multicenter drug-research study who had received a project diagnosis of schizophrenia. Of the 166, 44 (26.5%) met the Feighner criteria for schizophrenia. Along with the low coverage, Overall and Hollister found poor concordance of the Feighner definition with any of the other five definitions of schizophrenia.

The problems associated with the coverage of the Feighner criteria should not be surprising. Coverage and reliability are logically related concepts. Blashfield (1973) has argued that classificatory concepts with broad coverage are often unreliable because the patient populations associated with the two concepts are likely to overlap. On the other hand, the more precisely a classificatory concept is defined, the more likely it is to be reliable, but its precise meaning usually restricts the coverage of the patients to whom it is assigned. The use of prose definitions in the DSM-II meant that the coverage of its classificatory concepts could be as broad as subjective judgment would allow. Under these conditions, reliability could suffer. The Feighner criteria, in contrast, generally have more precise definitions, and apparently better reliability, but their coverage is altered. In fact, the definition of schizophrenia by Feighner *et al.* appears to be overly narrow and restrictive.

CONCLUDING COMMENTS

In this chapter, the rather considerable influence of the neo-Kraepelinian movement has been demonstrated through an analysis of the diagnostic criteria in the Feighner *et al.* (1972) paper. The influence of this paper, and therefore of the neo-Kraepelinian movement, was documented by (1) the strikingly large number of citations of the Feighner paper over a short period of time; (2) the adoption of diagnostic criteria modeled after those of Feighner *et al.* in the RDC and the DSM-III; and (3) the subsequent research, which has attempted to study the adequacy of the criteria.

The neo-Kraepelinian movement and its impact on classification provide a dramatic example of a point made by Deutsch (1966). He suggested that most changes in the language system of a science are brought about by a small group of scientists forming a collective body of related interests. When this group solidifies its ideas, it creates a jargon. This jargon often involves new classificatory concepts or new definitions of old concepts. As the power and prestige of this group increases within its scientific domain, its new jargon gains acceptance.

This process of change describes the growth of influence of the neo-Kraepelinians. The psychiatrists who favored a biological approach to psychiatry and who believed in the medical model began to form an "invisible college." Many of psychiatrists were at or had been associated with Washington University in St. Louis, the University of Iowa, and the New York State Psychiatric Institute. Because of their emphasis on science and research, this group became concerned with classification and description. As a result, they subsequently proposed a set of diagnostic criteria that could be used in research. The influence of these neo-Kraepelinians spread, and their diagnostic criteria gathered a great deal of attention, evidenced by frequent citations as well as incorporation within the newest official psychiatric classification (i.e., the DSM-III).

Despite widespread popularity, diagnostic criteria do not adequately resolve many of the apparent problems surrounding psychiatric classification. First, the use of diagnostic criteria does not guarantee high reliability. This point is dramatically illustrated by the low reliability associated with the Feighner criteria for *organic brain syndrome* (Helzer *et al.*, 1977). Second, the issue of coverage becomes a problem in studies using diagnostic criteria. For instance, the Feighner *et al.* definition of *schizophrenia* led to a very narrow interpretation of this category. In contrast, the definition of depression by Feighner *et al.* may be too broad.

SUGGESTED READINGS

Rosenhan, D. L. On being sane in insane places. *Science*, 1973, *179*, 250–258.

Weiner, B. "On being sane in insane places": A process (attributional) analysis and critique. *Journal of Abnormal Psychology*, 1975, *84*, 433–441.

Spitzer, R. L. On pseudoscience in science, logic in remission and psychiatric diagnosis: A critique of Rosenhan's "On being sane in insane places." *Journal of Abnormal Psychology*, 1975, *84*, 442–452.

Crown, S. "On being sane in insane places": A comment from England. *Journal of Abnormal Psychology*, 1975, *84*, 453–455.

Millon, T. Reflections on Rosenhan's "On being sane in insane places." *Journal of Abnormal Psychology*, 1975, *84*, 456–461.

Rosenhan, D. L. The contextual nature of psychiatric diagnosis. *Journal of Abnormal Psychology*, 1975, *84*, 462–473.

Farber, I. E. Sane and insane: Constructions and misconstructions. *Journal of Abnormal Psychology*, 1975, *84*, 589–620.

The journal articles listed above represent an interesting and varied collection of ideas about the diagnostic process and the criticism that labels are self-fulfilling prophecies. The stimulus for this set of articles was Rosenhan's paper in *Science* (1973). Rosenhan discussed two major topics in his original paper: (1) the validity of contemporary psychiatric diagnosis and (2) the depersonalization that the pseudopatients noted during their stays in the hospitals. Almost all responses to Rosenhan (1973) have focused on the first point.

Of the responses to the Rosenhan article, the most emotional is Spitzer's. He was determined to defeat Rosenhan and to demonstrate that contemporary psychiatric diagnosis is meaningful. Spitzer's attack, though often telling, is poorly organized. A more balanced critique of Rosenhan is Weiner's article, which uses attributional theory from social psychology to show the limitations of Rosenhan's approach. Rosenhan himself, in his 1975 article, wrote from a social psychological perspective and emphasized the contextual nature of the diagnostic process. His 1975 conclusions were not nearly as radical as the conclusions he had reached in the original 1973 article.

The last paper, by Farber, is the most thorough analysis of the Rosenhan paper. Farber explored ideas about disease and behaviorism, about reliability and validity, and about the implications of labeling. Farber concluded that the Rosenhan paper brought into relief some of the crucial philosophical issues concerning classification, but that the methodological limitations of the Rosenhan demonstration certainly did not warrant his conclusion that "we cannot distinguish the sane from the insane in psychiatric hospitals" (Rosenhan, 1973).

Price, D. S. *Little science, big science*. New York: Columbia University Press, 1963.

Price's short book is a collection of remarkable lectures given by him at the Brookhaven National Laboratory. In his preface, Price stated that his goal was to discuss how science, a fairly small part of society only two centuries ago, has become such a large part of modern society. In his sociological analysis of science, Price discussed the parameters of its growth, introduced the concept of invisible colleges, and discussed the politics of science. The book is quite readable and contains striking insights into the process of science.

A later work by Menard (1973), called *Science: Growth and Change*, is a more recent expansion of some of Price's ideas. The Menard book is less tightly written, yet it contains a number of intriguing thoughts. Also, Menard provided a number of examples of techniques that can be used in studies associated with the sociology of science.

Matza, D. *Becoming deviant*. Englewood Cliffs, N.J.: Prentice-Hall, 1969.

Of the three general criticisms of psychiatric classification that were being promoted in the 1950s and 1960s, the labeling perspective is the one that is the least thoroughly discussed in this book. Matza's monograph is not a good introduction to the labeling argument. A book by Scheff (1966), *Being Mentally Ill: A Sociological Theory,* is better as an introduction. However, Scheff's approach is disappointingly simpleminded and is likely to convince the sophisticated reader that the labeling perspective is of little value. Matza's monograph, though brief, is much richer and requires much more thought to absorb. It discusses three sociological models for the development of deviance: affinity, affiliation, and signification (labeling). Matza, from his antiscientific perspective, discussed all three, using modern sociological analysis to examine the depth of these models.

Woodruff, R. A., Goodwin, W., & Guze, S. B. *Psychiatric diagnosis.* New York: Oxford University Press, 1974.

Psychiatric Diagnosis is a book version of the Feighner *et al.* paper. The basic descriptive features (e.g., the presenting picture, the natural history, the epidemiology, and the clinical management) of 12 diagnostic categories are itemized; these 12 categories are largely the same as the categories defined in Feighner *et al.* (1972). After each chapter is listed a bibliography that contains much of the neo-Kraepelinian research on the category. There are two appendixes to the book: a slightly modified version of the Feighner criteria and a paper by Guze (1970) discussing the medical model. The book is intended as a primer for those interested in psychopathology, and it contains information that could be useful for training medical students (an interesting parallel to Kraepelin's textbooks).

Chapter 3

The Neo-Kraepelinian Movement and the Medical Model

The following was an informal conversation that occurred before group therapy on an inpatient unit.

PATIENT: What do you do?
CLINICIAN: I am a psychologist.
PATIENT: Oh really, that's interesting. You know, I've never understood the difference between a psychologist and a psychiatrist. What is the difference anyway?
CLINICIAN: Did you ever watch the Bob Newhart Show on TV?
PATIENT: Yes.
CLINICIAN: That's what I do.
PATIENT: (*Puzzled. After a long latency.*) I still don't understand the difference.

The previous chapter introduced the neo-Kraepelinian movement and discussed its impact on modern American psychiatry. This movement was presented through a detailed analysis of the Feighner paper.

In the next chapters, two basic themes that are frequently associated with the neo-Kraepelinian movement are discussed: the medical model and diagnostic reliability. Concerning the medical model, the views associated with the neo-Kraepelinian perspective will be examined and then criticized. The term *medical model* has invoked a sufficiently wide set of connotations and denotations so that there exist many discussions in the contemporary literature. Also the reader should be aware that discussions of the medical model are emotionally loaded. The rational reader who has no particular investment in one view or another may be surprised by the degree of emotionality and the frequency with which interesting, irrational, pseudo, enlightening, simplex, complex, and overtly crazy arguments are intermixed. This chapter will not do justice to the many issues associated with the concept of the *medical model.* Nor will it avoid the irrational complexities that have snared other writers.

THE MEDICAL MODEL

In relation to the medical model, especially its relevance to classification, two issues will be discussed (Sarason & Ganzer, 1968): (1) The meaning of *disease* is the first issue. A central thesis associated with the medical model is that mental disorders are diseases in the same sense that general medical disorders such as syphilis are diseases. (2) The second issue is the conflict over professional jurisdiction associated with the medical model. Manpower, acceptance, earnings, and power are topics that must be addressed as attempts are made to define professional territoriality. As a result, this issue involves the relative status of various mental health professions.

The definition of the term *disease* is relevant to the acceptance or the rejection of the medical model, and to the classification of psychopathology. Without the definition of disease, the boundary of psychiatric classification is fuzzy and unclear. Panzetta (1974) represented the problem of classification associated with the definition of the term *disease.* Suppose, Panzetta argued, that a scientist creates a classification of *zemptgorskies.* His colleagues might question him. "What is a zemptgorsky?" The scientist may respond that he is not certain how to give a precise definition. Perhaps he could say that a zemptgorsky is anything included in his classification.

Panzetta's story makes it clear that without a definition of the most inclusive concept in a classification, the boundaries of the classification are unclear. The scientist did not define a zemptgorsky. To say that a zemptgorsky is whatever is included in his classification does not help. Nor does it help for a clinician to say that a mental disease is anything included in psychiatric classification. Some definition of the concept of disease is necessary.

Despite the realization that the definition of mental disease is important, the neo-Kraepelinians have not found formulating an adequate definition easy. For instance, Samuel Guze, a coauthor of the Feighner paper, proposed three definitions during the 1970s. Each of these definitions is briefly analyzed below.

Definition #1 of Guze

Guze's first definition of disease was proposed in his paper "The Need for Toughmindedness in Psychiatric Thinking" (1970):

> In fact, any condition associated with discomfort, pain, disability, death, or an increased liability to these states, *regarded by physicians and the public as properly the responsibility of the medical profession,* may be considered a disease. (p. 662; italics in original)

In short, Guze was advocating that (1) disease is related to suffering and (2) disease is what physicians treat (Kendell, 1975). The first feature, disease as suffering, refers to the earliest meaning of the term. Historically, the term *dis-ease* referred to human conditions that caused pain, disability, and death. However, as Kendell (1975) pointed out, this historical approach to defining disease is no longer accepted. Clearly, there are persons with diseases who have no awareness of their condition, and they do not experience suffering. In addition, there are a number of persons who complain of pain, of disability, and of psychological distress without having a disease that requires medical treatment.

The other feature in Guze's first definition is that a disease is what physicians treat. Kendell (1975) presented a strong argument against this definition:

> The fact is that any definition of disease that boils down to "what people complain of" or "what doctors treat," or some combination of the two, is almost worse than no definition at all. It is free to expand or contract with changes in social attitudes and therapeutic optimism and is at the mercy of idiosyncratic judgements by doctors or patients. [In other words,] equating illness with other people's "therapeutic concern" implies that no one can be ill until he has been recognized as such, and also gives doctors, and society, free rein to label all deviants as ill, thus opening the door to all the inconsistencies and abuses Szasz has vividly conjured up. (p. 11)*

---

*Kendell was making a reference to Thomas Szasz, an American psychiatrist who has written frequently about the inappropriateness of the medical model as applied to psychopathology. In his first and best known book, *The Myth of Mental Illness* (1961), Szasz argued that psychiatrists have used the medical model to justify their status as moral gatekeepers of society in order to sanctify their value judgments about people. Szasz's writings are flamboyant in style, and his arguments are so charged with emotion-laden terms that separating the cognitive from the affective aspects of his arguments is difficult. By his contemporaries, Szasz is classified with a school of thought

In summary, Guze, in his first definition, argued that disease is related to suffering and that diseases are what doctors treat. Neither of these approaches to the definition of disease is adequate.

Definition #2 of Guze

Later Guze and his colleagues in the St. Louis group formulated another definition of disease. In 1974, Woodruff *et al.* published a book titled *Psychiatric Diagnosis.* The preface of this book contained the following definition:

> When the term "disease" is used, this is what is meant: a disease is a cluster of symptoms and/or signs with a more or less predictable course. Symptoms are what patients tell you; signs are what you see. The cluster may be associated with physical abnormality or may not. The essential point is that it results in consultation with a physician who specializes in recognizing, preventing, and, sometimes, curing diseases. (p. x)

This definition is similar to the syndrome definition of disease proposed by Thomas Sydenham. Sydenham was a great British physician of the seventeenth century who emphasized that physicians should carefully observe the clinical presentations of their patients. Physicians began to note that certain symptoms and signs were correlated and that patients with these observable "syndromes" often seemed to have about the same course. However, Sydenham's syndrome approach to the definition of disease was superseded in the 19th century by the approach of Rudolph Virchow and others who emphasized the discovery of causes for these syndromes through new sciences such as physiology and bacteriology. Since then, the definition of disease in general medicine has been popularly thought to refer to a disorder that has a recognizable clinical syndrome and that also has a definite etiology and pathophysiology and a definable course (Klerman, 1977). Nonetheless, physicians recognize the importance of isolating a clinical syndrome as an essential step in the scientific process of delineating a disease. After a syndrome has been clinically isolated, then the medical sciences can be rigorously applied to explain the underlying processes that led to this syndrome.

By using a syndrome approach to the definition of disease, Guze

---

called the *anti-psychiatrists.* Most mental health professionals are familiar with at least some of the writings by Szasz, and generally, he is perceived as a gadfly within the profession. Despite frequent attempts by many authors to deprecate Szasz's arguments, there exists a lingering suspicion that at least some of what Szasz so dramatically asserts is correct.

and his colleagues were proposing a definition that would not be considered acceptable when applied to general medical conditions. Different diseases can present with the same clinical syndrome. For instance, both paresis and schizophrenia can present with virtually the same clinical picture. Nonetheless, no one would argue that paresis and schizophrenia are the same disease.

Note also that, in the second definition, Guze reintroduced the notion that what physicians diagnose, prevent, and cure is important to any definition of disease. Again, this emphasis on "what doctors treat" does nothing to clarify the meaning of *disease.*

Definition #3 of Guze

Guze's third definition of disease appeared in a more recent paper, in which he defended the medical-model approach to psychopathology. This paper is titled "Nature of Psychiatric Illness: Why Psychiatry Is a Branch of Medicine" (1978). In this paper, Guze did not attempt to offer an explicit scientific definition of disease that would include both medical and psychiatric disorders. Instead, he proposed a semantic abstraction:

> Disease is a convention, a metaphorical concept. Because of limited and inadequate knowledge, all definitions of disease are ambiguous, and the limits of the concept cannot be sharply defined. On the contrary, new understanding and insight will certainly affect the limits. It may be said that there is no such thing as disease; there are only diseases. The concept of disease may be a "myth." But, if so, it is a myth when applied to all medical conditions and not just when applied to psychiatric disorders. (p. 298)

In brief, Guze was suggesting that disease cannot be accurately and adequately defined, a position that contradicts his initial stance on the subject. Nonetheless, he still believed that the concept of disease is important, and that an adequate definition would be a major advance in medicine and in the mental health field. Guze concluded:

> Only after defining disease in general can one expect to determine whether psychiatric disorders are diseases. If dictionary definitions are applied, most psychiatric conditions meet the criteria for disease. (p. 298)

Spitzer's Definition

Another member of the neo-Kraepelinian movement, Robert Spitzer, has also attempted to find an adequate definition. Spitzer differentiated between the terms *disease* and *mental disorder.* Spitzer argued that the term *disease* refers to disorders with observable pathophysiological processes. Thus, most of the organic brain syndromes and the mental retardation syndromes could be considered diseases within the DSM. Because the concept of disease does not fit, Spitzer preferred the concept of *mental disorder* to generically cover all categories in the DSM-III (Spitzer, Sheehy, & Endicott, 1977).

Spitzer offered a fairly detailed definition of *mental disorders:*

> 1. The manifestations of the condition are primarily psychological and involve alterations in behavior. However, it includes conditions manifested by somatic changes, such as psycho-physiological reactions, if an understanding of the cause and course of the condition is largely dependent on the use of psychological concepts, such as personality, motivation, and conflict.
>
> 2. The condition in its full blown state is regularly and intrinsically associated with subjective distress, generalized impairment in social effectiveness or functioning, or voluntary behavior that the subject wishes he could stop because it is regularly associated with physical disability or illness.
>
> 3. The condition is distinct from other conditions in terms of the clinical picture and, ideally, follow-up, family studies, and response to treatment. (Spitzer & Wilson, 1975, p. 829)

Spitzer and Wilson (1975) provided a reasonably lengthy defense of this definition, as well as an attempt to clarify concepts used in the definition. Interested readers should study the Spitzer and Wilson commentary. It is worth noting that this definition was formulated partly in response to the controversy within American psychiatry during the early 1970s on whether or not homosexuality *per se* is a mental disorder. Using this definition, Spitzer (1973) concluded that homosexuality is *not* a mental disorder. Homosexuality failed to meet Part 2 of the definition. That is, homosexuality is *not* regularly associated with subjective distress, with impaired social functioning, nor with attempts to terminate the condition because of a resulting disability. However, if a person is upset by his or her homosexual behavior, then the resulting dysphoric condition would be considered a mental disorder.

The Spitzer and Wilson definition of mental disorder, despite its complexity, is probably the best current definition of the term that has yet appeared in the literature. Nonetheless, even this definition has its problems. Consider, for example, the third part of the definition: "The condition is distinct from other conditions in terms of the clinical picture." Hardly any diagnostic category within the DSM can meet the requirement of distinctiveness. Most clinicians recognize that patients who clearly fit within one diagnostic category are less frequent than patients whose clinical picture seems to span a number of diagnostic categories. The lack of clear differentiation among traditional categories is shown by the postulation and the formalization of new diagnostic categories that represent the "border" between already recognized concepts. For example, there are some patients who present with the personality dynamics of character disorders and who also sporadically show the clinical symptoms of a psychosis. The patient on this border between psychosis and character disorder has now been placed in a popular new category called *borderline personality disorder.* Another border diagnosis is the concept *schizoaffective,* which refers to patients who have a mixed picture of schizophrenic and affective symptoms.

Kendell (1975) summarized the concern about how distinctive the traditional diagnostic concepts of psychopathology are:

> At present, there is little convincing evidence that the boundaries we have drawn between one syndrome and another lie at genuine "points of rarity," i.e. that interforms between adjacent syndromes are less common than patients exhibiting the uncontaminated symptoms of one or other syndrome. The frequency with which clinicians are driven to employing such labels as schizoaffective psychosis, borderline psychosis, anxiety depression, and pseudoneurotic schizophrenia, the widespread evidence that different schools of psychiatry draw the boundaries between one syndrome and another in different places, and the generally low reliability of psychiatric diagnoses all suggest that syndromes merge into one another without natural boundaries in between. (pp. 127–128)

To summarize, the neo-Kraepelinians have been concerned with the issue of defining disease. Achieving an adequate definition has not proved easy. Before devoting more attention to the problems of definition, this chapter now focuses on the issue of professional jurisdiction.

PROFESSIONAL JURISDICTION

There are a number of professions concerned with psychopathology and mental health. A partial list of these professions include clinical psychology, counseling, nursing, occupational therapy, psychiatry, recreational therapy, social work, and vocational rehabilitation. Throughout the history of the mental health movement, psychiatry has maintained the position as the dominant member of this collection of professions.

The rhetoric of the medical model has become popular in the competition among these professions. It is the issue of professional jurisdiction that is largely responsible for the emotional and irrational components of the arguments surrounding the medical model. To some extent, discussions about the meaning of the terms *medical model* and *disease* are better understood through an analysis of the political connotations of these terms rather than through an analysis of their scientific denotations.

To psychiatrists, advocacy of the medical model is a defense of psychiatry's dominance and a stimulus for the increased emphasis within psychiatry on its biological/medical heritage. If persons with mental disorders are ill, then they need diagnosis and treatment by physicians. Psychiatry is a specialty within medicine; thus, psychiatrists should be responsible for treating or, at a minimum, supervising other professionals in the treatment of patients.

To psychologists, social workers, nurses, and the other nonmedical professions, *medical model* has become a pejorative term to refer to their second-class citizenship and relatively low earning potential. They suggest that psychopathology is caused by social, cultural, and/or behavioral forces as well as by biological forces. Thus, these professions conclude that they should be as appropriate a choice as psychiatrists for the treatment of mental disorders.

The quotes that follow should give an indication of the political nature of discussions about the medical model. The first quotations are excerpted from Guze's 1978 article, which was mentioned earlier. In the conclusion to his paper, Guze argued:

> Biological processes play important, specific roles in the development of many psychiatric disorders, just as they do in general medical disorders. Medical training provides the best preparation for understanding etiology and pathogenesis in psychiatric conditions and for prescribing and supervising treatment.
>
> Diagnosis is the keystone of general medical practice; everything else that doctors do depends upon it. The same is true for psychiatry. The effects of our social and psychologic environments cannot be properly understood until they are studied in relation to different illnesses. (p. 306)

> Modern medicine is increasingly based on biomedical science. Psychiatry is beginning to do the same. . . .
>
> The medical model offers the best framework for approaching psychopathology because it provides for a balance of biologic and sociologic insights. (p. 307)

The opposing viewpoint is taken from an article by a psychologist, Herbert Rie, entitled "Psychology, Mental Health, and the Public Interest" (1977):

> Most obviously, the inclination to view extreme deviations in human

> adaptation—or psychopathology—as disease entities and hence as "medical" problems must finally be shed. This is neither a new nor a radical idea nor is it foreign to everyday practice. It is only at the policy levels, when acknowledgment of the realities of daily practice might require changes in prerogatives, in status, or in control, that the idea continues to meet resistance.
>
> Implicit in this change is the elimination of the tendency to transform a client into a "patient" by the fact of hospitalization. After all, the overwhelming majority of individuals with psychological problems suffer no medical disorder and may require protection, not aseptic isolation and physiological monitoring. . . .
>
> Implicit in this change is the final elimination of all vestiges of a hierarchic relationship between medicine and psychology and specifically between psychiatry and psychology. This would further imply the elimination of the transparent fiction that "psychotherapy" prefixed by the term *medical* and performed by a physician is somehow different from "psychotherapy" prefixed by the term *psychological* and performed by a psychologist. . . .*
>
> Implicit in this change is the acknowledgment that psychologists are at least as committed to their clients as are their medical colleagues and are surely no more likely to ignore their medical problems than are physicians to ignore their patients' psychological problems, and perhaps less so. (p. 3)

The dominance of psychiatry among the mental health professions has been exemplified in three ways: (1) psychiatrists are the highest paid of the mental health professionals; (2) psychiatrists have been the most numerous of the mental health professionals (except for nurses); and (3) psychiatry has been the primary mental health profession recognized by insurance companies. However, the dominance of psychiatry in these three areas is being challenged as the number of psychiatrists relative to other mental health professionals has steadily declined over the last 30 years. In order to better understand the competitive issues between the mental health professions, the topics of (1) earnings, (2) manpower, and (3) insurance will be addressed in turn.

Earnings

It is in the area of earnings that the dominance of psychiatry becomes most obvious. An attempt has been made to gather adequate earnings data to permit a national comparison of the earning potential of the various mental health professions. There was a surprising

* Rie was making reference to the fact that in the United States some health insurance companies will reimburse a person for receiving "medical psychotherapy" on an outpatient basis but not for psychotherapy administered by nonphysicians.

TABLE 5. Median Earnings in 1977 of Mental Health Professionals

Psychiatrists (M.D.)	$53,790
Psychologists (Ph.D.)	$27,375
Psychologists (M.A.)	$18,262
Social workers (M.S.W.)	$14,600
Psychiatric nurse specialist (M.A.)	$12,360

Note. These data were collected from Gottfredson and Dyer (1978); L. Jackson (1980); Morales and Sheafor (1980); and Mattera (1979). The datum for social workers is not an estimate of the median but is the minimum salary for 1977 as set by the National Association of Social Workers.

amount of difficulty in obtaining figures that were accurate and representative of the professions.

Table 5 contains a comparison of the median earnings for different professions in the year 1977. There was a clear difference in the average earnings of the various mental health professionals. Psychiatrists obviously earned a great deal more than did any other professional group. In fact, from these figures, it appears that a psychiatrist is worth two clinical psychologists with Ph.D.'s, or three M.S.W.'s, or four nurses with advanced degrees.

Concerning earnings, it is also relevant to note that, of the medical specialties, psychiatry is one of the most poorly paid. Psychiatry, family practice, and pediatrics have consistently been found to have the lowest earnings among physicians. For instance the median net earnings of all physicians in 1977 was $65,400. Surgical specialties were generally the highest paid and had earnings of $78,200 in 1977 (Mattera, 1979).

Notice also from Table 5 that even if psychiatry were deleted from the earnings data, there would still be a clear pecking order among the remaining mental health professionals. Clinical psychology is the dominant member of the nonmedical professions. Even psychologists who hold only master's degrees generally have greater earnings than the other professionals with the same educational degrees.

Two other features of the earnings data are worthy of comment. First, educational background as shown by the college degree is obviously related to monetary gain. All psychiatrists are M.D.'s, though not all physicians who call themselves psychiatrists have completed an approved psychiatric residency, and a large percentage are not board-certified. Among clinical psychologists, the standard degree is a

Ph.D. Psychologists who have completed only a master's degree cannot become full members of the American Psychological Association, but they can have an associate member status. Also, in some states, only persons who have attained a Ph.D. in psychology are permitted to call themselves psychologists.

Among social workers, the standard degree is an M.S.W. (master of social work). The proportion of social workers with a Ph.D. is relatively small. Specifically, 2% of the membership of the National Association of Social Workers in 1977 held Ph.D.'s, whereas 73% of the membership of the American Psychological Association did. There is also a reasonably large number of persons with bachelor's degrees in various social sciences and humanities who work as "caseworkers" in state and federal government positions. These caseworkers are often titled *social workers,* and since 1973, they have been able to join the National Association of Social Workers.

Among nurses in the mental health field, the base degree has been an R.N. In addition, there have been a number of L.P.N.'s and nurse's aides who work under the generic title of *nurse* in inpatient settings. Over the last two decades, there has been a movement to increase the educational sophistication of nurses who are interested in specializing in psychiatry. As a result, a master's degree has been offered that entitles the holder to use the title *psychiatric nurse specialist.* There are relatively few nurses in the mental health field with the degree.

The second relevant feature concerning the earnings data (and the implicit pecking order of power) is the sex ratios within the professions (Albert, 1978). Nursing is virtually entirely a female profession (96% of all nurses are women). In contrast, psychiatry is heavily dominated by males (82% of psychiatrists are males, and 93% of all physicians are males). Clinical psychology is more evenly split (55% males). Social workers tend to be women (69%). In addition, the social work field has the highest proportion of any of these professions in racial and cultural minorities.

Manpower

During the 1950s, Hollingshead and Redlich (1958) published a series of important analyses of the mental health field in the New Haven, Connecticut, area. Their data show that in 1950 the largest number of mental health professionals were nurses (53%), with psychiatrists being the second largest of the four professions (26%). It is worth noting that nurses were primarily employed in inpatient settings and were under the direct supervision of the psychiatrists. Although these nurses were working with mentally disturbed patients, most had had relatively little educational background for doing so. In fact, it was 1955 before the American Nursing Association recognized psychiatric nursing as a specialty within the nursing field.

By 1975, 25 years later, there was a distinct growth in the number of mental health professionals, with the most dramatic growth occurring among psychologists and social workers. The data reported by Redlich and Kellert (1978) are shown in Table 6.

There are other data that demonstrate the relative numerical decline of psychiatry among the major mental health professionals. For instance, psychiatry is now the smallest of the four professions. As of 1974, a National Institute of Mental Health survey showed that psychiatrists constituted only 7% of all full-time professional staff. In contrast, 9% were psychologists; 18% were social workers; and 29% were nurses (R.N.'s). The remainder of the full-time professionals in outpatient clinics or inpatient settings were vocational rehabilitation counselors, school teachers (B.A. and above), occupational therapists, and other mental health professionals with a B.A. or higher (National Institute of Mental Health, 1976).

Another set of data also attests to the relative numerical decline of psychiatry. The national memberships of three professional organizations—the American Psychiatric Association, the American Psychological Association, and the National Association of Social Workers—have shown steady growth over the last two decades. However, the rates of growth have been quite different. The number of psychiatrists has doubled once every 22 years, whereas the number of psychologists has doubled once every 14 years. The number of social workers has grown even faster, doubling once every 12 years.

The final indication of the relative decline of psychiatry in terms of numbers is the change in the professional staffing of community mental health centers (CMHCs). Data on staffing patterns are shown

TABLE 6. Approximate Number of Mental Health Professionals in the New Haven Area

	1950	1975
Psychiatrists	50	200
Psychologists	20	140
Social workers	20	140
Nurses	100	300

Note. Data from Redlich and Kellert (1978).

TABLE 7. Average Number of Full-Time Staff in Community Mental Health Centers

	1970	1975
Psychiatrists	3.1	2.4
Psychologists	3.8	7.2
Social workers	8.1	10.7
Other professionals	7.3	12.2

Notes. Data from Winslow (1979).

in Table 7 (Winslow, 1979). These data show an increase in the average number of professionals per community mental health center for all professions except psychiatry. In this short period of time, psychiatry dropped from representing 14% of the professional staff to only 7%. When presenting these data in the *American Journal of Psychiatry,* Winslow expressed concern about the loss of psychiatrists in both clinical and leadership roles within CMHCs. He suggested that organized psychiatry should attempt to reverse this trend.

The manpower issues concerning psychiatry do not involve just the relative number of psychiatrists in practice. Concern has also been expressed in psychiatry about its ability to recruit from the available pool of medical students. Nielsen (1980) showed that during the 1950s about 7% of all U.S. medical students chose psychiatry as a specialty. In contrast, the projected figures for 1980 suggested that only 2.5% of graduating American medical students would select psychiatry. A study by Eagle and Marcos (1980) found that the medical students who choose psychiatry were below average in class rank. An additional source of concern to some psychiatrists has been the fact that the number of foreign medical school graduates in psychiatry residency programs has increased from 28% to 38% over the last five years. In short, psychiatry is currently having difficulty recruiting new members to the profession. In contrast, for instance, clinical psychology has experienced no recruitment problem. During the 1970s, the number of applicants to a clinical psychology graduate program was at least 10 times the number who were finally selected for admission.

What the data on manpower suggest is that psychiatry has not been growing nearly as rapidly as psychology or social work. Moreover, the data suggest that the relative numerical decline of psychiatry will continue. As a result, the law of supply and demand suggests that there will be increasing competition among the professions within the marketplace of mental health services.

Insurance

In the preceding sections, it has been shown that psychiatrists have suffered a drop in numbers relative to the other mental health professionals. Data have also been presented that show that psychiatry has a much higher earning potential than the other professions. The result is that psychologists, in particular, have attempted to upgrade their status to a level more commensurate with that of psychiatry.

From its inception, the American Psychological Association was controlled by psychologists who worked in academia. The training of psychologists, even clinical psychologists, had been oriented toward teaching graduate students to become faculty members. During the 1970s, however, control of the American Psychological Association has gradually been taken over by practicing clinicians. The effect has been to promote an emphasis on psychology more as a profession than as a science.

Associated with the internal changes in the American Psychological Association has been the enacting of licensing laws in all 50 states. The emphasis on licensing has been meant to show that the psychological profession can monitor itself, can ensure a uniform meaning for the term *psychologist,* and can remove charlatans from its midst.

Another major advance in the professional emphasis within psychology has been the successful passage of freedom-of-choice laws in 29 states (Dorken & Webb, 1980). These laws, in effect, state that a client should be able to choose among different professional groups for psychotherapy. In order to ensure a fair choice, psychiatrists and licensed clinical psychologists are considered equal in terms of receiving compensation—including compensation from private insurance companies—for their services.

As the Congress of the United States has slowly moved toward the enactment of a national health-insurance program, clinical psychologists have been very active in attempting to be certain that their services, as well as the services of psychiatrists, will be included. Interest in this topic has been sufficiently strong so that the *American Psychologist,* the main professional journal of the American Psychological Association, devoted an entire issue to the topic (September 1977). The consensus of the authors who published articles in this journal issue was that psychology must be included in any national health-insurance program if it is to survive as a profession:

> Our profession is at a major crossroads. If we are able to obtain independent inclusion under national health insurance, then we will be assured of continuous training funds, financial security, professional challenges, etc.

> If not, I am afraid that our training institutions will soon be phased out, and we will find ourselves as very expensive specialists for whom no one wants to pay the bill. (DeLeon, 1977, p. 751)

The available evidence suggests that the inclusion or the exclusion of clinical psychology from health insurance will have a distinct impact on this profession. For instance, Dorken (1977) demonstrated that when licensed psychologists were included as providers of psychotherapy under CHAMPUS (Civilian Health And Medical Program of the Uniformed Services), the hourly fees of psychologists quickly approached the hourly fees of psychiatrists. Dorken and Webb (1980) showed that there had been a substantial increase in the number of clinical psychologists in private practice in those states where freedom-of-choice (FOC) laws had been passed. States with FOC laws, in effect, permit psychologists to compete with psychiatrists, without any major financial favoritism being shown to psychiatry.

Professional Jurisdiction and Classification

What does this conflict among the mental health professions have to do with classification? The answer is that psychiatrists, especially those who advocate the medical model, have emphasized classification and diagnosis as a cornerstone of their approach to psychopathology. Thus, the other mental health professionals have become concerned with traditional classificatory systems because these systems are *psychiatric* classifications.

This point was clearly stated by two psychologists, Schact and Nathan (1977), in their commentary on the DSM-III. These psychologists reviewed and criticized the available information about the new edition of the DSM that was being created. Generally, they applauded the increased emphasis in the DSM-III on precision and clarity of definition. However, at the end of their paper, Schact and Nathan concluded:

> In our judgment, DSM-III has the potential to be very bad for psychologists, despite the real and potential advances in scope, diagnostic reliability, and diagnostic logic it represents. Notwithstanding its clarification of important aspects of the diagnostic process, the extensive pilot testing which is planned for it, and the inclusive nature of the diagnostic categories it includes, DSM-III has sufficiently important shortcomings that organized psychology, in our judgment, ought to be most concerned at its possible adoption. Its professional and legal significance for psychology appear to be especially troublesome. It is entirely possible, for example, that promulgation of DSM-III as an official action of the American Psychiatric Associa-

> tion will carry sufficient weight to call it to the attention of insurers and legislators who will see it in quasi-official recognition of the primacy of physicians in the diagnosis and treatment of the disorders categorized by DSM-III. Since these disorders are inclusive of virtually every psychological ill to which men and women are heir, little will be left for psychologists and other mental health professionals to work with. Even less will be left if legislators and third-party payers conclude that, since these conditions are all "medical disorders," they must first be evaluated by physicians, who will then decide whether they or ancillary mental health professionals (e.g., the psychologists) will treat the disorder. (p. 1024)

Organized psychology did respond to the concern about *psychiatric* classification. At the 1976 annual meeting of the American Psychological Association, there was a symposium devoted to an analysis of the proposed DSM-III and its perceived threat to psychology. The result was that the American Psychological Association sponsored a committee to form a *psychological* classification that could compete with the DSM-III. This committee was given the name Task Force on Descriptive Behavioral Classification. This task force attempted to draw on existing knowledge about behavior; knowledge about classification; knowledge about the psychological, social, and environmental factors affecting deviant behavior; and knowledge about scientific approaches to description and assessment.

In 1978, this task force was dismantled. The members of the task force had also become aware of the enormity of their undertaking and had been requesting increased support if they were to continue. Critics had argued that creating a new classification of mental disorders would be an economically staggering undertaking that probably would not yield useful results in the near future.

During 1979, however, the pressure to form a psychological classification was still present among the leaders of the American Psychological Association. A decision was made to create a new committee. However, this new committee was to focus only on the classification of childhood disorders. The reason for this focus was that the area of childhood disorders had been a popular topic of many descriptive studies in psychology during the 1960s and 1970s, and because it was believed that these studies had yielded sufficiently convergent results to make the proposal of a new classification into a manageable undertaking. As of this writing, members of the task force have received independent funding to proceed with the development of a measurement instrument that can be used to describe the psychopathology of children. Their future work will draw heavily on factor-analytic approaches to description (see Chapter 7).

THE MEDICAL MODEL REVISITED

Wing's Two Definitions of Mental Illness

Stereotypes are a cognitive device commonly used to help us understand other human beings. People are complex creatures who defy easy conceptualization. We use stereotypes that allow us to simplify by grouping people in order to bring into relief their similarities and differences. However, stereotypes have the unfortunate property of often seeming more real in our minds than the complexities of the human beings we initially wanted to understand. Thus, stereotypes can be caricatures that fail to represent people. Not all professional football players are dumb; not all male homosexuals speak in a falsetto voice and have limp wrists; not all Republicans voted for Richard Nixon in 1972; not all psychiatrists believe that psychiatry should be dominant among the mental health professions; not all neo-Kraepelinians eschew psychoanalysis and perform research on the biological factors of psychopathology; not all clinical psychologists believe that organized psychiatry is trying to destroy them.

This chapter, by its simple arguments, has suggested that all thinking about the medical model is simpleminded. Unfortunately, much of the literature has been as simple as this chapter suggests. But not all. A case in point is Wing's book *Reasoning about Madness* (1978). John Wing is a contemporary British psychiatrist who has been actively interested in the social forces that affect psychopathology. His initial work involved the study of mental institutions and the effect of institutionalization on psychiatric patients (Wing & Brown, 1970). He also created the Present State Examination, which is a structured interview with a well-defined scoring system (Wing, Cooper, & Sartorius, 1974). Other major interests for Wing have been the study of schizophrenia and the international differences in the use of classificatory systems.

Wing has distinguished between two meanings of the term *mental illness.* The first is the broad meaning and encompasses all types of deviations and abnormalities that may lead to becoming the patient of a mental health professional. In effect, this meaning of *mental illness* results from the conjunction of two processes: statistical deviance and social attribution. Statistical deviance means that the individual is behaving in an unusual or "ab-normal" fashion when compared with most persons from his or her cultural group. However, statistical deviance by itself is not considered sufficient to label a person as mentally ill. Social attribution is the second important ingredient. Society must consider the deviant behavior a norm violation. If so, then the labeling process (Scheff, 1966) comes into play, and various social forces gradually converge to result in the individual's becoming a patient.

The second meaning of *mental illness* is much more specific and refers to the concept of *disease.* Wing argued that the concept of *disease* has two essential parts: (1) a disease is associated with a clinical syndrome; and (2) there is an underlying biological disturbance. Wing particularly emphasized the importance of a clinical syndrome because historically the recognition of clinical syndromes predates the discovery of underlying biological mechanisms. Also, it is at the level of defining clinical syndromes that the science of psychopathology is still struggling.

It was noted earlier, when discussing Guze's second definition of *disease,* that to define a disease as a recognizable clinical syndrome was not sufficient. Wing would agree. In fact, he believed that to call a clinical syndrome a disease was to formulate a hypothesis that then required testing. Could an underlying biological mechanism (or set of mechanisms) be discovered that is essential to an understanding of the syndrome? Until an affirmative response is given, the clinical picture cannot be identified as representing a mental illness in the narrow sense of the term (i.e., as a disease). Wing also pointed out that even if a clinical syndrome is eventually discovered to be a disease, social factors can still play a very important role in the development and maintenance of the deviant behaviors.

Wing suggested that mental illness, in the "disease" sense, refers to only a small portion of what generally is included under the rubric of psychopathology. It is through the broad meaning that mental illness covers the range of individuals who are identified as patients by mental health professionals. More specifically, for instance, the entire general heading of the personality disorders could not be claimed to be diseases. Sufficient evidence to support the presence of biological mechanisms is not available. More important, Wing argued that recognizable clinical syndromes had not been identified for what are generally called *personality disorders.* Therefore, personality disorders have not even achieved the first necessary stage for fitting a disease model of psychopathology.

In contrast, Wing believed that schizophrenia is a mental illness in the disease sense:

> With one crucial exception, the evidence in favour of an explanation of the central syndromes of schizophrenia in terms of a disease theory or

theories is very similar to that for *diabetes mellitus*. The most florid clinical forms can be reliably recognized by trained clinicians. The aetiology is unknown, but there is some evidence in favour of multigene inheritance, with all that implies by way of the likelihood of "latent," "sub-clinical," and "asymptomatic" cases. A good deal is known about somatic and social precipitating factors. Methods of pharmacological treatment are highly effective in suppressing and preventing the acute manifestations, and techniques of long-term management have been worked out. Where a disease theory of schizophrenia is thoroughly incomplete is in the absence of a firm link between the clinical syndrome and an underlying homeostatic mechanism, equivalent to that which controls carbohydrate metabolism and which is known to be defective in diabetes. Of course, it is clear that the pathophysiology of diabetes itself is by no means understood, but if we could get so far with schizophrenia it would be a very long step.

All that can be claimed, therefore, is that a disease theory can be postulated, at least for the large central group of the schizophrenias, and that it seems to be useful in both of the ways mentioned earlier. It is helpful to be able to make the diagnosis, first of all, because this will lead in a large number of cases to the reduction of impairments. Secondly, making the diagnosis also allows us to further the scientific investigation of this condition. Progress to date is highly suggestive that further advances are likely to be made. Such advances, in turn, are likely to benefit our patients. (Wing, 1978, pp. 123–124)

The Demarcation Problem

The importance of defining the concepts of *disease* and *mental disorder* is that their definition will help to resolve the *demarcation problem*. In other words, these definitions will help draw a boundary between persons who should be seen by mental health professionals and persons who should not. In addition, the demarcation problem involves the question of what proposed disorders actually should be subsumed under the classification of psychopathology.

The demarcation problem is a serious scientific issue for the classification of psychopathology (Morse, 1978). For instance, the DSM-III contains a category (under the administrative codes) for persons who see a psychotherapist because of marital discord. As any practicing clinician knows, marital discord is a major social issue that is often involved in the reasons that someone wants to see a clinician. But does this mean that marital discord should be a category of psychopathology? Is it a mental disorder? Does the *de facto* importance of marital discord in clinical practice show that the traditional approaches to classification are too narrow because of their focus on the individual? Or does it reflect the failure of clinicians to make an important distinction between a reason for referral and a category of psychopathology?

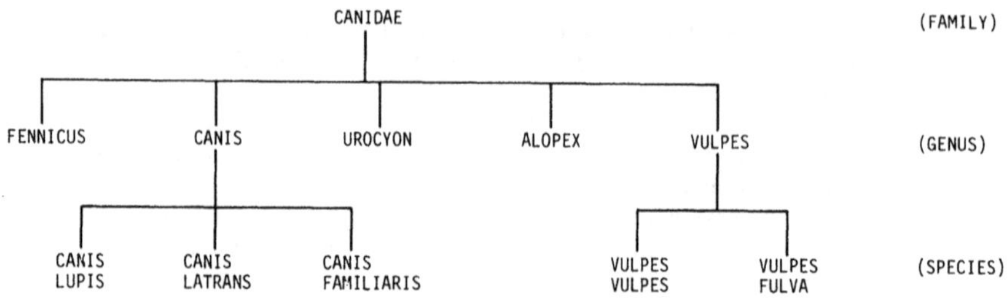

FIGURE 5. Hierarchical structure of the biological classification of dogs.

To understand the importance of the demarcation problem in the classification of psychopathology, a digression is made here into biological classification. This digression involves the presentation of ideas about the hierarchical structure of biological classification.

Within the biological sciences, classification has been recognized as a fundamental process. Classificatory systems serve to organize almost all known information about living organisms. There is a dictum in botany, for example, that states, "The name of a plant is the key to its literature."

The hierarchical organization of biological classification, as proposed by Linnaeus, can be understood by examining Figure 5. This figure shows the classification of dogs. The major categories in this classification are represented by names such as *Vulpes* (fox) and *Canis* (dog). The hierarchical organization is shown by the superordinate-subordinate relationships among the categories. Thus, categories like the wolf (*Canis lupis*), the coyote (*Canis latrans*), and the common dog (*Canis familiaris*) are relatively specific categories, whereas the other categories, such as mammal and animal (not shown in Figure 5), are relatively inclusive terms.

The hierarchical arrangement of classificatory categories had been recognized before Linnaeus. What Linnaeus added was a new set of terms: *kingdom, phylum, family, genus, species,* and so on. Some of these terms are shown in the far right column of Figure 5, and they correspond to rows among the classificatory categories. Thus, *species* refers to the most specific categories in the classificatory system, and *kingdom* refers to the most inclusive term.

The reason that Linnaeus's innovation became so important was that classifiers within biology realized that the definition of *species* would be crucial to biological classification. In other words, the definition of *species* could be used to decide whether a newly proposed category of living organisms should be added to biological classification. Before Linnaeus, biological classification had been in a relative state of anarchy because different classifiers had proposed a large variety of classificatory categories, a number of which overlapped, and deciding which of these categories should be kept in a scientific classification and which should not proved difficult. For instance, should the "red fox" found in Europe and the "red fox" found in North America be classified in the same category? What about the "black fox" found in the United States? Or the "corsac fox" found in European Russia?

The definition of the concept *species* allowed biologists to resolve questions like those above. As currently defined, the concept of *species* contains any category of living organisms that have a common gene pool (i.e., they interbreed), that exist in a reasonably well-defined geographic and ecological segment of the world, and that have similar morphological characteristics. By applying this definition, one can show that the "red fox" of Europe and the "red fox" of North America, despite their common popular name, represent different species (*Vulpes vulpes* is the technical name for the European red fox and *Vulpes fulva* for the North American). The "corsac fox" is also a separate species (*Vulpes corsac*). On the other hand, the "black fox" of North American simply represents a color phase of the red fox and thus does not signify a separate species.

To summarize, Linnaeus, by emphasizing the concept of species, made biologists aware of the demarcation problems. Also, the terms he proposed for organizing the rows in a classificatory hierarchy allowed classifiers to formulate orderly principles by which to structure biological classification. In addition, it is important to note that an acceptable definition of *species* did not appear until Darwin had postulated his theory of evolution. Thus, an adequate resolution of the demarcation problem required the proposal of a general theory of biology.

To understand the relevance of this digression about biological classification to the demarcation issue in psychiatric classification, the reader should examine Figure 6. This figure displays some of the categories from the DSM-II. Notice that, like the organization among biological categories, the DSM-II categories are arranged in a hierarchy and that some terms are more inclusive than others. However, unlike in biological classification, there are no names in the far right column to serve to name the rows in the DSM-II hierarchy. That is, traditional psychiatric classification contains no terms that are analogous to the concepts of *kingdom, phylum,* or *species.*

One hypothesis is that the term *disease* is analogous to the terms *species* in biological classification, in that *disease* refers to the class of the most specific categories in psychiatric classification. If so, the definition of *disease* is as crucial to psychiatric classification as the defini-

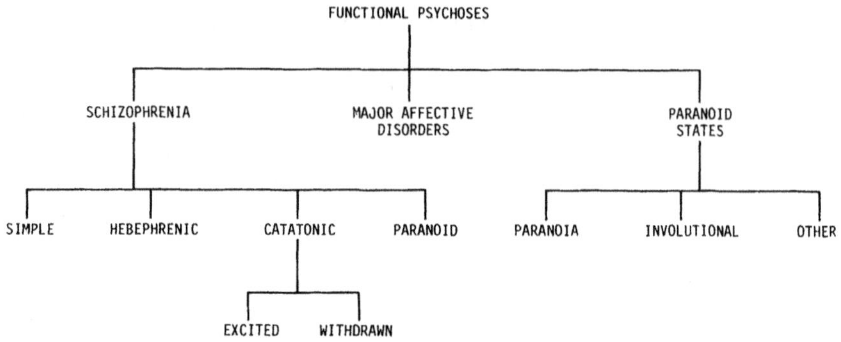

FIGURE 6. Hierarchical structure of the classification of functional psychoses.

tion of *species* was to biological classification, because both definitions are needed to resolve the demarcation problem. In addition, if *species* and *disease* are analogous terms, the failure to define *disease* as it applies to psychopathology should not be considered a major deficiency, as adequate definitions of *species* did not appear until Darwin's theory of evolution was formulated, that is, nearly 150 years after Linnaeus. In order to adequately resolve the scientific side of the demarcation problem, it will be necessary to develop a comprehensive theory of psychopathology.

As noted in the beginning of this book, classificatory systems and theories are necessarily intertwined. From a scientific perspective, the issues about the medical model have revolved around the type of theory that will best yield a final, detailed framework for explaining psychopathology. However, no scientifically adequate theory of psychopathology exists. Nor is one likely to exist within this century. As our understanding of psychopathology develops, better theories will evolve and increasingly precise definitions of mental disease will become possible. But without an adequate theory, no scientific resolution of the demarcation problem will occur.

The other side of the demarcation problem is its inherent political implications. This chapter has attempted to demonstrate that an important political issue associated with the medical-model approach to the demarcation problem is professional jurisdiction. Like the scientific side, the political side of the demarcation issue is not trivial and will not be easily resolved. For instance, consider the economic forces that feed the controversy over professional jurisdiction. Psychiatry is clearly growing less rapidly than the other professions. The result will probably be an increase in professional competition. In addition, the striking differences in current earnings will be an ongoing source of antagonism.

The political aspects of the demarcation problem are not limited to professional competition. A good example is the debate concerning whether homosexuality should be included as a mental disorder in the official psychiatric classification. In the DSM-I and the DSM-II, homosexuality was a recognized category. However, during the late 1960s and the early 1970s, gay activist groups began a systematic attempt to persuade psychiatrists that to keep homosexuality as a mental disorder category was unwarranted discrimination. The resulting political encounter is documented in a book by Bayer (1981) entitled *Homosexuality and American Psychiatry: The Politics of Diagnosis.* The result was that homosexuality *per se* was eliminated as a category, but "sexual orientation disturbance" was introduced in order to handle individuals who were distressed by their sexual preferences.

To conclude, this chapter attempted to discuss some of the issues associated with the medical model. These issues represent serious dilemmas whose resolution will have major effects on the study of psychopathology and on future classifications of mental disorders. In particular, there has been an emphasis on defining *disease,* on the politics of professional jurisdiction, and on the demarcation problem.

The "medical model" has attracted a great deal of attention over the last three decades. As Klerman (1977) suggested, the concept does not exist in isolation; instead, it has been enveloped in a variety of emotionally laden issues:

> Like "law and order" and "neighborhood schools," "medical model" has become a slogan for use in public debate—more an epithet for attacking one's enemies or a rallying cry for gathering one's followers than a term of precision for furthering intellectual discussion or scientific discovery. To M.D. psychiatrists it has become a standard around which to gather to protect benefits threatened by insurance companies who propose to reduce coverage for mental health services. To behavior therapists, who are followers of Skinner, it designates the adherents of psychoanalysis, whose practices are based on the postulates of an "underlying illness" rather than dealing directly with manifest behaviors. To black militants in the urban ghetto, the term refers to the futile efforts of the local community mental health center to deal with deep social ills, by "applying band-aids to the social casualties and the walking wounded." To psychologists and social workers in mental health settings, it is a term of contempt to use in the struggle over whether medical degrees are necessary for positions of greater authority or higher salary. To sociologists, learned in "labeling theory," it refers to the narrow reductionism of conventional psychiatric thinking that refuses to see the social forces in the generation and perpetuation of social deviance. To civil libertarians, it is the basis for the unwillingness of the mental professions to acknowledge the extent to which mental institutions are agents of social control. To psychotherapists—medical and nonmedical—it is a synonym for biological treatment, especially "shock" and psychosurgery, which threaten to destroy the minds and the souls of their

> patients. To biological psychiatrists, proud of the advances in new drugs, it is a summary call to psychiatrists to return to scientific medicine and the mainstream of medical practice. (pp. 220–221)

SUGGESTED READINGS

Caplan, A. L., Englehardt, H. T., & McCartney, J. J. (Eds.), *Concepts of health and disease: Interdisciplinary perspectives.* Reading, Mass.: Addison-Wesley, 1981.

The topics of the medical model, disease, and health are important topics for the entire medical profession. Caplan *et al.* have put together a useful collection of articles, essays, position papers, and so on that reflect the wide range of opinions about the medical model and its impact on society. In structuring their collection, they commented that it is in the area of the mental disorders (especially the addictions, sexual behavior, and criminal behavior) that the meaning of *disease* has been most frequently questioned. An example of a paper included in this book is an article by Englehardt about "The Disease of Masturbation," which reviews the three centuries of arguments about whether this behavior should be considered a disease. Authors whose writing are included in the book are Thomas Sydenham, Rudolph Virchow, Walter Cannon, Abraham Maslow, Eduard von Hartmann, Robert Kendell, Thomas Szasz, David Mechanic, Horacio Fabrega, and Jacob Moreno.

One of the most informative papers in the book is by Veatch, entitled "The Medical Model: Its Nature and Problems." Veatch defined the medical model as an integrated system of concepts that is used to understand a large class of deviances. Veatch further argued that a deviancy is placed within the scope of the medical model if it has four characteristics: (1) the deviancy is nonvoluntary; (2) there is a demonstrable organic (i.e., anatomical, biochemical, or physiological) change associated with the deviancy; (3) the physician is seen as the most qualified to handle the deviancy; and (4) the deviancy has been noted by society as falling outside some limit of acceptability.

Clare, A. *Psychiatry in dissent: Controversial issues in thought and practice.* London: Tavistock, 1976.

This is an intriguing little book that addresses some of the most vexing issues in the contemporary mental health field: the concept of mental illness, the antipsychiatrists, psychosurgery, electroconvulsive shock, schizophrenia, and classification. Clare was able to inject a sense of clarity about these emotion-laden topics and to give the reader a sense of the powerful political and social forces that influence our thinking about abnormal behavior and about the human beings we call *patients.*

Kraupl-Taylor, F. *Psychopathology: Its causes and symptoms.* Baltimore: Johns Hopkins University Press, 1979.

This textbook about psychopathology was written from a European perspective, which is not often presented clearly in the American literature. Unlike the writers of most American textbooks in psychiatry and clinical psychology, Kraupl-Taylor did *not* organize his chapters around headings based on the various categories of mental disorders. Instead, his book is organized around the various headings concerning classes of symptoms (perception, intellect, affect, anxiety, mobility, memory, and so on). Thus, Kraupl-Taylor's book is, in a fundamental sense, more concerned with the description of psychopathology than are most American texts.

Of the most relevance to this chapter is Kraupl-Taylor's discussion of the concept of disease. He has provided a simple, but powerful, discussion of the classic meaning of this concept. Disease entities, stated Kraupl-Taylor, "are symptoms complexes which are abnormal by population and/or individual standards, are directly observable, and cause suffering and concerns to the patients and their social environment" (p. 279). Later in the book, Kraupl-Taylor admitted that this definition is inadequate but argued that it comes as close as is currently possible to what is meant by the concept.

Farrell, B. A. Mental illness: A conceptual analysis. *Psychological Medicine,* 1979, *9,* 21–35.

This paper contains an interesting and insightful philosophical analysis of the meaning of the phrase *mental illness.* The author has started with the sentence "Smith is mentally ill" and then analyzes potential translations of this sentence, such as "Smith's condition is a deviant one (in a statistical sense)," or "Smith's body is in a certain specifiable state, which it is correct to call abnormal, and which abnormality contributes to (his abnormal behavior)." After his detailed commentary, Farrell has concluded that psychiatry has made a mistake in trying to search for a definition of *mental illness.* As a science, psychiatry simply has not formulated a sufficiently complex and integrated body of knowledge to define a simple concept.

Chapter 4

Diagnostic Reliability

> A preoccupation with description and classification is itself portrayed as a symptom, an abnormal psychological phenomenon whereby psychiatrists defend themselves against their own inadequacy and fear by emphasizing the "objectivity" inherent in their arid statistical and classificatory approaches and by hiding behind a smokescreen of measurements and "scientific" jargon.
>
> Anthony Clare, 1976, p. 83

A major area of research concerning classification has focused on the reliability of conventional psychiatric systems. The neo-Kraepelinians have emphasized this research because they have been promoting a scientific approach to psychopathology. They have come to believe that if a psychiatric classification is not reliable, the development of a science about mental disorders will not be possible:

> Classification systems such as diagnosis have two primary properties, reliability and validity. Reliability refers to the consistency with which subjects are classified; validity, to the utility of the system for its various purposes. In the case of psychiatric diagnosis, the purposes of the classification system are communication about clinical features, aetiology, course of illness and treatment. A necessary constraint on the validity of a system is its reliability. There is no guarantee that a reliable system is valid, but assuredly an unreliable system must be invalid. (Spitzer & Fleiss, 1974, p. 341)

Because of the concern about reliability, many of the improvements fostered by the neo-Kraepelinians were directed at improving reliability. For instance, the use of explicit diagnostic criteria in the Feighner *et al.* (1972) paper was clearly intended to increase interclinician agreement when assigning diagnostic categories. In addition, Spitzer and his associates devoted a great deal of effort to creating structured interviews and to using computers in diagnosis. These researchers hoped to overcome, thereby, the vagueness and the subjectivity inherent in the traditional diagnostic process.

This chapter is devoted to an analysis of reliability and its relevance to psychiatric classification. The perspective adopted in the chapter is historical. That is, research on the issue of reliability over the last three decades is analyzed chronologically so that we may discuss how the different themes surrounding reliability have developed.

HISTORICAL REVIEW OF DIAGNOSTIC RELIABILITY

Pre-1950s Research

The earliest studies concerning diagnostic reliability were performed in the 1930s. For instance, Masserman and Carmichael (1938) studied 100 patients at the time of admission to an inpatient unit and one year later during a systematic follow-up. They found that 40% of the patients were assigned to different categories on follow-up. If the diagnostic process were completely reliable, it would be expected that the percentage of altered diagnoses would be small.

In another study, Boisen (1938) contrasted the frequency of different diagnoses across comparable hospitals. For instance, he found that in one state hospital in Illinois 11% of all newly admitted schizophrenics were hebephrenic. At a different Illinois state hospital, 76% of the admitted schizophrenics were called hebephrenic. Although it was possible for the incidence of hebephrenia to show some variation, it seemed likely to Boisen that such a large amount of variation was associated with nonsystematic aspects of the diagnostic processes.

Ash's Study of Interclinician Agreement

A potential concern about both Boisen (1938) and the Masserman and Carmichael (1938) studies is that neither study directly examined how well different clinicians would agree when diagnosing the same patients. The first major study to focus exclusively on interclinician agreement was performed by Ash (1949) when he was a graduate student in industrial psychology. He had three psychiatrists collectively interview 52 male patients being seen at an outpatient clinic. Ash found that all three psychiatrists agreed on only 20% of the specific diagnoses. However, on nearly half the cases (48.6%), two of the three psychiatrists agreed. Ash noted that if rates of agreement were computed for major categories instead of for specific categories (e.g., "psychosis" instead of "hebephrenic schizophrenia"), the values were higher. Thus, all three psychiatrists agreed on the major category in nearly half the cases (45.7%), and at least two of the three psychiatrists agreed on another 51.4% of the cases.

Ash concluded from his results that (1) psychiatrists have poor levels of interclinician agreement and (2) agreement is better on "major" categories than on specific diagnoses. Although the Ash study has been consistently criticized for its methodological limitations, these two conclusions have gained general acceptance.

A major criticism of the Ash study has been its sample of patients. Ash himself noted that his sample was quite small; hence, his study needed replication. A more serious criticism was pointed out by Matarazzo (1978). In the Ash study, 75% of the assigned diagnoses were "normal with a predominant personality characteristic." Thus, specific category agreement required that the psychiatrists note the same personality features, such as unethical, egocentric, and selfish; or uninhibited, dull, and anhedonic. In short, the Ash study was better characterized as an analysis of interclinician agreement about personality descriptors than as a study of the diagnosis of psychopathological conditions.

Beck (1962) also criticized the Ash study because it used a pre-DSM classification, permitted categories that were not mutually exclusive, and allowed the psychiatrists to use "vague" terms such as the personality traits noted above. Matarazzo (1978) suggested that, given these methodological problems, the most impressive finding in the Ash study was that the psychiatrists showed any agreement at all. It is interesting to note that despite the consistent negative comments on its methodology, the study by Ash (1949) has been frequently cited as demonstrating the poor reliability of psychiatric classification.

Studies from the 1950s

Three major research studies on diagnostic reliability appeared during the 1950s. The first of these was by Hunt, Wittson, and Hunt (1953). These researchers examined the stability of diagnoses on 794 naval enlisted men who had been identified in boot camp as unsuitable for active duty for psychiatric reasons. These enlistees were assigned diagnoses by the clinicians examining them when they entered the military. Then, the enlistees were transferred to a naval hospital and were independently assigned diagnoses (i.e., the recruiting-station diagnoses were not sent to the hospital). In 54.1% of the cases, there was agreement on the major category, and 32.6% agreement occurred for the specific diagnoses. Hunt and his colleagues performed other research studies concerning the process of diagnosis as used in the navy and, especially, focused on the predictive validity of diagnostic decisions. These studies were summarized by Hunt (in 1955).

A second study was carried out by Schmidt and Fonda (1956). As in the Ash study, these researchers focused on agreement among clinicians. However, Schmidt and Fonda used inpatients at a state hospital. The patients were diagnosed by one of eight psychiatry residents and also by one of three psychiatrists. There was some variation in the amount of information available to the two groups. This was the first reliability study to use the first edition of the DSM as the classification system. The overall rate of agreement found by Schmidt and Fonda was 55% for specific diagnoses, and there was an average of 84% agreement on major categories. Schmidt and Fonda concluded that, when the new DSM-I was used, psychiatric classification has "satisfactory reliability demonstrated for some psychiatric diagnoses." Also like Ash (1949), Schmidt and Fonda emphasized that reliability appeared to be better for major categories than for specific diagnoses.

In a third study from the 1950s, Norris (1959) used a large sample of over 6,000 patients. This study had a procedure similar to that used by Hunt *et al.* (1953). Norris analyzed the diagnoses assigned to patients at subsequent institutional settings. Unfortunately, however, the clinicians making the second diagnoses were not blind to the diagnoses made at the initial observation units. The overall rate of agreement for specific diagnoses was relatively high (about 60%, according to Beck, 1962). Agreements were particularly high for many, but not all, of the specific organic brain syndromes, plus some of the functional psychoses. The reliability values estimated for the neuroses and the personality disorders were lower.

Kreitman's Review

The first general review of the reliability literature was written by Kreitman (1961) and was published in the *Journal of Mental Science.*

Kreitman discussed all of the studies mentioned above as well as studies by Doering and Raymond (1934), Foulds (1955), and Goldfarb (1959).

The review by Kreitman was organized into three major sections: (a) an analysis of the variables of concern in reliability studies; (b) a summary of previous research; and (c) a discussion of the results. Kreitman submitted that there were five variables of relevance to diagnostic reliability:

(1) Variables relating to the psychiatrists (p. 876)
(2) Variables relating to the psychiatric examination (p. 877)
(3) Variables relating to the nomenclature and reporting (p. 877)
(4) Variables relating to patients (p. 878)
(5) Manner of analysis (p. 879)

Second, Krietman argued that the best of the previous studies was the one performed by Norris (1959) because she provided the most detail about the disagreements that had occurred. Like later reviewers, Kreitman was not impressed by the Ash study (1949) because most of the subjects did not clearly have a mental disorder and because a systematic classification was not being studied.

The third part of the review concerned Kreitman's conclusions. Kreitman did *not* conclude that his review showed psychiatric classification to be unreliable. Instead, two novel comments were made: (1) Kreitman believed that the apparent lack of reliability in psychiatric diagnosis had been exaggerated; and (2) he suggested that the importance of reliability in analyzing psychiatric classification had been overstated. These two comments should be kept in mind while the reader is studying the remainder of this history.

Three Articles by Beck *et al.*

In 1962, Beck and his associates at the University of Pennsylvania published three journal articles about reliability. The first article was a review of the previous literature (Beck, 1962); the second was a research study of interclinician agreement using four experienced clinicians (Beck, Ward, Mendelson, Mock, & Erbaugh, 1962); and the third commented on the reasons for the disagreements that had occurred in the reliability study (Ward, Beck, Mendelson, Mock, & Erbaugh, 1962).

The first of these articles, Beck's review (1962), covered most of the same literature as had Kreitman's earlier review. Beck made five recommendations for future studies of diagnostic reliability:

1. The most recent version of the DSM should be analyzed in any new studies.
2. The psychiatrists responsible for making diagnoses should be adequately trained and have sufficient clinical experience.
3. The diagnostic process should be uniform for all patients and all clinicians.
4. Diagnostic interviews should either be viewed jointly or the time interval between interviews should be kept to a minimum.
5. The research diagnosis should be kept separate from the diagnosis appearing on a patient's clinical record since administrative issues may affect the latter.

These five recommendations by Beck were generally followed in the research studies that appeared in the 1960s and 1970s.

For the second article, Beck and his colleagues had four board-certified psychiatrists independently interview and diagnose patients in pairs. There were 153 patients, who had been randomly chosen from new referrals to two outpatient clinics in Philadelphia. The two interviews were done back to back and within one week of the patient's initial screening at the clinic. After completing an interview, each clinician made a diagnosis, assigned a degree-of-certainty rating to the diagnosis, and rated the patient's depression on a 22-item scale (probably the Beck Depression Inventory or its predecessor). After completing their interviews of a patient, the two psychiatrists compared their diagnoses and assigned reasons for any disagreements they noted.

The portion of this research focusing on interclinician agreement was published in the paper by Beck *et al.* (1962). The major conclusions from this study were: (1) the overall agreement was statistically significant ($p < .001$) and higher than that obtained in other comparable studies; (2) agreement was high when both clinicians rated their certainty as high (82%); and (3) the best agreement occurred for neurotic depression (63%) and the worst for personality trait disturbance (38%).

The methodology of this reliability study had serious defects. First, because the clinicians discussed their diagnoses after each pair of interviews, the agreement among the psychiatrists could be expected to improve over time as the clinicians learned about each other's preferences. Thus, the methodology was biased toward relatively good rates of agreement.

The finding of maximal agreement among clinicians on the category of "neurotic depression" is inconsistent with the results of other reliability studies. One potential reason for this finding is that, because the psychiatrists were also completing a depression rating scale on the patients, it is likely that they were sensitive to this diagnosis. Another possible reason for the high rate of agreement on the diagnosis of neurotic depression was that this category was frequently assigned to the outpatients being studied. In other words, the base rate of this diagnosis was high (Meehl & Rosen, 1955). Because percentage-of-agreement statistics are affected by the base rates of the diagnoses, using the percentage of agreement as the statistical estimate of reliability could have led to an artificially high estimate (see later comments by Spitzer & Fleiss, 1974).

The third journal article from this research was the paper by Ward *et al.* (1962). This article focused on the attributions that the clinicians made for the sources of disagreement. The four psychiatrists believed that there were three possible reasons for disagreement. These reasons and the relative frequency with which they were endorsed were:

> (a) inconstancy on the part of the patient (2 disagreements: 5%); (b) inconstancy on the part of the diagnostician (13 disagreements: 32.5%); (c) inadequacies of the nosology (25 disagreements: 62.5%). (p. 199)

Ward *et al.* concluded that the major changes needed for improved reliability were changes in the DSM classification system.

The last article, by Ward *et al.*, had a marked impact on subsequent thinking about diagnostic reliability. It blamed interclinician disagreements on the flaws inherent in the DSM-I classification. This attribution has generally been accepted by other researchers and clinicians when they comment on both the DSM-I and DSM-II. Nonetheless, the Ward *et al.* article and its conclusions had serious methodological problems. There were three possible objects of blame for any disagreement that had occurred in the Beck *et al.* reliability study: the patient, the clinicians, or the classification system. The fact that for three of every five disagreements Ward *et al.* decided to blame the system is not surprising. Certainly most persons, including psychiatrists, find it easier to blame a system for their problems rather than themselves or other human beings. For instance, suppose one clinician felt that part of the reason for a disagreement about a diagnosis was that the other clinician had performed a sloppy interview. The first psychiatrist may have felt unwilling to make this attribution because it would insult a colleague. An easier solution would be to claim that the definition of a particular diagnostic category was too vague.

In short, the research by Beck, Ward, and colleagues did not conclusively demonstrate that the major cause of diagnostic disagreement was the DSM system. Nevertheless, many commentators have agreed with the conclusions of Ward *et al.* Thus, this study has been frequently cited despite its limitations. The phenomenon of accepting the conclusions of one unreplicated study, even though a critical analysis of the study should generate suspicions, is not unusual in the social sciences. Samelson (1980), for instance, documented a similar, uncritical acceptance of the famous "little Albert" case published by J. B. Watson (Watson & Rayner, 1920).

Sandifer, Pettus, and Quade

Sandifer *et al.* (1964) performed a reliability study in three North Carolina state hospitals. This research on interclinician agreement was methodologically one of the best of the studies that have been performed. Sandifer *et al.* used 13 clinicians in their research, of which 11 were psychiatrists who had completed a residency training program. Of the 13, 1 was a state hospital superintendent; 2 were assistant superintendents; 6 were clinical directors; and 4 were ward physicians. The length of psychiatric experience of the diagnosticians ranged from 4 to 37 years.

The sample consisted of 91 patients, all of whom were first admissions and who were diagnosed within three weeks of admission. Any patient who had a history of alcohol abuse or who was over 60 years old was excluded. Fifty-five of the patients were female.

The information on each patient was presented in a diagnostic conference at which there were at least 10 clinicians. The presenting problem, the social history, the medical workup, the laboratory tests, and the psychological test results were presented by a member of the hospital staff. Then, the patient was interviewed by one psychiatrist in the presence of the clinicians. After the interview, the clinicians independently assigned diagnoses, using the DSM-I system. The entire diagnostic process required 30–45 minutes.

The results are shown in Table 8. In this table, the columns represent the diagnoses assigned by one clinician; the rows represent the diagnoses assigned by a second clinician. Thus, of the patients given a diagnosis of acute brain syndrome by one clinician, 46% were called acute brain syndrome by a second clinician. The main values of interest occur along the diagonal of the table and are enclosed in small boxes. If the psychiatrists agreed completely in the assignment of all diagnoses, there would be values of 100% down the diagonal of the table. It should be obvious that complete interclinician disagreement

TABLE 8. Agreement Data from Sandifer *et al.* (1964) (in Percentages)

	1	2	3	4	5	6	7	8	9	10	11	12
1. Acute brain syndrome	46									12	30	
2. Chronic brain syndrome		66		6							9	9
3. Schizophrenia			74							6	7	
4. Involutional psychosis		6	10	26	7	12				32		
5. Manic-depressive			7	9	36					31	11	
6. Psychotic depression			11	21		22				33		
7. Paranoid reaction	13		40	9			13			7	9	
8. Psychosis, other		8	17					17				50
9. Psychophysiologic									40	33	16	
10. Psychoneurosis				8	6					56	13	
11. Personality disorder	6		5							14	66	
12. Mental deficiency		12	7					5				73

did not occur. If the psychiatrists were totally unreliable, one would expect values of approximately 8% in *all* cells of Table 8. This did not occur either. Instead, disagreement appears to have occurred in systematic ways. The most striking disagreements are demonstrated by the values that are circled. For instance, if one clinician diagnosed a patient as either involutional psychosis, manic-depressive, or psychotic depression, it was quite likely that another clinician would diagnose the same patient as neurotic (probably neurotic depression). Sandifer *et al.* commented that the overall interclinician agreement was 57%. If disagreement concerning the depressive disorders could have been resolved, the rate of agreement in the study would have jumped to 84%.

The Sandifer, Pettus, and Quade study is not beyond criticism. Two major concerns can be expressed. First, the statistic used in this study (percentage of agreement) does not permit comparison with other studies. Zubin (1967) demonstrated how the statistical formulas for the calculation of "percentage of agreement" varied from one study to another. In addition, the meaning of *percentage of agreement* changes as a function of the base rates with which different diagnoses are used. Suppose, for example, that only two diagnostic categories are considered for a set of patients. Also, suppose that Category A is assigned to 90% of the patients, and Category B to 10%. A reliability study of interclinician agreement is performed and shows 88% agreement. Although 88% seems to be a high rate of agreement, its meaning pales when one recognizes that there would be 82% agreement if both clinicians randomly assigned Diagnosis A to 90% of the patients. In other words, 88% agreement is only slightly higher (6% higher, to be exact) than what might be expected by chance in this hypothetical study.

The other comment made about the Sandifer *et al.* research was that state hospitals tend to have relatively severely disturbed patients, and that diagnostic agreement is generally greater concerning more disturbed patients (Kendell, 1975; Zubin, 1967). Thus, the sampling by Sandifer *et al.* may have led to a somewhat greater agreement percentage than if an outpatient sample had been used. On the other hand, the study does seem representative of the types of patients, the clinicians, and the diagnostic process used in state hospitals. Diagnoses can have important effects for both administrative and therapeutic decisions in a state hospital, whereas the utility of diagnosis for outpatients is less clear.

Zubin's Review

In 1967, Zubin published a review of the classification of psychopathology in the *Annual Review of Psychology.* A major portion of Zubin's review was devoted to his analysis of reliability studies. Zubin subdivided journal articles containing research on diagnostic reliability into three groups. First, there was a set of studies that focused on *interclinician agreement.* The research by Ash (1949), Beck *et al.* (1962), and Sandifer *et al.* (1964) fit in this group. The second type of study analyzed *diagnostic consistency over time.* The studies by Masserman and Carmichael (1938), Hunt *et al.* (1953), and Norris (1959) fell into this group because diagnoses on the same patients from different points in time were being compared. Finally, there were studies that focused on *diagnostic frequencies across comparable samples.* The only study mentioned here so far that belongs to this group is the research by Boisen (1938).

Since Zubin's review and his useful classification of pre-1965 articles, the main focus concerning diagnostic reliability has been on interclinician agreement. Reliability among diagnosticians is essential if a psychiatric classification is to be used as a basis for communication (Spitzer & Fleiss, 1974).

Comparison of British and American Diagnosticians

During the late 1960s and the early 1970s, a series of studies was performed comparing the diagnoses made by British and American psychiatrists (Sandifer, Hordern, Timbury, & Green, 1968; Katz, Cole,

& Lowery, 1969; Kendell, Cooper, Gourlay, Copeland, Sharpe, & Gurland, 1971; Cooper, Kendell, Gurland, Sharpe, Copeland, & Simon, 1972). The major initial stimulus for these studies was the realization that Americans consistently diagnosed a large percentage of inpatients as schizophrenic, whereas Britains used the diagnosis of manic-depressive psychosis more frequently than Americans. Did these differences reflect different frequencies of these two disorders in the two countries? Or did American and British clinicians have different interpretations of what the terms meant? To answer the latter question, most of the studies showed the same videotaped interviews to psychiatrists in both countries and asked them to assign diagnoses.

The best known of the cross-cultural comparisons was performed by Kendell *et al.* (1971). These investigators selected eight videotaped interviews of psychiatric patients. Five of the patients were selected to emphasize suspected diagnostic differences between the British and the Americans.

Kendell and his associates showed the videotapes to groups of psychiatrists in Great Britain. Five of the tapes were seen by a group of 40 psychiatrists at three British institutions. The other three tapes were seen by about 200 psychiatrists at a British national meeting.

Interestingly, Kendell *et al.* had distinctly more difficulty recruiting American psychiatrists to view the tapes. It seemed that these researchers had trouble obtaining psychiatrists in the United States, and thus American diagnosticians were recruited on a "catch-as-catch-can" basis. The number of American psychiatrists seeing each videotape varied from 27 to 133. The vast majority of the psychiatrists were from the New York City area.

Kendell *et al.* found that the Americans most frequently assigned the diagnosis of schizophrenia to all eight videotaped patients. However, the British preferred the diagnosis of manic-depressive for some patients, schizophrenia for others, and personality disorder for others. Kendell *et al.* argued that the Americans had an overinclusive conception of schizophrenia. In other words, what the American psychiatrists mean by *schizophrenia* is what the British call *mania, depression, neurosis, personality disorder,* and *schizophrenia.*

The Kendell *et al.* study did have its limitations. First, because the videotapes were selected to emphasize diagnostic differences between British and American psychiatrists, the cross-cultural variation among clinicians was not surprising. Also, given the problem in recruiting American psychiatrists, it is difficult to decide how generalizable this sample was. Despite these methodological limitations, the other studies comparing American and British views of schizophrenia also showed Americans as having a relatively broad interpretation of this disorder. Kendell believed that these studies had a major impact on the thinking of American psychiatrists: "It is my impression that one of the factors which led to a dramatic change in attitude towards classification and diagnostic criteria by American psychiatrists earlier in this decade was the realization that they were out of step with the rest of the world" (Kendell, personal communication, 1981).

Another perspective on the studies comparing British and American psychiatrists is that these studies demonstrated how important variables reflecting differences among clinicians (e.g., training, socioeconomic class, and race) are in interclinician agreement (Kreitman, 1961). American psychiatrists appear to have been taught a very general concept of schizophrenia, whereas the British, like most other European-trained psychiatrists, have a much narrower meaning of the term. Thus, the differences in the training of these diagnosticians have had a systematic effect on interclinician agreement.

Blashfield and Draguns (1976a) also pointed out that when clinicians have cultural backgrounds different from those of their patients, the clinicians are more likely to assign severe diagnoses. For instance, Simon, Fleiss, Gurland, Stiller, and Sharpe (1973) compared *research diagnoses*, carefully made after studying a wide range of information, with the *hospital diagnoses* of the same patients. They found that, for blacks, the hospital diagnoses had a higher proportion of the schizophrenic label than the research diagnoses.

Review by Spitzer and Fleiss

The review of the reliability literature by Spitzer and Fleiss (1974) clearly and emphatically states the position of the neo-Kraepelinians concerning diagnostic reliability. These authors argued that (1) reliability is a necessary feature of any adequate classification system; (2) studies performed before 1975 showed psychiatric classification to be unreliable; and (3) the reason for the poor reliability was the inherent vagueness in the DSM-I and the DSM-II:

> The reliability of psychiatric diagnosis as it has been practiced since at least the late 1950's is not good. It is likely that the reasons for diagnostic unreliability are the same now as when Beck et al. (1962) studied them. They found that a significant amount of the variability among diagnosticians was due to differences in how they elicited and evaluated the necessary information, and that an even larger amount was due to inherent weakness and ambiguities in the nomenclature. (p. 345)

An important feature of the Spitzer and Fleiss paper was the use of a statistic called *kappa* as a measure of interclinician agreement. The formula for kappa is:

$$\kappa = \frac{P_o - P_c}{1 - P_c}$$

where P_o is defined as the proportion of observed agreement among clinicians and P_c is the proportion of agreement expected by chance. Spitzer and Fleiss advocated the use of kappa in studies of interclinician agreement because it corrects for the different base rates at which diagnostic categories are used.

Spitzer and Fleiss reanalyzed the results from all previous studies that provided sufficient detail and calculated the kappa values for various diagnoses. The results after recomputation are shown in Table 9. These results suggest that the reliabilities for organic brain syndromes and alcoholism were relatively good. In addition, the reliability for schizophrenia appears "fair," even though obvious variability existed. For other disorders, such as the neuroses, the personality disorders, and the affective disorders, the level of interclinician agreement was inadequate according to Spitzer and Fleiss.

Notice that the use of kappa did affect the interpretation of the studies. For instance, Beck *et al.* (1962) concluded that their study found higher levels of interclinician agreement than previous research and also that neurotic depression was the most reliable category that they had examined. From Table 9, neither of the conclusions of Beck *et al.* appears to be correct. First, the kappa values from the Schmidt and Fonda study were uniformly higher than for the Beck *et al.* study. Second, sociopathic disorders, not neurotic depression, had the highest reliability in the Beck *et al.* study once a correction was made for base rates using kappa.

Kappa

During the 1970s, the statistic kappa became standard in almost all reliability studies. The next dozen pages focus on the details concerning kappa and the implicit assumption that this statistic makes about reliability. As a result, these pages contain a foray into the world of statistics in the hope that it will clarify the meaning of reliability and the notions about "error" that are inherent in the use of reliability statistics. Once the statistical aspects of reliability have been dis-

TABLE 9. Kappa Measures of Diagnostic Reliability (Computed by Spitzer and Fleiss, 1974)

	Study[a]							
					V			
Category	I	II	III	IV	New York	London	VI	Mean
Mental deficiency				.72				.72
Organic brain syndrome	.82	.90					.59	.77
Acute brain syndrome				.44				.44
Chronic brain syndrome				.64				.64
Alcoholism					.74	.68		.71
Psychosis	.73	.62		.56	.42	.43	.54	.55
Schizophrenia	.77		.42	.68	.32	.60	.65	.57
Affective disorder					.19	.44	.59	.41
Neurotic depression			.47		.20	.10		.26
Psychotic depression				.19	.24	.30		.24
Manic-depressive				.33				.33
Involutional depression			.38	.21				.30
Personality disorder or neurosis	.63			.51	.24	.36		.44
Personality disorder			.33	.56	.19	.22	.29	.32
Sociopathic			.53					.53
Neurosis		.52		.42	.26	.30	.48	.40
Anxiety reaction			.45					.45
Psychophysiologic reaction				.38				.38

[a]Study I, Schmidt and Fonda (1956); Study II, Kreitman (1961); Study III, Beck *et al.* (1962); Study IV, Sandifer *et al.* (1964); Study V, Cooper *et al.* (1972); Study VI, Spitzer, Endicott, Cohen, and Fleiss (1974).

cussed, this chapter returns to the fundamental conceptual issues related to this topic.

The Spitzer and Fleiss (1974) article was not the first to advocate the use of kappa. Cohen first proposed this measure in 1960 as an estimate of nominal scale agreement. Independently, Scott (1955) also suggested the same statistic. Krippendorf (1970, 1971) also discussed generalizations of the kappa statistic. Spitzer, Cohen, Fleiss, and Endicott (1967) were the first to specifically advocate the use of kappa as a measure of reliability in studies of diagnostic agreement among psychiatrists.

Prior to the use of kappa, reliability studies were often likely to be subject to one or more statistical errors (Bartko & Carpenter, 1976):

(1) The computational formula for estimating the percentage of agreement varied across reliability studies (Zubin, 1967). Thus, a direct comparison of the results was very difficult. (2) The use of percentage-of-agreement statistics did not correct for the amount of agreement expected by chance. In other words, the different base rates were ignored. (3) The statistical properties of such measures as the percentage of agreement were unknown. Thus, it was not possible to perform tests of significance on any results.

Bartko and Carpenter demonstrated that kappa had the advantage of overcoming all three of the problems mentioned above: (1) Kappa had a well-defined computational formula. The statistic could be used when there were a number of clinicians, when more than one diagnosis was assigned to a patient, and when certain diagnostic disagreements were much more serious than other disagreements. (2) Kappa, by definition, was designed to correct for chance. Hence, the meaning of kappa should be consistent regardless of base rates. (3) The statistical theory behind kappa was relatively well described, and significance tests could be performed (Fleiss, 1973; Hubert, 1977, 1978). During the 1970s, kappa became the standard statistical measure of reliability in the studies of interclinician agreement. The various percentage-of-agreement indexes fell into disrepute and were no longer used in this literature.

Maxwell's Criticism of Kappa

Although kappa became the statistical measure of choice for reliability studies, it was not without its critics. The most effective criticism was made by Maxwell (1977). He argued that the manner in which chance was assumed to operate when using kappa may not always be plausible.

To understand Maxwell's criticism, consider Table 10. This table presents the hypothetical results when two clinicians diagnose 100 patients for alcoholism. In 4 cases, both clinicians agree that the patients are alcoholic, and in another 64 cases, both clinicians agree that the patients are not alcoholic. Therefore, the percentage of agreement is 68%:

$$68\% = \frac{4 + 64}{100}$$

In contrast, the *specific percentage of agreement*, as defined by Helzer,

TABLE 10. Agreement between Two Clinicians in the Diagnosis of Alcoholism

		Clinician 1	
		Alcoholic	Not alcoholic
Clinician 2	Alcoholic	4	4
	Not alcoholic	28	64

Clayton, Pambakian, Reich, Woodruff, and Reveley (1977), is only 11%:

$$11\% = \frac{4}{4 + 28 + 4}$$

Remember that the latter statistic, as discussed on page 50 in Chapter 2, considers only those cases in which at least one clinician identifies the case as alcoholic. Thus, the 64 clearly negative cases are ignored.

The kappa statistic can also be computed for these data:

$$P_o = \text{Proportion of observed agreement}$$
$$P_o = \frac{4 + 64}{100}$$
$$P_o = .68$$

Notice that P_o has the same value as the most common interpretation of the percentages of agreement statistic.

$$P_c = \text{proportion of agreement expected by chance}$$
$$P_c = P_1 Q_1 + P_2 Q_2$$

where P_1 is defined as the proportion of cases called alcoholic by the first clinician; P_2 is the proportion of cases called not alcoholic by the first clinician; and Q_1 and Q_2 are the analogous proportions for the second clinician.

$$P_1 = \frac{4 + 28}{100} = .32$$
$$P_2 = \frac{4 + 64}{100} = .68$$
$$Q_1 = \frac{4 + 4}{100} = .08$$
$$Q_2 = \frac{28 + 64}{100} = .92$$

Notice that there is clear disagreement between the two clinicians in terms of the number of alcoholics present in the sample. Clinician 1 assigns the diagnosis of alcoholism to almost one of every three patients. In contrast, Clinician 2 believes that alcoholics are infrequent and uses the diagnosis in only 1 of every 12 cases:

$$P_c = (.32) \times (.08) + (.68) \times (.92)$$
$$P_c = .6512$$

In other words, given the base rates with which these two clinicians use the diagnosis of alcoholism, their expected agreement by chance is 65%. As a result:

$$\kappa = \frac{P_o - P_c}{1 - P_c}$$
$$\kappa = .08$$

This value of kappa is low and suggests that these two clinicians are not very reliable when making a diagnosis of alcoholism. Notice that a different inference might be made if the researcher performing this study reports instead that there is a 68% agreement between the two clinicians.

As noted earlier, Maxwell was concerned with the manner in which kappa implicitly assumes chance to be operating. The value of P_c in the kappa statistic is dependent on the base rates by which the two clinicians use the diagnosis. Clinicians 1 and 2 have distinctly different base rates: $P_1 = .32$, and $Q_1 = .08$. These different base rates are used in the computation of kappa.

Maxwell argued, however, that if the diagnoses being made by the two clinicians are truly random events, then

$$P_1 = Q_1$$

In other words, the fact that the two clinicians use this diagnosis with different frequencies suggests a systematic difference in the way the clinicians assign the diagnosis. For instance, Clinician 2 may use the diagnosis only in those cases in which there is physiologic evidence of addiction (e.g., blackouts, withdrawal after abstinence, difficulty with gait, and liver or gastric symptoms). In contrast, Clinician 1 may assign the diagnosis to anyone whose alcohol intake is reported as a fifth of liquor per week or more. Maxwell would consider such a difference between the two clinicians as informative and as not being due to chance. If diagnostic assignment were a purely random event, such as being controlled by the flip of a coin, one would expect the proportion of cases called "alcoholic" to remain relatively constant.

Maxwell suggested an alternative approach to reliability. Suppose that A_1 is defined as the proportion of patients that both clinicians agree, *without any doubt,* are alcoholic, and A_0 is the proportion that they are certain are not alcoholic. Then, the remaining cases (G) are those about which the two clinicians have guessed. If PA is defined as the observed proportion of cases for which the clinicians agreed, then according to Maxwell:

$$PA = A_1 + \frac{G}{4}$$

$$PA = A_1 + \frac{1 - A_1 - A_0}{4}$$

$$\text{since } A_1 + A_0 + G = 1$$

In other words, Maxwell was suggesting that the proportion of cases on which both clinicians agree (which is .04 in the hypothetical example) is the sum of the proportion for which they agree without any doubt plus one-quarter of the questionable cases. Thus, the questionable cases are spread equally among all four cells of the agreement matrix shown in Table 10, a plausible occurrence if the questionable cases are randomly distributed. A similar argument could lead to the following estimate for the observed proportion of negative cases (NA):

$$NA = A_0 + \frac{G}{4}$$

$$NA = A_0 + \frac{1 - A_1 - A_0}{4}$$

Solving for A_1 and A_0,

$$A_1 = 1.5PA + .5NA - .5$$
$$A_0 = .5PA + 1.5NA - .5$$

From Table 10,

$$PA = .04$$
$$NA = .64$$

After substituting the observed values in the two equations for NA and PA above, the equations can be solved for A_1 and A_0:

$$A_1 = -.12$$
$$A_0 = .48$$

Maxwell commented that the negative values for the proportion of positive agreements that occurred with certainty should lead a researcher to be seriously concerned about using the four patients who are called "alcoholic" by both clinicians. Their judgments in this diagnosis are suspect. On the other hand, they appear to be in relatively close agreement about the diagnosis of cases who are not alcoholic.

In addition to the ideas mentioned above, Maxwell proposed a new statistic called the *random error coefficient of agreement* (*RE*). The estimate of the statistic for the hypothetical example is:

$$RE = A_1 + A_0$$
$$RE = -.12 + .48$$
$$RE = .36$$

Notice that the value of $RE = .36$ is distinctly higher than kappa (.08), thereby suggesting greater interclinician agreement but still showing agreement to be much lower than optional (i.e., lower than $RE = 1.00$).

Janes (1979), expanding on the criticism by Maxwell (1977), provided a further comparison of the *RE* statistic and kappa. Janes demonstrated that the most striking difference between the two occurs when the base rates of the clinicians are relatively different. When the base rates are nearly equal, *RE* and kappa yield comparable results. Like Maxwell, Janes criticized the implicit definition of chance that is used when kappa is the statistical measure of reliability.

The Overemphasis on Reliability

Carey and Gottesman's Critique of Reliability Studies

The final paper to be discussed in this historical review is by Carey and Gottesman (1978). These two authors presented what they considered four important misunderstandings about reliability:

1. *High reliability is necessary if one is to have at least a moderately valid classificatory system.* Carey and Gottesman argued that this assertion is incorrect. And they presented a hypothetical example in which the diagnosis of schizophrenia appears to be low ($\kappa = .36$) but the validity estimate is high ($\phi = .93$). Moreover, if these same clinicians diagnosed a different sample of patients, the reliability would decrease to $\kappa = .26$ but the validity would become nearly perfect, $\phi = .981$. Because this argument by Carey and Gottesman is important and is counterintuitive, it is developed subsequently in more detail.

2. *Diagnostic reliability, when statistically estimated by means of kappa, is not affected by changes in base rates.* Carey and Gottesman asserted the converse: base rates do effect estimates of reliability. They gave an example in which a *strict definition* of "bipolar affective

disorder" is contrasted with a *loose definition*. A strict definition is one that assigns a tight meaning to the concept in order to avoid including any questionable cases. A loose definition, on the other hand, is relatively broad and attempts to be sufficiently wide so that any true member of the diagnostic category will be included. On a sample in which the base rate of persons with a bipolar disorder was low (e.g., one of every seven patients), the reliability of the strict definition was $\kappa = .88$, whereas for the loose definition, it was $\kappa = .57$. However, on a sample where bipolars occurred with a high base rate (e.g., two of every three patients), the strict reliability of the definition would be $\kappa = .72$ and the loose definition would appear to be more reliable at $\kappa = .86$. Thus, the choice of a strict or a loose definition on the basis of which was more reliable would be affected by the base rate within the sample being examined.

Comment: For readers who are concerned about this example because kappa appears to correct for base rates, remember that kappa corrects for the base rates at which the clinicians use the diagnoses. Kappa does not correct for the base rates of the actual disorders among the patient sample.

3. *In order to improve the classification of mental disorders, efforts should be directed at increasing reliability.* Carey and Gottesman cautioned that after a certain point, improving reliability can actually lead to decrements in validity. This is the *attenuation paradox*, which has been recognized in psychometric theory (Loevinger, 1954). This caution should not be interpreted as claiming that improvements in reliability are a waste of time. Generally, such improvements are useful. However, definitions of diagnostic categories can become overly precise; and even though reliability would be improved, the result could be a decrease in validity. The attenuation paradox illustrates the need for estimates of validity as well as estimates of reliability in the empirical analysis of classificatory systems.

4. *The errors in diagnosis are random.* Carey and Gottesman, like Maxwell (1977) and Janes (1979), differentiated between those cases in which the clinicians assigned diagnoses "without doubt" and the questionable cases in which diagnoses were assigned randomly. Maxwell assumed that the questionable cases would be evenly and randomly distributed across the four cells of the agreement matrix.

Carey and Gottesman countered that much of the disagreement associated with the diagnostic process may, in fact, be systematic and not random. For instance, in the example of the two clinicians who were screening for alcoholism, it was suggested that one clinician used a relatively strict definition, whereas the other used a relatively loose definition. The disagreement on the tightness of their definitions was

TABLE 11. Diagnoses of All Actual Alcoholics by Both Clinicians

		Clinician 1	
		Alcoholic	Not alcoholic
Clinician 2	Alcoholic	3	2
	Not alcoholic	21	4

not random, and reliability estimates based on random error models do not adequately account for these systematic sources of disagreement. This argument led Carey and Gottesman to assert that "reliability and agreement are not important in some circumstances where validity can be examined directly" (p. 1454).

As noted earlier, Carey and Gottesman argued that high reliability is not always necessary for adequate validity. This claim is important, and further discussion will be devoted to this argument. In order to clarify what Carey and Gottesman were saying, an example will be presented in some detail.

This example is based on the hypothetical study of alcoholics whose reliability results were presented in Table 10 (see pp. 98–102). In this study, two clinicians assign the diagnosis of alcoholism to 100 patients. Suppose that, among these 100 cases, 70% are actually not alcoholic and 30% are. Suppose also that the distribution of actual alcoholics occurs as shown in Table 11 and of nonalcoholics as shown in Table 12. In these tables, the four cells of the clinician-agreement matrix are presented, and the distributions of the actual alcoholics as well as the actual nonalcoholics are shown for the four cells. From Table 11, one will note that 3 of the 4 cases that both clinicians call alcoholic actually are alcoholics. Table 12 shows that 1 nonalcoholic is

TABLE 12. Diagnoses of All Actual Nonalcoholics by Both Clinicians

		Clinician 1	
		Alcoholic	Not alcoholic
Clinician 2	Alcoholic	1	2
	Not alcoholic	7	60

TABLE 13. Validity When Using Consensus Nonalcoholics in Research Study

		Actual	
		Alcoholics	Nonalcoholics
Consensus of two clinicians	Possibly alcoholic	26	10
	Not alcoholic	4	60

called alcoholic by both clinicians. Also from Table 12, of the 64 cases called not alcoholic by both clinicians, 60 are really not alcoholics.

Suppose that the diagnosis of alcoholism were conducted as part of the screening process for a biochemical study of patients with probable affective disorders. The screening for alcoholism is being conducted because the biochemical measures would be invalid for patients with a history of alcoholism. Thus, the researchers have decided to include in their biochemical study only the 64 cases identified as not alcoholic by both clinicians. Table 13 presents a 2 × 2 matrix from which the validity of the consensus decision by the two clinicians can be estimated. From this table, it can be noted that 60 of the actual nonalcoholics have been called not alcoholic; 4 of those called not alcoholic are, in fact, alcoholics; 10 of the cases discarded as possibly alcoholic are, in fact, nonalcoholics; and 26 of the discarded cases have been correctly discarded because they are alcoholics.

A statistic called *phi* can be used to calculate the validity of the consensus decision concerning which patients are not alcoholic. From these data, the value of ϕ is .69. This value is distinctly larger than might be expected from the low reliability for these clinicians, as $\kappa = .08$. Thus, it appears that the clinicians have very low reliability, yet the validity of their decisions is fairly high! This example demonstrates that high levels of reliability are not always necessary for adequate levels of validity.

Notice in comparing Tables 11, 12, and 13 that the reason for the moderately high validity is that only the patients called not alcoholic are important. The cases called alcoholic by both clinicians are lumped together with the 32 cases over which the two clinicians disagree.

The validity estimate would not be so high if the decision were made to use the patients called alcoholic by both clinicians. Suppose, for example, that the intent of the study is to research the family

TABLE 14. Validity When Using Consensus Alcoholics in Research Study

		Actual	
		Alcoholics	Nonalcoholics
Consensus of two clinicians	Alcoholic	3	1
	Possibly not alcoholic	27	69

histories of patients with probable affective disorders who have also become alcoholics. If the consensus diagnosis were used, only 4 of the 100 total patients would be kept for the family history research (and 1 of these 4 cases is actually a nonalcoholic).

Table 14 presents the validity matrix for the family history study. If phi is used as the validity index, the validity is low: $\phi = .20$. This level of validity would be considered unsatisfactory. Under these circumstances, the researchers should use different clinicians or an improved diagnostic process for detecting alcholism in this sample of patients.

The results described above are quite consistent with the comments made by Carey and Gottesman and by Maxwell. First, this example demonstrates, as did Carey and Gottesman, that high reliability is not necessary for high validity. In other words, the assertion by Spitzer and Fleiss (1974) that "There is no guarantee that a reliable system is valid, but assuredly an unreliable system must be invalid" is not correct. An apparently unreliable system can be valid.

Maxwell's analysis of the meaning of reliability when applied to interclinician agreement helps to explain how the apparent paradox of low reliability and high validity can occur. Maxwell divided agreement into two types: agreement about positive cases for a diagnosis (called A_1) and agreement about negative cases that do not have a disorder (called A_0). In the cases of two clinicians diagnosing alcoholics, the value of A_0 is moderately good ($A_0 = .48$). Thus, the moderately good validity index of $\phi = .69$, when only agreed-upon negative cases are used, is expected. In the same way, Maxwell's analysis showed these two clinicians to have low agreement over positive cases of alcoholism ($A_1 = -.12$). In this context, the low validity when only positive cases are used is also not surprising.

CONCLUDING COMMENTS

The issue of reliability became an important focus of empirical studies on classification under the aegis of the neo-Kraepelinians. The clearest statement of the position by the neo-Kraepelinian movement is contained in the Spitzer and Fleiss (1974) review. These authors asserted:

1. "There is no guarantee that a reliable system is valid, but assuredly an unreliable system must be invalid" (p. 341).
2. Studies performed before 1974 showed traditional psychiatric classification to be unreliable.
3. The reason for the unreliability was the inherent vagueness in definitions contained within the DSM-I and the DSM-II.

As a result of these assertions, the neo-Kraepelinians believed that the DSM-III should use definitions modeled after the Feighner *et al.* (1972) criteria in order to increase the preciseness and, therefore, the reliability of this classification.

1. *"Assuredly an unreliable system is invalid."* As demonstrated by Carey and Gottesman, this claim is incorrect. This demonstration undermines the emphasis on reliability by the neo-Kraepelinians and suggests a focus on other aspects of classification.

Readers who are familiar with the recent changes in the concept of reliability in psychometric theory (Cronbach, Rajaratnam, & Gleser, 1973) are probably not surprised by the Carey and Gottesman demonstration. Reliability is *not* an inherent property of a classification system (nor of a psychological test) (cf. Sanson-Fisher & Martin, 1981). Any definition of reliability involves some implicit or explicit decision about what types of variance are undesirable. Depending on the uses of a classification system, different interpretations of error variance may be appropriate. In other words, a particular diagnostic category may be used reliably for one purpose, but the same category may be "unreliable" if used for another purpose.

2. *Studies before 1974 showed traditional psychiatric classification to be unreliable.* This assertion is not correct. Admittedly, interclinician agreement is less than optimal. But a review of the reliability literature concerning medical diagnosis (Koran, 1975a,b) showed that all clinicians show less than perfect agreement when making decisions about patients. And psychiatrists are no exception to this rule.

The failure to obtain perfect agreement, however, should not be interpreted as demonstrating unreliability. As shown in Table 8 (from

Sandifer *et al.*, 1964), diagnostic disagreements generally occur in an organized fashion. Clinicians do not assign diagnoses randomly. There exist certain biases that influence the diagnostic process and how clinicians use classificatory concepts.

3. *The reason for reliability was the inherent vagueness in traditional category definitions.* To date, only one study (Ward *et al.*, 1962) has provided support for this conclusion; however, it had a number of methodological shortcomings that make its central conclusions suspect. Without further research, the assertion that vague definitions were a major source of error variance in reliability studies cannot be regarded as adequately proven.

Given that the conclusions suggested by the neo-Kraepelinians seem unwarranted, what conclusions do seem reasonable? First, interclinician agreement when assigning diagnoses is important and valuable, even though it is not always necessary. The example contained in this chapter demonstrates that reliability is not logically necessary in order for a classification to have utility. Nonetheless, interclinician agreement is extremely important. In particular, interclinician agreement is valuable if a classification system is to be useful as a basis for communication among clinicians. If clinicians do not have generally accepted meanings associated with diagnostic concepts, the clinical value of a classification system will be nil. In addition, improvements in a classification system that are designed to increase its reliability generally increase the fidelity of the diagnostic measurements, thereby making diagnosis more useful for developing a science of psychopathology.

Second, an important area for future research will be an analysis of the reasons for diagnostic disagreement. Additional studies that demonstrate whether psychiatric classification is reliable or unreliable would not be particularly informative at this time unless the researchers were studying a brand-new classification system. The numerous reliability studies that have been reviewed in this chapter and elsewhere have all shown that diagnostic disagreement is common. The important issue becomes, What are the reasons for this disagreement? Potential sources of disagreement appear to be related to (1) the perspective on classification that the clinicians learned while in training; (2) the type and amount of information that the clinicians have; (3) the classification system that is being used; and (4) the particular diagnostic concepts that are most likely for a given case. These comments about the sources of disagreement are frankly speculative. A great deal of research is needed to clarify the reasons for diagnostic disagreement so that the precision of the diagnostic process can be increased.

Third, another important area for future research on classification will be the analysis of the *validity* of classificatory concepts. Admittedly, these studies will be more complicated and more difficult to complete than studies of reliability. Nonetheless, until well-conceived validity studies are performed, the utility of classificatory concepts will not be adequately assessed, regardless of how reliable the concepts have been shown to be.

The conclusions reached by this review of the literature on the reliability of psychiatric classification are very much in agreement with the conclusions, two decades ago, of Kreitman (1961) in his initial review of the same literature:

> The importance of diagnosis for psychopathology, epidemiology, therapeutic assessment, clinical psychology, and administration needs no repeated emphasis here. Even though there is scant evidence that the present state of diagnosis is as uniformly bad as it is often reported to be, there is clearly room for improvement. Such progress must depend not only on diagnostic studies themselves but also upon developments in a number of other fields. The problem of diagnosis is ultimately one of "validity," which implies the use of some other criteria, such as the natural history of pathology of an illness against which the procedures of clinicial diagnosis can be checked. Hence the contribution which studies of reliability can make is necessarily limited, nor is the relation of reliability to validity always clear-cut. (p. 884)

SUGGESTED READINGS

Matarazzo, J. D. The interview: Its reliability and validity in psychiatric diagnosis. In B. B. Wolman (Ed.), *Clinical diagnosis of mental disorders: A handbook*. New York: Plenum Press, 1978.

This 50-page chapter is an analysis of the psychiatric interview as a diagnostic device. Contained within the chapter is a historically organized review of the studies on diagnostic reliability. Matarazzo's review is recommended because he analyzes many of the same studies discussed in this chapter, but he often generates conclusions different from those drawn here. Thus, reading Matarazzo will help the reader to see this literature from a different perspective.

Meehl, P. E. *Psychodiagnosis: Selected papers*. New York: Norton, 1977.

Meehl has been an influential and important commentator on the diagnostic process. Meehl, a psychologist with a behavioral orientation whose early research had important effects on the Minnesota Multiphasic Personality Inventory, is knowledgeable in the philosophy of science and has become a psychoanalytically oriented therapist. This book contains many of the papers about diagnosis for which Meehl has become well known, and it discusses topics such as base rates, clinical versus statistical prediction, construct validity, and the process of clinical case conferences. For anyone interested in studying how clinicians approach the diagnostic process and how this process might be improved, this remarkable book is a must.

Wiggins, J. S. *Personality and prediction: Principles of personality assessment.* Reading, Mass.: Addison-Wesley, 1973.

A major argument in this chapter has been that the study of reliability *per se* has been overemphasized in the literature on psychiatric classification. Instead, future research would be more profitably devoted to understanding the diagnostic process (i.e., how a clinician decides to which category a patient should be assigned). The Wiggins book is a challenging and penetrating summary of much of the research on human judgment that was performed during the 1960s and 1970s. The ideas and research on this topic of Lewis Goldberg, Robin Dawes, Kenneth Hammond, Paul E. Meehl, and other psychologists are explored in detail.

Chapter 5

DSM-III

DSM-III: Diagnostic Delight or Nosological Nightmare?

Peter Nathan, 1979, p. 477

The third edition of the *Diagnostic and Statistical Manual of Mental Disorders,* otherwise known as the DSM-III, was published in 1980. As its authors noted, this new classification system for American psychiatry was met with a variety of reactions including "interest, alarm, despair, excitement and joy." The publication of the DSM-III represented the culmination of the work on classification by members of the neo-Kraepelinian movement. Their concern about diagnostic reliability, their advocacy of psychiatry as a viable branch of scientific medicine, and their belief that description is an important step in the process of understanding psychopathology are all reflected in the DSM-III. Because of the importance of the DSM-III, this chapter is devoted to a discussion and criticism of this new system.

The most obvious difference between the DSM-III and the two previous editions is that the DSM-III is much longer. The DSM-I consists of 130 pages and approximately 34,000 words. The DSM-II is virtually the same size (134 pages and 34,000 words). In contrast, the DSM-III is about five times longer, with 494 pages and 177,000 words.

In format, the DSM-III is more like a textbook. The majority of the DSM-III is subdivided according to major diagnostic categories, rather like an abnormal psychology textbook. The DSM-III has a table of contents, an index, and an appendix that defines the technical terms used in the body of the book. Also like a textbook concerned with psychopathology, the DSM-III contains a great deal of descriptive information about the categories. For instance, material is presented about the prevalence, the complications, the familial patterns, sex ratios, and predisposing features for most disorders. As noted in Chapter 1, Kraepelin did not propose classification systems *per se.* Instead, he wrote textbooks, and his classifications were extrapolated from the table of contents in his books. By being structured like a textbook, the DSM-III is more like Kraepelin's system than is either the DSM-I or the DSM-II.

Other differences exist between the DSM-III and DSM-I/DSM-II. The authors of the DSM-III suggested that this new classification made four major changes when contrasted with the earlier editions: (1) the use of diagnostic criteria; (2) a multiaxial approach to patient evaluation; (3) expanded descriptive information about the categories; and (4) a reorganization of the diagnostic categories. Each of these innovations is discussed here in turn.

INNOVATION 1: DIAGNOSTIC CRITERIA

One difference between the DSM-III and the two earlier editions of the DSM is the use of diagnostic criteria. The form of these definitional criteria is modeled after the criteria used in the Feighner *et al.* paper (1972). Many authors have described the DSM-III diagnostic criteria as "Chinese menu definitions." That is, when using a Chinese menu, one must choose so many foods from Category A, a certain number from Category B, and so on. In the same way, a DSM-III definition requires that a patient demonstrate two symptoms from Category A, one from Category B, and two from Category C, for example.

To give the reader some sense of the difference between the DSM-III criteria and the prose definitions from earlier DSM editions, three definitions of mania are presented below: the first is from the DSM-I; the second is the definition listed in the DSM-II; and the third is the diagnostic criteria as listed in the DSM-III:

DSM-I Definition

000 - x11 MANIC DEPRESSIVE REACTION, MANIC TYPE

This group is characterized by elation or irritability, with overtalkativeness, flight of ideas, and increased motor activity. Transitory, often momentary, episodes of depression may occur, but will not change the classification from the manic type of reaction. (p. 25)

DSM-II Definition

296.1 MANIC-DEPRESSIVE ILLNESS, MANIC TYPE (MANIC DEPRESSIVE PSYCHOSIS, MANIC TYPE)

This disorder consists exclusively of manic episodes. These episodes are characterized by excessive elation, irritability, talkativeness, flight of ideas, and accelerated speech and motor activity. Brief periods of depression sometimes occur, but they are never true depressive episodes. (p. 36)

Note that in the DSM-II definition, six symptoms are mentioned:

1. Excessive elation
2. Irritability
3. Excessive talkativeness
4. Flight of ideas
5. Accelerated speech
6. Accelerated motor activity

The DSM-II does not tell the clinician how many of the six symptoms are sufficient to make a diagnosis of mania or if any of the symptoms are more informative than others.

DSM-III Definition

DIAGNOSTIC CRITERIA FOR A MANIC EPISODE

A. One or more distinct periods with a predominantly elevated (MA1), expansive (MA2), or irritable mood (MA3). The elevated or irritable mood must be a prominent part of the illness and relatively persistent, although it may alternate or intermingle with depressive mood.

B. Duration of at least one week (or any duration if hospitalization is necessary), during which, for most of the time, at least three of the following symptoms have persisted (four if the mood is only irritable) and have been present to a significant degree:

 (1) increase in activity (either socially, at work, or sexually) [(MB1a)] or physical restlessness [(MB1b)]

 (2) more talkative than usual [(MB2a)] or pressure to keep talking [(MB2b)]

 (3) flight of ideas [(MB3a)] or subjective experience that thoughts are racing [(MB3b)])

 (4) inflated self-esteem [(MB4a)] (grandiosity, which may be delusional [(MB4b)])

 (5) decreased need for sleep [(MB5)]

 (6) distractibility, i.e., attention is too easily drawn to unimportant or irrelevant external stimuli [(MB6)]

 (7) excessive involvement in activities that have a high potential for painful consequences which is not recognized, e.g., buying sprees, sexual indiscretions, foolish business investments, reckless driving [(MB7)]

C. Neither of the following dominates the clinical picture when an affective syndrome is absent (i.e., symptoms in criteria A and B above):

 (1) preoccupation with a mood-incongruent delusion (MC1a) or hallucination [(MC1b)]

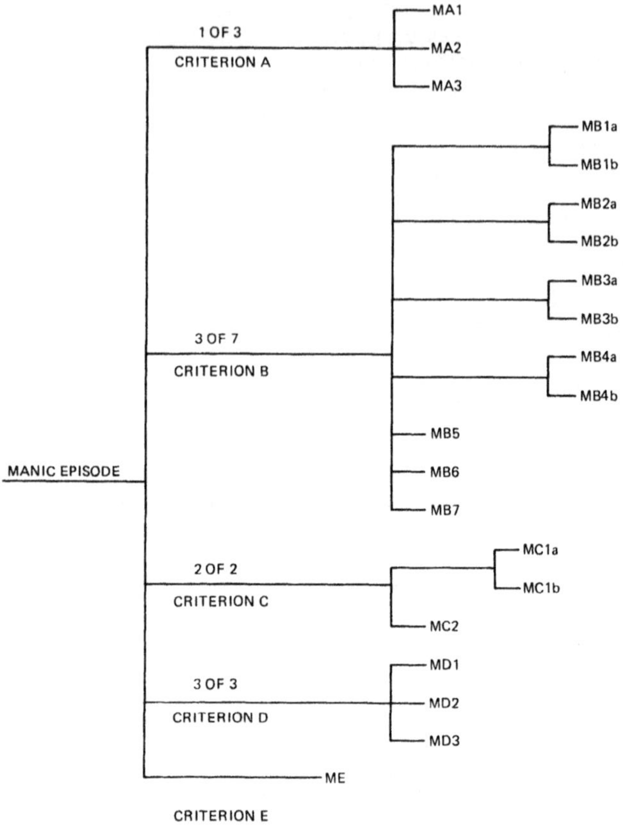

FIGURE 7. Implicit information structure in DSM-III criteria for manic episode.

(2) bizarre behavior [(MC2)]

D. Not superimposed on either Schizophrenia [(MD1)], Schizophreniform Disorder [(MD2)] or a Paranoid Disorder [(MD3)].

E. Not due to any Organic Mental Disorder, such as Substance Abuse [(ME)]. (pp. 208–209)

Note that there are 21 separate symptoms listed in the DSM-III definition. Each symptom is denoted by a code listed in parenthesis in the definition. Figure 7 displays how these symptoms are combined to form a diagnosis. Compared with the 6 symptoms listed for mania in the DSM-II, 21 are listed in the DSM-III. Also the DSM-III notes which symptoms are the most important for a diagnosis. For instance, an elevated, expansive, or irritable mood is considered essential according to the authors of the DSM-III. Although two of these mood-related symptoms are listed in the DSM-II definition, they are not emphasized as being more crucial to the diagnosis of mania than the other symptoms.

The major reason for using diagnostic criteria is to increase the reliability among clinicians. Spitzer, Williams, and Skodol (1979) have suggested that there are three major sources of variance that account for most unreliability in the diagnostic process: (1) information variance; (2) observation/interpretation variance; and (3) criterion variance. *Information variance,* according to Spitzer, refers to the differences among clinicians in the information that they elicit from patients. For instance, one clinician may learn a great deal about the patient's family history during an interview, whereas another may focus on the symptoms of the patient. The second source, *observation/ interpretation variance,* concerns the variability among clinicians in what observations they make about a patient and how those observations are interpreted. If two clinicians watch the same videotape of a patient, the clinicians are presented with the same information. But one clinician may make more observations about the nonverbal behavior of the patient or may make a different interpretation of the patient's glances about the room. The last of the three is *criterion variance.* This variance is associated with the different definitions that clinicians may associate with a diagnosis. For instance, one clinician may have been taught that a patient with mania will not display a thought disorder, whereas another clinician may diagnose a patient as manic even if he or she has a thought disorder because the mood changes predominate in the patient's clinical picture.

Spitzer and his colleagues have frequently cited the Beck *et al.* (1962) study as showing that criterion variance (i.e., lack of precision in definitions) is responsible for the largest percentage of variations among clinicians when assigning diagnoses. As noted in Chapter 4, the methodology of the study by Beck and his colleagues, however, was inadequate to support the conclusion that most problems with diagnostic reliability are associated with criterion variance. Nonetheless, the neo-Kraepelinians believe that when diagnostic criteria are being used, criterion variance should cease to be a major problem. Thus, the use of diagnostic criteria should make the DSM-III a better classification because of the associated improvement in reliability. The data reported by Spitzer, Forman, and Nee (1979) demonstrated that the DSM-III criteria improved the reliability of mania from .61 with the DSM-I/DSM-II definitions (averaged across five studies by Spitzer and Fleiss, 1974) to .87 with the DSM-III diagnostic criteria (as calculated from the field studies).

In a useful paper that discusses how classificatory systems can be created for psychopathology, Pfohl and Andreasen (1978) discussed two approaches to the formation of category definitions: (1) an intuitive approach and (2) an empirical approach. The intuitive approach had been the standard means of creating definitions in the DSM-I and the DSM-II. These classifications were written by small committees of 7–10 psychiatrists who were considered experts in psychopathology. On the basis of their clinical and scientific knowledge, the experts intuitively created what they considered the best current definition of the categories. The empirical approach, in contrast, uses statistical procedures such as discriminant analysis and logical decision-tree analysis to determine empirically which symptoms are most informative concerning a diagnostic category. For instance, Wing *et al.* (1974) had used the logical decision-tree method to analyze the relationships among a collection of symptoms and a group of diagnostic categories. The symptoms were all items from a structured interview developed by Wing *et al.* called the Present State Examination (PSE). This analysis led to the creation of a computer program called CATEGO, which can take the symptoms as elicited on the PSE and postulate the diagnosis for the patient.

The diagnostic criteria used in the DSM-III were created by the intuitive approach. In 1974, the American Psychiatric Association appointed a 19-person DSM-III Task Force to write this new classification. The chairperson of the task force was Robert L. Spitzer. Of the 19 persons on the task force, 10 were from universities previously noted as being associated with the neo-Kraepelinian movement. The task force decided to appoint 12 advisory committees, each of which was to focus on the classification and the definition of specific diagnoses in one or more of the major diagnostic categories. For instance, one advisory committee of 10 members (Nancy Andreasen, Paula Clayton, Jean Endicott, Joseph Lipinski, Michael Mavroidis, Harrison Pope, Robert Spitzer, Janet Williams, Robert Woodruff, and Lyman Wynne) was concerned with the schizophrenic, paranoid, and affective disorders. This committee was responsible for forming the diagnostic criteria used to define the disorders conventionally known as the *functional psychoses.*

Even though the criteria were created by means of an intuitive approach, the criteria can be examined empirically. Morey (1979) performed a study to empirically analyze the diagnostic criteria used to define dementia, mania, and schizophrenia in the DSM-III. He found that a total of 91 different symptoms were included in the three criteria. He asked psychiatrists and clinical psychologists to rate these 91 on a 5-point scale according to their informativeness for diagnosing mania, dementia, and schizophrenia. From these ratings, Morey detected those symptoms that were not sufficiently discriminative to be included in the criteria. For instance, one symptom listed under the diagnostic criteria for schizophrenia is "marked impairment in role functioning as a wage-earner, student or homemaker." This symptom was given an informativeness rating of 4.07 for dementia, 3.20 for schizophrenia, and 2.97 for mania (where a rating of 5 meant "highly informative" and a rating of 1 meant "not informative"). Thus, the clinicians believed the symptoms of impaired role functioning to be more indicative for dementia than for schizophrenia. This judgment suggests that the symptom is not of use in the differential diagnosis of schizophrenia. Morey found that 5 of the 21 symptoms in the diagnostic criteria for mania were not sufficiently discriminative to separate mania from schizophrenia or dementia. For instance, one of the symptoms is defined as "Distractibility, i.e., attention is drawn too easily to unimportant or irrelevant external stimuli" (Symptom MB6 from Figure 7). The clinicians rated this symptom as moderately informative for *both* dementia and mania (dementia = 3.43, and mania = 3.70). Thus, the symptom of distractibility could not be used to differentiate between these two major diagnoses.

Another approach to studying the symptoms to be included in diagnostic criteria is exemplified by a study of Andreasen (1979a,b). She was interested in examining the various symptoms that are associated with the concept of a thought disorder: thought blocking, derailment, clanging, neologisms, perseveration, tangentiality, and so on. First, Andraesen proposed definitions for these various symptoms:

> *Distractible Speech.*—During the course of a discussion in an interview, the patient repeatedly stops talking in the middle of a sentence or idea and changes the subject in response to a nearby stimulus, such as an object on a desk, the interviewer's clothing or appearance, etc. *Example.*—"Then I left San Francisco and moved to . . . . Where did you get that tie? It looks like it's left over from the 50's. I like the warm weather in San Diego. Is that a conch shell on your desk? Have you ever gone scuba diving?" (1979a, p. 1318)

Andreasen demonstrated that when these definitions were used, most symptoms had good reliability (i.e., the kappa estimate of reliability for distractible speech was .78). Then interviews of 113 patients were rated by means of these definitions. The patients represented three different diagnoses as defined by the Research Diagnostic Criteria: mania, depression, and schizophrenia. Andreasen (1979b) found that many of the symptoms associated with a thought disorder were, in fact, more frequent among manics than among schizophrenics. For instance, distractible speech was noted in 10 of 32 manic patients but in only 1 of 45 schizophrenics. The thought disorder symptoms that occurred with about equal frequency among schizophrenics and manics were tangentiality, derailment (loosening of associations), incoherence, illogicality, and loss of goals. It was the "negative" formal thought-disorder symptoms that were most indicative of schizophrenia (i.e., poverty of speech and poverty of content of speech). In contrast, some "positive" formal thought-disorder symptoms, such as distractible speech and pressure of speech, were more characteristic of mania. Andreasen also showed that some symptoms commonly associated with schizophrenia, such as neologisms and thought blocking, occurred so rarely that they were of little diagnostic value.

Andreasen's study was notable for its careful definition of the symptoms being analyzed. Her results suggest that the presence of a formal thought disorder in the clinical picture of the patient is not sufficient to discriminate schizophrenia from mania. Instead, only certain of the symptoms commonly associated with the concept of a thought disorder are indicative of schizophrenia (i.e., poverty of speech), whereas other symptoms are actually more indicative of mania (e.g., distractible speech and pressure of speech).

Notice that Morey (1979) had found that distractibility was not considered a good differential symptom by clinicians. In contrast, Andreasen (1979b) showed distractible speech to be useful in differentiating mania from schizophrenia. This apparent inconsistency of results is worth exploring. First, Morey found that his clinicians rated distractibility as being significantly more informative for mania (mean rating for mania = 3.70 on a 5-point scale) than for schizophrenia (mean rating = 2.70). The problem with distractibility as a symptom, according to Morey, was that it did not differentiate between mania and dementia (mean rating = 3.43). Second, the DSM-III defines distractibility in terms of general problems with attention. In contrast, Andreasen was concerned only with speech patterns. Interestingly, general attentional difficulties have often been implicated in the psychological deficit studies of schizophrenics (Maher, 1966). In a different study attempting to isolate the symptoms that can differentiate schizophrenia from primary affective disorder, Fowler, Liskow, Tanna, Lytle, and Mezzich (1980) found that a behavioral rating for inattention ("preoccupied, inattentive, fails to understand questions") was significantly more frequent among schizophrenics. In sum, these studies are consistent in suggesting that "distractibility" can be an important diagnostic indicator. However, exactly what this symptom indicates depends on how the symptom is defined. Since the DSM-III uses symptoms such as distractibility in its diagnostic criteria, precise definitions of these symptoms are necessary if the diagnostic criteria are to have precise meaning.

At a more general level, the studies by Morey (1979) and by Andreasen (1979a,b) demonstrated a major implicit issue associated with diagnostic criteria. The use of these criteria assumes (1) that the meaning and definition of individual symptoms is clear to diagnosticians and (2) that the symptoms can be used to differentiate the diagnostic category that they define from other categories. As regards any particular category, either or both assumptions could be false. This suggests that a fruitful area for descriptive research would be to examine the relations between various symptoms and the diagnostic categories that they define.

INNOVATION 2: MULTIAXIAL DIAGNOSIS

The second major innovation in the DSM-III was the use of a multiaxial system. A complete diagnosis according to the DSM-III involves five axes:

Axis I—Clinical Syndrome
Axis II—Personality Disorder (adults) or
Axis II—Developmental Disorder (children)
Axis III—Medical Disorder
Axis IV—Severity of Psychosocial Stressor
Axis V—Highest Level of Adaptive Functioning

Therefore, the complete diagnosis for a patient might read:

Axis I—Alcohol abuse, episodic [305.02]
Axis II—Passive-aggressive personality disorder [301.84]
Axis III—Malnutrition, hepatitis, pancreatitis
Axis IV—Moderate (chronic marital discord)
Axis V—Good (functioning in two jobs, but poor home life)

Interest in multiaxial systems for psychiatric classification has a fairly recent history. The earliest proposal was made by the Swedish psychiatrist Essen-Moller (1961, 1971). He suggested that a psychiatric diagnosis should occur along two axes: (1) the clinical syndrome and (2) the etiology. He made this suggestion because, like Stengel (1959), he believed that a major source of confusion and controversy in traditional psychiatric classification was the nonsystematic admixture of etiological and syndrome-oriented diagnoses. For instance, "paresis" is a diagnosis that has clear etiological implications. However, paresis can present via a number of clinicial forms, including a paranoid syndrome, a depressive syndrome, or a grandiose syndrome. None of these clinical syndromes of paresis are noted as subtypes in the DSM-I or the DSM-II. In contrast, "psychotic depressive reaction" is defined entirely in terms of the clinical syndrome. The etiology of psychotic depression is unknown. In other words, paresis is a diagnosis that focuses on etiology but ignores the clinical syndrome. Psychotic depression is defined in terms of a clinical syndrome, but the etiology is unclear. In order to remove this confusion, Essen-Moller suggested that the clinical syndrome and the etiology both be diagnosed on separate axes. The idea of using "axes" to represent different types of information about a patient was attractive, and soon, other clinicians were suggesting additional axes (Rutter, Lebovici, Eisenberg, Sneznevskij, Sadoun, Brooke, & Lin, 1969; Rutter, Shaffer, & Shepherd, 1975; Wing, 1970; Ottosson & Perris, 1973; Helmchen, 1975; Strauss, 1975). These various suggestions have been reviewed by Mezzich (1979) and by Helmchen (1980).

There are three major advantages to the use of a multiaxial system. First, a multiaxial diagnosis provides more information than does a standard single-category classification. As Helmchen (1980) pointed out, "The psychiatrist with human sensitivity for the particular individuality of each of his patients may hesitate to code a patient as a case in a system" (p. 43). To diagnose a patient by assigning her or him to one category oversimplifies the knowledge of a clinician about the complexity of the patient. Although a multiaxial system may also oversimplify the issues concerning a patient, the axes permit the representation of various facets about the patient that are important in his or her treatment.

The second advantage of a multiaxial system is that it provides a framework for acquiring more complete and useful information about mental disorders. For instance, Rutter and his colleagues (1969, 1975) have suggested that mental retardation should be coded on an axis separate from the clinical syndrome when diagnosing children. Rutter believed that this separation was important because clinicians often either diagnose the clinical syndrome or diagnose the mental retardation. As a result, the meaning of statistics on the prevalence of either mental retardation or the clinical syndromes is uncertain. Rutter *et al.* (1975) cited an example at a World Health Organization seminar on child psychiatry in which the clinicians were presented with the case of a mentally retarded, epileptic girl who was also psychotic. Rutter *et al.* found that even though all the psychiatrists were aware of the three conditions affecting this girl, most assigned only one diagnosis, and the choice of the one diagnosis among three possibilities varied across the psychiatrists.

The third advantage of a multiaxial system is its use in education. Students who are learning about psychopathology are generally taught to use conventional classificatory categories to aid them in forming general abstract concepts about patients. One major effect of a classification system is that it serves to focus the attention of students on the important features that patients can present. The traditional single-category classification leads primarily to a focus on the patient's self-reported complaints and his or her unusual behaviors. The multiaxial system used in the DSM-III also requires that a student examine the personality of the patient, his or her medical condition, the psychosocial environment, and the patient's previous role functioning.

Although the advantages of a multiaxial system are impressive, it does have two major drawbacks. First, it complicates the work that a clinician must do before making a diagnosis. A multiaxial system involves a time investment for clinicians who are likely to be already overburdened with report writing, the completion of insurance forms, and so on. Even though the additional time will not necessarily cause clinicians to reject the multiaxial system in the DSM-III, it will mean that clinicians must be convinced that the positive benefits of a multiaxial system will outweigh the time required to use it.

The second disadvantage of the contemporary multiaxial systems is that these systems have been proposed only recently and the measurement issues associated with the various axes have not been successfully resolved. More specifically, although the DSM-III has five axes, in actuality it has only two meaningful axes: the clinical snydrome and the medical disorder. For instance, Axis II was included in order to present information about the *personality style* of the patient. However, Axis II contains categories only for personality *disorders* and does not permit the adequate recording of information about the normal personality structure of the patient. Also, the personality disorders, as defined in the DSM-III, have been found to have low interclinician reliability despite the use of explicit diagnostic criteria. Therefore, by their own standards, the authors of the DSM-III are not satisfied with the personality disorder axis.

The last two axes of the DSM-III (psychosocial stressor and highest level of adaptive functioning) are particularly disappointing. Both axes suggest a recognition of the importance of psychosocial environments in the development and maintenance of abnormal behavior. However, the measurement of these two axes involves crude rating scales that contain very little specific information. For instance, Axis IV, psychosocial stressor, is rated along a 6-point rating scale from "None" to "Severe." What does it mean when a specific patient is rated as having a "moderate" stressor on Axis IV? It would be much more useful to know that the patient developed disturbed behavior after learning that her husband was having a homosexual affair or that the patient had just been fired from the first steady job that she had managed to hold in six years or that the patient was being ostracized by a group of good friends after they had learned that she had had an abortion two years before. To use a 6-point rating scale to summarize information about the social, cultural, occupational, and physical environments that form the context for a patient's abnormal behavior is ridiculous. It is rather like replacing Axis I (the clinical syndrome axis) with a 6-point rating scale that ranges from "no mental disorder" to "severely disturbed." Although the inclusion of Axes IV and V is admirable in principle, their measurement by crude rating scales reduces their clinical utility to worthlessness.

Other criticisms can be directed at the multiaxial system as adopted in the DSM-III. Rutter and Shaffer (1980), for instance, were disturbed that mental retardation was not included as a separate axis for children. The reason given by the authors of the DSM-III for *not* incorporating mental retardation as a separate axis was that the categories of mental retardation (mild, moderate, severe, and profound retardation) refer to clinical syndromes. The DSM-III advocates the multiple diagnosis of clinical syndromes on Axis I. Hence, the mental retardation disorders did not need to be kept as a separate axis. In response, Rutter and Shaffer (1980) noted that if the same logic were applied to personality disorders, then personality disorders should be included under Axis I. Nonetheless, the authors of the DSM-III decided to define personality disorders as representing a separate axis. Rutter *et al.* concluded:

> The decision [to not include mental retardation as a separate axis] seems curious in that the studies undertaken [by the World Health Organization] indicated that the greatest need for a multiaxial system was precisely in those cases in which a disorder of emotions or behavior coexisted with intellectual retardation. (p. 380)

INNOVATION 3: DESCRIPTIVE INFORMATION ABOUT CATEGORIES

The third innovation in the DSM-III is the amount of information provided about the various diagnostic categories. In the earlier editions of the DSM, the only information provided about a diagnostic category is the few prose sentences that are used to define the category. For instance, the DSM-II definition of schizophrenia is completed in four *sentences*. In contrast, the DSM-III discusses schizophrenia for 10 *pages*. Included in these 10 pages is information under the following headings:

Essential features	Impairment
Deterioration from previous level of functioning	Complications
	Premorbid personality
Characteristic symptoms involving multiple psychological processes	Predisposing features
	Prevalence
	Sex ratio
Associated features	Familial pattern
Age at onset	Differential diagnosis
Course	Diagnostic criteria

This is a striking increase in the scope of information presented within a DSM and leads to the comparison of the DSM-III with a textbook about psychopathology.

As noted earlier, there are structural similarities between the DSM-III and a textbook. Both are physically about the same size. Both have a table of contents, a glossary, and an index. Both are organized into chapters associated with the major categories of psychopathology.

In addition to the similarities, there are distinct differences between the DSM-III and a textbook. First, the DSM-III does not contain information about suggested or standard treatment approaches to the various mental disorders. A possible reason for this omission was to avoid the controversy that would result from whatever recommendations might have been made. A second difference is that the DSM-III does not attempt to summarize the research literature related to the various disorders. Nor should it. Major textbooks about psychopathology, such as the *Comprehensive Textbook of Psychiatry,* are enormous undertakings. For instance, the CTP-III (Kaplan, Freedman, & Sadock, 1980) is over 3,300 pages long and has almost 250 contributing authors. Such monumental summaries of current information about psychopathology are best left to textbook authors and should not be undertaken by official committees who create classificatory systems. A third difference between the DSM-III and a textbook is that the DSM-III does not contain any references. Without references, the accuracy of the information presented in the DSM-III cannot be checked, nor is there any indication of what sources the authors of the DSM-III used to guide their decision-making processes.

INNOVATION 4: DIAGNOSTIC CATEGORIES

Since Kraepelin's sixth edition, there has been a steady increase in the number of categories recognized officially in psychiatric classifications. The Standard Classified Nomenclature had 97 categories; the DSM-I contains 108 categories; and the DSM-II has 182. The DSM-III, consistent with this trend of expansion, has a total of 265 categories. Extrapolating from these data, the expectation is that the DSM-IV will contain about 350 categories.

In addition to an increase in the number of categories, the DSM-III has also significantly reorganized the relationships among the categories. For instance, the following outline contains all the DSM-II categories in which depression is a major feature:

- I. Psychoses not attributed to physical conditions
  - A. Major affective disorders
    - 1. Involutional melancholia
    - 2. Manic-depressive illness
      - a. Depressed type
      - b. Circular
  - B. Other psychoses
    - 1. Psychotic depressive reaction
- II. Neuroses
  - A. Depressive neurosis

In contrast, the DSM-III organization of affective disorders is very different. No longer is there a major psychotic/neurotic distinction. Nor is there a distinction between endogenous and reactive depression. Instead, the unipolar/bipolar dimension has become prominent in the DSM-III. Below is an outline of the DSM-III categories in which depression would be a major feature.

- I. Affective Disorders
  - A. Major affective disorder
    - 1. Bipolar disorder
      - a. Mixed
      - b. Depressed
    - 2. Major depression (i.e., unipolar depression)

a. Single episode
b. Recurrent

B. Other affective disorders (affective personality disorders)
1. Cyclothymic disorder
2. Dysthymic disorder

C. Atypical affective disorder
1. Atypical bipolar disorder
2. Atypical depression

The reorganization of the affective disorders within the DSM-III is in significant contrast to the DSM-II. One reason for this change is that the reliability studies had shown variability in the classification of depression to be a significant source of clinician disagreement (Sandifer, Pettus, & Quade, 1964). In addition, a number of biochemical and family pedigree studies have emphasized the validity of the unipolar versus bipolar distinction (Winokur *et al.*, 1969). Because the DSM-III contains such a dramatic reorganization of diagnostic categories, a comprehensive analysis of the changes will not be attempted. A review of these changes can be found in an article by Spitzer, Williams, and Skodol (1979). As the reader who compares the DSM-II and the DSM-III can see, most changes have occurred in the areas of organic brain syndromes, the childhood disorders, the affective disorders, the sexual deviations/dysfunctions, and the neuroses. In the remainder of this chapter, the changes concerning schizophrenia and childhood disorders are explored in detail. Schizophrenia was selected because it is a topic of major clinical and research interest in the mental health field. The changes in the area of childhood disorders are important because the expansion of these categories reflects the increased interest among professionals in child psychopathology.

Schizophrenia

Changes in the classification of schizophrenia as proposed within the DSM-III have attracted a relatively large amount of attention (Fenton, Mosher, & Matthews, 1981; Fox, 1981; Lipkowitz & Idupuganti, 1981; Singerman, Stoltzman, Robins, Helzer, & Croughan, 1981; Endicott, Nee, Fleiss, Cohen, Williams, & Simon, 1982). As noted in Chapter 1, schizophrenia has been a historically important disorder in psychopathology. Part of the reason for Kraepelin's rapid achievement of international fame was his advocacy of the controversial concept of dementia praecox. Even the name of this category has been controversial.

The 1933 Standard Classified Nomenclature called the disorder "Dementia Praecox (Schizophrenia)," thereby reflecting the hesitancy of official American psychiatry to incorporate the Bleulerian term. In 1952, the DSM-I used the name "Schizophrenic Reaction." The word *reaction* was used in order to reflect the heritage of Adolf Meyer, who had suggested that clinical syndromes represented behavioral reactions by patients to the changes in their internal environments (e.g., body sensations caused by biochemical abnormalities) as well as the demands of external environments (e.g., the emotional demands made by parents). Meyer's influence apparently was less pervasive in 1968, and the DSM-II used the simple name "Schizophrenia." At least one commentator (Menninger, 1969) suggested that this choice of a name reflected the simpleminded belief that schizophrenia is a unitary disease. The DSM-III returned to a two-word name: "Schizophrenic Disorders." The term *disorder* does not carry strong connotations; hence, this name change is expected to stimulate less controversy than did previous alterations in the title of this disorder.

Another major change associated with the DSM-III is the selection of subtypes. Kraepelin's sixth edition noted only three subtypes:

Hebephrenic
Catatonic
Paranoid

In Bleuler's classic book, *Dementia Praecox, or the Group of Schizophrenias* (1950), four subtypes were noted:

Hebephrenic
Catatonic
Paranoid
Simple

These four subtypes of Kraepelin/Bleuler were the only subtypes in the Standard Classified Nomenclature and are generally considered the classic subtypes associated with schizophrenia. However, by the early 1950s, the DSM-I had been expanded to contain nine subtypes:

Hebephrenic
Catatonic
Paranoid
Simple
Schizo-affective
Acute undifferentiated
Chronic undifferentiated

Residual
Childhood

The number of subtypes hit a peak with the DSM-II, which contains 14 subcategories of schizophrenia:

Simple
Hebephrenic
Catatonic
  Catatonic type, excited
  Catatonic type, withdrawn
Paranoid
Acute schizophrenic episode
Latent
Residual
Schizo-affective
  Schizo-affective, excited
  Schizo-affective, depressed
Childhood
Chronic undifferentiated

In contrast, the DSM-III reduced the number of subtypes to five:

Disorganized (hebephrenic)
Catatonic
Paranoid
Undifferentiated
Residual

Notice that the first three subtypes in the DSM-III are the classic subtypes of Kraepelin, whereas the other two subtypes are "wastebasket" categories that represent clinical pictures that are either mixed (i.e., undifferentiated) or not currently psychotic (i.e., residual).

The reasons for the reduction in the number of subtypes are many. The authors of the DSM-III defended their choice of subtypes by listing the reasons that many of the subtypes from the DSM-II should be deleted. For instance, Spitzer, Williams, and Skodol (1979) argued that "latent" and "simple" were eliminated because these subtypes referred to nonpsychotic forms of schizophrenia that could be included in certain personality-disorder categories such as "schizotypal" or "borderline." If this explanation is correct, why was "residual" kept as a viable subtype? *Residual* refers to a form of the disorder in which the patient is not psychotic and is manifesting a relatively chronic and stable characterological style. The authors of the DSM-III also decided to eliminate "acute schizophrenic episode" and "chronic undifferentiated" because these subtypes focus on the course of the disorder. They decided that it would be better to code the chronicity of the disorder separately (i.e., chronicity is a separate axis to be applied to schizophrenic disorders). Four types of course are possible: chronic without acute exacerbation, chronic with acute exacerbation, subchronic without acute exacerbation, and subchronic with acute exacerbation.

One additional alteration in the subtypes of schizophrenia does deserve some recognition. This alteration concerns the subtype "schizoaffective." This category became a topic of significant vacillation on the part of the DSM-III Task Force. The concept of a schizoaffective psychosis was first proposed by Cobb (1943). Clinicians had noted that there are patients who present with significant affective symptoms who, under close observation, can be noted as showing clear indications of a thought disorder. Because the presence of a thought disorder had become the primary American indication of schizophrenia, it was suggested that these schizoaffective patients represent a new variety of schizophrenia. The concept was quickly adopted by Americans but was not popular in Europe.

The initial drafts of the DSM-III kept schizoaffective as a subcategory of schizophrenia. However, an important review paper about the concept *schizoaffective* (Procci, 1976) was published in the influential journal *Archives of General Psychiatry.* From his detailed review, Procci concluded that some, perhaps even many, schizoaffectives have family histories of affective disorders and respond to drug therapies as if they had a major affective disorder. Procci's review clearly challenged the inclusion of schizoaffective as a subtype of schizophrenia. As a result, the 1978 draft of the DSM-III included "schizoaffective disorders" as a new major category that appeared at the same hierarchical level as schizophrenic disorders, affective disorders, and paranoid disorders. However, the authors of the DSM-III became dissatisfied with this resolution. As a result, the final 1980 draft of the DSM-III kept the category of schizoaffective disorder but included it as a subtype of the general wastebasket category, "psychotic disorders not elsewhere classified." Also, the authors of the DSM-III took the unusual action of not proposing diagnostic criteria for this disorder. They believed that there was insufficient consensus on its definition, and they suggested that a great deal of research was needed to clarify this concept.

After removing the subtypes in the DSM-II that they considered questionable, the DSM-III authors left only five subtypes. This reduction in subtypes also probably reflects the current interest in tightening the meaning of *schizophrenia*. However, this reduction of subtypes should not be taken as meaning that the resulting five subtypes are the most valid or the most useful subtypes available.

Carpenter and Stephens (1979) carefully reviewed the subtype literature concerning schizophrenia. They concluded that the four classic subtypes—simple, hebephrenic, catatonic, and paranoid—have questionable usefulness. They asserted that these categories had been demonstrated to lack adequate reliability (Blashfield, 1973) and acceptable validity (Katz, Cole, & Lowery, 1964; Stone, Hopkins, Mahnke, Shapiro, & Silverglate, 1968; Hay & Forrest, 1972). Instead, Carpenter and Stephens suggested that the use of a multiaxial system including the course of the disorder (acute vs. chronic) and prognosis (good premorbid vs. bad premorbid) would be an improvement. These axes have been well established in experimental studies of schizophrenia.

To summarize, a major goal of the authors of the DSM-III was to narrow the concept of schizophrenia in order to bring it more into line with the manner in which the concept is used in most other nations. In accomplishing this goal, they decided to reduce the number of subtypes of schizophrenia to five by removing all questionable subtypes. What remained were three classic subtypes plus two wastebasket categories. Carpenter and Stephens have noted that the evidence for the reliability and the validity of even these subtypes is questionable. Instead, the use of a multiaxial system concerning course and prognosis would more likely be of value than the classic subtypes.

Childhood Disorders

Another major change in the transition from the DSM-II to the DSM-III concerned the disorders of childhood. There are a total of 45 specific categories in the DSM-III concerned with childhood disorders, more than four times the number of specific categories concerning children in the DSM-II. These 45 DSM-III categories are all grouped under a heading with the title of "*Disorders usually first evident in infancy, childhood or adolescence.*" As implied by this heading, a patient of *any* age can be diagnosed by means of the categories under this heading. The disorders listed under the general heading usually begin in childhood but may still be evident in adults. For instance, if an adult is mentally retarded, he would be diagnosed under the childhood disorder categories for mental retardation. Also, if a 22-year-old woman developed anorexia nervosa, she would be diagnosed under the heading of childhood disorders because most cases of anorexia nervosa begin during adolescence.

The 45 categories in the DSM-III that concern disorders of childhood can be subdivided into five major headings: intellectual, behavioral, emotional, physical, and developmental. There are five specific categories under disorders of *intelligence.* All of these categories concern mental retardation (i.e., mild, moderate, severe, profound, and unspecified mental retardation). The different subtypes are defined in terms of the level of intellectual impairment. The concept of "borderline mental retardation" from the DSM-II was not kept because the authors of the DSM-III believed that it had been frequently misused and often overinterpreted. As with the organic brain syndrome, any known or suspected physical etiology of mental retardation is not coded in this section of the classification. Instead, the appropriate medical syndrome, such as Down's syndrome (a term used to refer to children with a particular chromosomal aberration), is coded on Axis II.

The second group of childhood disorders is the *behavioral* disorders. These contain three specific categories under the general concept of "attention deficit disorder" and five specific categories under "conduct disorder." *Attention deficit disorder* is a new name for a disorder that has become very popular in both professional and public discussions of child psychopathology: minimal brain dysfunction.

The third major heading under childhood disorders is the disorders of *emotion.* The seven specific categories listed under this heading are subdivided into "anxiety disorders" and into miscellaneous categories including "schizoid," "oppositional disorder," and "elective mutism."

The fourth major heading is the *physical* disorders, which include such categories as "anorexia nervosa," "bulimia," "stuttering," "enuresis," "tics," and "sleepwalking." Many of the categories under this heading were included in the DSM-II under the odd title of "special symptoms not elsewhere classified."

The last collection of disorders is *developmental disorders.* These are organized under two headings: "pervasive developmental disorders" and "specific developmental disorders." The major specific category under the heading of "pervasive developmental disorder" is "infantile autism." This disorder was previously titled "childhood schizophrenia" in both the DSM-I and the DSM-II. However, there was little evidence that this clinical state was really a particular form of schizophrenia (Rutter & Schopler, 1978). Hence, including it as a subtype of schizophrenia seemed unwarranted.

A number of criticisms can be made of DSM-III classification of the mental disorders of childhood. The most serious general criticism was noted by Rutter and Schaffer (1980), who argued that there was insufficient evidence for either the reliability or the validity of many specific disorders to justify their inclusion in the DSM-III. In other words, the fourfold expansion in the number of categories appeared unreasonable.

To substantiate their criticism, Rutter and Shaffer discussed a number of questionable specific categories. For instance, they cited Garmezy (1978), who had criticized the category of "oppositional disorder." Garmezy noted that features used to define "oppositional disorder" (which is a category associated with disturbances of emotion) are "violations of minor rules, temper tantrums, argumentativeness, provocative behavior and stubbornness." All of these alleged symptoms are common among children in general. Rutter and Shaffer argued that there is no good reason to suppose that a child with this collection of characteristics has a sufficiently severe and handicapping condition to warrant being diagnosed as having a mental disorder.

Another set of disorders criticized by Rutter and Shaffer was the three specific categories under "anxiety disorders" ("separation anxiety disorder," "avoidant disorder of childhood or adolescence," and "overanxious disorder"). Mattison, Cantwell, Russell, and Will (1979) found these disorders to show low levels of interrater reliability. Also Cantwell, Russell, Mattison, and Will (1979) designed case histories to represent various specific categories of childhood disorders from the DSM-III. After the case histories were diagnosed by selected clinicians, Cantwell *et al.* found that the percentage of raters agreeing with the actual diagnoses for "anxiety disorders" was also low (15%–55%). Barrios, Hartmann, and Shigetomi (1981) also argued that problems exist in the classification of anxiety disorders. At least one subcategory was included simply because of its popularity in contemporary thinking rather than because there was evidence to suggest a discrete syndrome. "Separation anxiety" is a currently popular concept in the child psychopathology literature because of the considerable research on attachment behavior. However, Barrios *et al.* suggested that there is no good reason currently to suppose that "separation anxiety" refers to a constellation of symptoms different from "noncompliance, enuresis, encopresis, pouting, refusal to go to bed at scheduled times and the like."

Another critique of the disorders of childhood as classified in the DSM-III was made by Achenbach (1980). He is a psychologist who has been actively involved in research using multivariate statistical techniques to empirically explore the classification of childhood disorders (Achenbach, 1966, 1978; Achenbach & Edelbrock, 1979). Achenbach (1980) noted that there is some correspondence between the DSM-III classification of childhood disorders and the syndromes empirically derived from factor-analytic studies. The major result of the factor-analytic studies was a consistent distinction between (1) *internalizing* syndromes, which include anxiety, social withdrawal, overcontrol, and depression as major features, and (2) *externalizing* syndromes, which emphasize features such as delinquency, school problems, aggressive acts, and hyperactivity. This major distinction is somewhat like the DSM-III distinction between emotional disorders ("anxiety disorders" plus the miscellaneous categories of "schizoid," "elective mutism," and so on) and the behavioral disorders (such as "conduct disorders" and "attentional deficit disorders"). However, not all emotional disorders in the DSM-III refer to internalizing syndromes (e.g., "oppositional disorder"). Achenbach also commented that some specific categories recognized under the behavioral and emotional heading in the DSM-III have not been found in empirical research. For instance, factor analysis studies have not consistently noted a factor that is similar in meaning to the DSM-III category of "attentional deficit disorder without hyperactivity." Nor do all specific syndromes empirically noted in the factor-analytic research appear in the DSM-III. For instance, the DSM-III does not contain a category titled "depressive disorder of childhood." Achenbach (1980) reported that at least seven factoring studies have empirically discovered a depressive syndrome among children. In addition, childhood depression has been the focus of at least one book (Schulterbrandt & Raskin, 1977) and a quickly enlarging journal literature (Costello, 1981). Admittedly, childhood depression is a controversial concept (Lefkowitz & Burton, 1978), and good reasons could be offered for its exclusion from an official classification. Nonetheless, one wonders why certain controversial concepts were excluded from the DSM-III, whereas other concepts such as "oppositional disorder," which had virtually no supporting data, were included.

CONCLUDING COMMENTS

There are four areas of innovation in the DSM-III: (1) the use of explicit diagnostic criteria; (2) the inclusion of multiaxial dimensions; (3) the amount of information presented about the categories; and (4) the increase and reorganization of diagnostic categories. These changes represent a major attempt to restructure psychiatric classification and to create a document that is consistent with the science of psychopathology in the 1980s.

If the DSM-III has been successful in stimulating research on psychiatric classification, the hope is that each of the major innovations mentioned above will be subjected to careful and rigorous analysis. For example, the diagnostic criteria for virtually all categories were created by a committee process. However, any set of symptoms in the diagnostic criteria for a specific category should have certain statistical associations between the symptoms and the category. Suppose that Category X is defined by six symptoms: a, b, c, d, e, and f. Also suppose that the diagnostic rule is that, if a patient has four of these six symptoms, then Diagnosis X can be applied to the patient. By proposing these symptoms in the diagnostic criteria for Category X, the DSM-III is hypothesizing the following: (1) First, the DSM-III is asserting that these six symptoms tend to co-occur. That is, these symptoms are correlated and form a syndrome. (2) Second, the DSM-III is also assuming that these symptoms are relatively unique to Category X and generally are not observed among other diagnostic categories. (3) By requiring four of six symptoms, the DSM-III is asserting that this "cutting score" of four symptoms provides an optimal amount of coverage. For instance, if a cutting score of only one of six symptoms were used, it is likely that too many patients would be diagnosed as belonging to Category X. On the other hand, if a cutting score of six symptoms were used, the coverage of Category X might be so small that very few patients would be assigned to the category. Further discussions of the cutting-score issue can be found in Finn (1982).

Important research regarding the DSM-III should not be limited to studying the details of diagnostic criteria. Other important topics of research are (1) an analysis of what categories in the DSM-III are actually applied in clinical practice, as it is unlikely that practicing clinicians will use the entire range of the 265 categories in the DSM-III; (2) a study of the measurement properties (reliability and validity) of the different axes in the five-axis system proposed by the DSM-III; (3) a critical analysis of the impact and utility of the DSM-III for teaching new clinicians about clinical diagnosis; and (4) evaluative studies about the diagnostic validity of the new categories proposed in the DSM-III, for example, the schizophreniform disorder (Coryell & Tsuang, 1982).

Before completing this chapter, a comment should be made about the general impact of the group that was responsible for the DSM-III—the neo-Kraepelinians. As noted in Chapter 1, Kraepelin was a major innovator. Since Kraepelin, classification has become an object of benign neglect among mental health professionals. Even though the literature of science grows exponentially, and scientific and technological knowledge has increased dramatically since 1900, the literature on psychiatric classification has been sparse. Relatively few positive changes have occurred. Whether the DSM-III is evaluated by history as a good classificatory system or not, the rebirth of interest in classification has been an important development directly related to the neo-Kraepelinian movement.

SUGGESTED READING

Foulds, G. A., & Bedford, A. Hierarchy of classes of personal illness. *Psychological Medicine*, 1975, *5*, 181–192.

This chapter presented the DSM-III as the culmination of the neo-Kraepelinian view of classification. However, other approaches to traditional psychiatric classification do exist. For instance, there is the approach taken by Foulds and Bedford to the hierarchical nature of traditional classification. Hopefully, this discussion of their views will provide the reader with some idea about possible alternatives to the neo-Kraepelinian perspective.

Foulds and Bedford are two British clinicians who have tried to improve psychiatric classification by creating a new system that has a coherent internal logic. The most important feature of their classification is its *hierarchical* structure. Foulds and Bedford realized, as did other clinicians before them, that there is a "pecking order" among major diagnostic categories. For instance, the first major category to be considered when diagnosing a patient is the general category of "organic brain syndrome." If sufficient evidence is found to warrant this diagnosis, the diagnostic process is stopped even if the patient has some other syndrome—depression, for example. The second major category to be considered is "schizophrenia." The hallmark of this category is a thought disorder. If clear evidence of a thought disorder is isolated (for instance, if a Schneiderian first-rank symptom is noted), then again the diagnostic process is stopped regardless of whatever other symptoms are noted. The third step is to consider the other major psychotic states: "mania," "depression," and "paranoia." Finally if no other diagnosis has been made, "neurosis" and "personality disorder" are considered.

Besides commenting on the implicit order by which diagnoses are made, Foulds and Bedford have also suggested that a patient who fits an early or "higher" category is likely to have at least some of the symptoms in categories further down in the order. Thus, a patient who is diagnosed as schizophrenic can also have the symptoms of mania and of neurosis. On the other hand, a neurotic should not have the symptoms of schizophrenia or of an organic brain syndrome.

Foulds and Bedford presented the major outline of their classificatory system in an article published in *Psychological Medicine* (1975). Earlier, Foulds (1964, 1965, 1971) had discussed the general structure of the system. Foulds and Bedford call their classification a "Personal Illness Model." This system contains 12 categories and 4 classes. The categories and classes are outlined below.

Class 1—DYSTHYMIC STATES (DS)

This class contains those categories that refer to mood states that may be distressing to the patient or the persons around the patient. These mood states can occur in normal persons. However, all persons in this class are *disturbed.* The three categories included as mood states are *Anxiety* (sA), *Depression* (sD), and *Elation* (sE). Foulds and Bedford noted that these mood states occur in almost all psychiatric patients; hence, being diagnosed at this class level has little explanatory power.

Class 2—NEUROTIC SYMPTOMS (NS)

This class contains five categories, which are likely to have a more serious effect on the lives of patients. Patients in these categories tend to have relatively high levels of subjective distress. In Foulds and Bedford's terminology, a cardinal feature of patients in this class is *dissonance.* The categories in this class are *Pithiatic* (conversion symptoms) (Ps), *Dissociative* (Ds), *Fear* (phobic) (Fs), *Compulsive* (Cs), and *Ruminative* (Rs).

Class 3—INTEGRATED DELUSIONS (ID)

The key concept for this class is *distortion.* Patients who belong to this class are generally perceived as "crazy" by others around them. The ability of these patients to handle the stress and complexity of everyday life is seriously compromised. These categories fit into the ID class: *Persecution* (dP), *Grandeur* (dG), and *Contrition* (dC).

Class 4—DELUSIONS OF DISINTEGRATION (DD)

This is the most pathological of the classes and is denoted by the concept of *disintegration.* Patients in this category have lost their self-identity and often have distorted body images. These patients are unable to interact successfully with others and will probably require hospitalization. Foulds and Bedford did not subdivide this class into separate categories.

Foulds and Bedford have also proposed a measurement instrument to permit assignment to these 12 categories and 4 classes. The instrument is the Delusion-Symptoms-States Inventory (DSSI), which contains 84 self-report items. These 84 items consist of 12 groups of 7 items, each group representing a given category. A patient rates each item on a frequency-of-occurrence scale from none (0) to nearly always (3). Example items are: "Recently, for no good reason, I have had feelings of panic." (Item indicative of Anxiety, or sA.) "Recently my feelings have been taken over by someone else." (Item associated with Delusions of Disintegration, or DD.) A score of 4, summed across items, is considered sufficient for assignment to a category and to a class.

Foulds and Bedford asserted that a patient whose scores are sufficient to

TABLE 15. Hierarchical Arrangement of Symptoms as Proposed by Foulds and Bedford

	Dysthymic states	Neurotic symptoms	Integrated delusions	Delusions of disintegration
Class 4–DD	X[a]	X	X	X
Class 3–ID	X	X	X	NO
Class 2–NS	X	X	NO	NO
Class 1–DS	X	NO	NO	NO

[a]X—patients in the class on this row will have symptoms from this category.

meet Class 4 should have scores that also fit Classes 3, 2, and 1. That is, the hierarchical arrangement among classes should ensure that only four possible patterns of class membership will occur. These patterns are shown in Table 15. A patient who fits Class 4 (delusions of Disintegration) should also have integrated delusions (Class 3), neurotic symptoms (Class 2) and mood states (Class 1). In contrast, a patient who fits Class 1, dysthymic states, should *not* have any symptoms from the other three classes.

The major research emphasis of Foulds and Bedford has been to show that the hierarchical structure that they have postulated among the classes is correct. Foulds and Bedford (1975) and Bedford and Presly (1978) found results that they interpreted as supporting the postulated hierarchical relationships. Other positive studies were done by McPherson, Antram, Bagshaw, and Carmichael (1977) and by Bagshaw and McPherson (1978). However, a recent study published by Surtees and Kendell (1979) was inconsistent with the Foulds and Bedford model. Surtees and Kendell concluded: "Our findings do not support Foulds' hierarchy model despite the fact that its requirements are fulfilled overall by at least 75 per cent of the patients. . . . The incompatible findings are that at least 64% of patients with Delusions of Disintegration do not possess the Neurotic Symptoms they are required to have" (p. 443).

The notion of a pecking order among diagnostic classes is a useful idea. Although this idea had been proposed earlier by others, it is the work of Foulds and Bedford that has helped to popularize it. The major attractiveness of the idea is that it fits an implicit clinical approach to diagnosis. Flow charts (Spitzer & Endicott, 1968; Wing *et al.*, 1974) of how clinicians function show the same basic pattern.

However, the particulars of the Foulds and Bedford system do appear to have problems. First, the system is based on the Delusions-Symptoms-States Inventory of 84 self-report items. The DSSI has not been extensively normed, nor has it demonstrated the psychometric properties required of good psychological tests. In particular, it is important to decide if a response style dimension accounts for a large proportion of variance among DSSI items. Response style dimensions, which are commonly noted in self-report inventories, can have a major effect on the meaning of a test. If response style does significantly contribute to variance on the DSSI, a likely possibility, then the DSSI would be a poor instrument to test the hierarchical model postulated by Foulds and Bedford.

Another comment on the Foulds and Bedford approach concerns their use of the term *hierarchy*. Historically, psychiatric classification has had a hierarchical structure. For instance, in the DSM-II, paranoid schizophrenia is a subset of schizophrenia, which is a subset of functional psychosis. This means that a patient who is assigned to the paranoid schizophrenia category also has to be assigned to the categories of schizophrenia and of functional psychosis. This is the type of hierarchical structure that is used in biological classification, in which hierarchical relations involve category memberships.

In contrast, Foulds and Bedford have used the term *hierarchy* to refer to a relationship structure that is more like the hierarchy of rank in the military. That is, a major outranks a captain, who outranks a lieutenant, and so on. In addition, the military rank structure assumes that a major will know and understand the duties of the lieutenant. However, the lieutenant is not expected to know and understand the job and the decisions of the major. It is this second sense of the term *hierarchy* to which Foulds and Bedford refer.

A final comment about Foulds and Bedford is that these authors are entertaining writers who frequently use interesting quotations. Although their writing style does not mean that their arguments are correct, it does help to attract attention to their work—much in the way that Kraepelin's excellent case descriptions attracted attention to his ideas on psychopathology. An example of the Foulds and Bedford prose will demonstrate their writing style and their ideas about other topics discussed in this book: "With regards to psychologists, our own attempts may be less important than the need to overcome their misplaced apathy and pessimism. The diagnostic situation is neither as gloomy nor as intractable as they believe. Even though there is scant evidence that the present state of diagnosis is as bad as it is often reported to be, there is clear room for improvement (Kreitman, 1961). Nowadays, this article (by Kreitman), often mis-referenced, is taken as demonstrating that psychiatric diagnosis is unreliable, and, even more erroneously, that it is inappropriate. 'O Lord I believe. Help thou my unbelief' " (Foulds & Bedford, 1974, p. 14).

Part II

Quantitative Approaches to Classification

The next three chapters of this book focus on the quantitative approaches to classification. These approaches have been favored by psychologists because they are empirically based and because they support the creation of classificatory systems through research. A common theme in most of the writings by the quantitatively oriented psychologists is the rejection of traditional psychiatric classification as represented by Kraepelin and the neo-Kraepelinians, a rejection that is ironic, given Kraepelin's interest in experimental psychology and his behavioral analysis of patients. Nonetheless, many contemporary clinical psychologists, perhaps for the political reasons outlined in Chapter 3, have suggested that traditional psychiatric classification should be completely eliminated. They further propose that new classification systems should be created using the sophisticated statistical techniques of factor analysis and cluster analysis. In others words, these psychologists are rejecting old classificatory concepts whose existence appears to be justified only by tradition. Instead, they recommend the statistical examination of large numbers of descriptive data in order to discern empirically the natural patterns of symptoms.

The discussion of quantitative approaches to classification is organized in three chapters. The first of these chapters, Chapter 6, presents an overview of multivariate statistical methods and models that can be applied to classification issues. Chapter 7 focuses on a family of techniques called *factor analysis,* which has a 50-year history of applications to psychology and has become a relatively well-enunciated multivariate methodology within psychology. The last chapter, Chapter 8, focuses on the newer multivariate method called *cluster analysis*. Although this family of procedures has attracted attention in the past 20 years, its statistical basis is less well understood than that of factor analysis.

All of the chapters in this section on quantitative approaches to classification involve discussions of statistical methodology. An attempt is made to present this methodology clearly so that a novice can understand the material. Nonetheless, statistical issues are unavoidable, and clinicians who are intimidated by formulas and statistics may find parts of these chapters challenging to read.

Chapter 6

An Overview of Multivariate Models and Methods

Statistics is neither an art nor a science, but a dodge.

Author unknown

Statistical procedures are tools that researchers use to help them understand their data and to decide what hypotheses are most consistent with the data. Multivariate statistical techniques are relatively complicated tools. Knowing when to use the tools, to apply them appropriately, and to understand how they affect the results that appear in a solution requires sophisticated training. Multivariate statistical procedures can be exceedingly useful: however, their misuse can produce distorted results, distortions that may not be apparent to the researcher who has an insufficient understanding of the procedures.

This chapter is intended to be a general introduction to the multivariate statistical procedures that are relevant to the topic of classification. Before these tools became available, the only method by which a researcher could create and test a classificatory system was through a laborious subjective analysis of the entities being classified. For instance, in the biological sciences, it was not uncommon for a researcher to devote his or her entire professional life to the detailed study of a set of organisms and, at the end of his or her career, to publish a *magnus opus* setting forth his or her classification.

Currently, multivariate statistical tools, coupled with the high-speed computer, have vastly reduced the time required for classification research. A major advantage of these quantitative approaches to forming classificatory systems is that they are highly *efficient*. Modern computers are extremely fast. Performing a factor analysis on a data set with 40 variables once required weeks, perhaps months; now a typical computer program can perform a more accurate factor analysis of the same data in a few seconds.

Another advantage of using multivariate statistical procedures to create a classification is their *objectivity*. When a researcher searches by hand through any large amount of information, there is a tendency to see only what he or she expects to see. In other words, a classification formed by one person is a product of both the available data *and* the theoretical biases of the classifier. In contrast, quantitative statistical procedures are objective in that these methods "see" the same structural aspects of a common data set regardless of the theoretical biases of the researchers who use the procedures.*

In introducing the reader to multivariate statistics, no attempt will be made to teach the reader to perform these methods. Despite their general availability, the statistical theory behind most of these tools is complex, and the reader should be cautioned about applying these tools without being educated in their strengths and weaknesses. An interested reader is referred to the growing number of multivariate textbooks, which describe the details of these techniques (Harris, 1975; Thorndike, 1978).

MODELS OF CLASSIFICATION

Before describing any of the particular methods that can be applied to the study of classification, the conceptual models that govern the use of these methods should be presented. As used in this context, the term *model* refers to a general perspective about the relationship among the concepts in a classificatory system. Although the models

* As the reader will learn, the "objectivity" of multivariate statistical techniques is more apparent than real. This comment is particularly appropriate for cluster analysis and factor analysis, which, as exploratory statistical procedures, involve a number of "subjective" decisions. Nonetheless, the lure of objectivity is often a major reason for choosing to use these methods.

being described below can be elaborated into detailed statistical frameworks, they will be considered only at a general level.

Skinner (1981) has suggested that there are five basic models that can be used for understanding psychopathology, all of which have been utilized in empirical attempts to create classificatory systems. These five models are (1) categorical; (2) dimensional; (3) hierarchical; (4) circumplex; and (5) hybrid. Because a controversy has developed over the comparative superiority of the dimensional model and the categorical model, the discussion of these two models will be relatively detailed.

Categorical Model

A *categorical model* assumes that assigning patients to diagnostic categories will summarize the basic information about a patient. Typically, categories are formed so that patients with the same, or nearly the same, symptom patterns will belong to the same category. When using a categorical model, treatment decisions are made on an all-or-none basis. That is, if a patient is assigned to diagnostic Category X, the patient should receive Treatment Y. Traditionally, categorical classifications were created by clinicians, who, after seeing a large number of patients, inferred the presence of categories of patients with similar symptom patterns who probably had the same etiology and needed the same treatment. More recently, cluster analysis methods have been used to create categorical classifications.

Dimensional Model

The second model is the *dimensional model,* which contrasts with the categorical model in that the dimensional model does not assign patients to categories on an all-or-none basis. Instead, this model is concerned with finding a small set of dimensions that can be used to assess and to visualize the relationships among patients. These dimensions are usually estimated through linear equations combining the observed variables. The linear equations are created through the application of procedures such as factor analysis and discriminant analysis.

An example of a particular dimensional model of psychopathology may help the reader to understand this model and how it contrasts with the categorical model. Hans J. Eysenck is a British psychologist in the quantitative tradition who has performed a great deal of research on both psychopathology and personality theory. His extensive factor-analytic research led him to propose three basic dimensions of relevance to psychopathology. These dimensions are *psychoticism, neuroticism,* and *introversion–extraversion.* This three-dimensional model of psychopathology had its historical antecedents in the ideas of Carl Jung and Ernst Kretschmer (Levy, 1970). Jung had suggested that hysterical neurosis was the extreme form of extraversion, whereas the extreme of introversion was an obsessive-compulsive neurosis. For his part, Kretschmer had proposed a continuum running from schizophrenia to manic-depressive psychosis, with normal falling in the middle of this continuum. Eysenck's factor-analytic studies of personality with psychiatric patients led him to reject the ideas of both Jung and Kretschmer and, instead, to postulate his three-dimensional model of psychopathology. Thus, Eysenck found that hysterics were not just persons who were extremely high on the extraversion dimension; they were also high on neuroticism. Normals who were low on neuroticism could be high on extraversion without being hysteric. Also, he found that psychoticism was a separate dimension.

Contrast of Dimensional and Categorical Models

Notice that Eysenck's dimensional model is superficially similar to a three-category model of psychopathology that recognizes psychotics, neurotics, and normals. However, Eysenck's dimensional model appears to be more powerful than the categorical model. In the dimensional model, no person is assigned to just one concept, as would occur in the categorical model. In other words, by means of the categorical model, a patient might be classified as being psychotic. Once she or he is classified, an assumption is made that this patient is similar to all other psychotics and should be treated in the same way that all psychotics are treated. In contrast, the dimensional model represents patients in terms of all dimensions included in the model. Thus, a patient who is high on the psychoticism dimension might also be found to be high on the neuroticism dimension and to be somewhat in the introverted direction on the introversion–extraversion dimension. In conventional diagnostic terms, this pattern might be associated with what is commonly called *psychotic depression.* Thus, an important positive feature of the dimensional model is that it retains more information in the classificatory process than the categorical model does.

A second advantage of the dimensional model is that particular classifications based on a dimensional model are generally tied to sophisticated measurement systems for psychopathology. That is, when particular dimensional models of psychopathology have been proposed, they are almost always associated with a particular assessment device or devices that can be used to determine where patients fall on these dimensions. Thus, Eysenck's three-dimensional model of psychopathology was associated with a particular test that he developed called the *Maudsley Personality Inventory.* In the same way, Maurice Lorr's dimensional model of psychosis was associated with a behavioral rating system that he developed for clinical interviews called the Inpatient Multidimensional Psychiatric Scale. (See Chapter 7 for more details on Lorr's dimensional model and his research on the IMPS.)

Categorical models of psychopathology, on the other hand, are rarely tied to particular assessment devices. As a result, there is a certain inherent vagueness about how patients are to be assigned to categories. For instance, the occurrence of auditory hallucinations is a symptom that is often used in a definition of the category of schizophrenia. The measurement of this symptom is usually not specified. Yet, there is clearly a difference between observing a patient talking to a nonexistent person in the room and having the patient respond "Yes" to the question, "Have you heard voices that were not real?" Behavioral observation and self-report during an interview are different forms of measurement—a fact that all clinicians recognize but that categorical systems of classification generally ignore.

Also, if categorical classifications are viewed as measurement systems, they are rather crude forms in which assessment can occur only in a binary form (either a patient is a member of a category, or she or he is not). In contrast, dimensional models imply measurement systems that are extended beyond being binary to some type of continuous scale. In other words, instead of a person's being either introverted or extraverted, a dimensional model permits the representation of exactly where the patient falls on the introversion–extraversion dimension.

In the preceding discussion, the dimensional model was presented by contrasting it with the categorical model. Traditional psychiatric classification is categorical in its structure. Writers who have favored the dimensional model have offered a number of arguments about why the latter is superior. As a result, a debate has ensued over the superiority of the two models. Like many debates in science, this one has generated much heat and, occasionally, some light. Specifically, four issues have appeared in this debate: (1) professional jurisdiction; (2) the concept of discontinuity; (3) the statistics of the methods used to realize the models; and (4) the structure of the categorical model. Each is discussed in turn.

Professional Jurisdiction

The first of these issues introduces political overtones into the debate. In general, the categorical model has been associated with traditional psychiatric classification, whereas the dimensional model has been embraced by psychologists. Thus, the debate about the dimensional versus the categorical model has often contained the same political implications as in the turf battle between psychiatry and psychology discussed in relation to the medical model. Because these issues of professional rivalry were discussed in some detail in Chapter 3, they will not be reviewed again here. Readers who are interested in how professional rivalry has been involved in the debate over the categorical versus the dimensional model should refer to Kendell (1975).

The Concept of Discontinuity

Kendell (1975) has been one of the major advocates of the concept of discontinuity. Kendell believed that if a categorical model is to be the preferred model, then there should exist boundaries or areas of discontinuity between different categories: "In terms of the familiar aphorism that classification is the art of carving nature at the joints, it should indeed imply that there is a joint there, that one is not sawing through bone" (p. 65).

A visual representation of the concept of discontinuity and how it could be relevant to the debate between categorical and dimensional models is shown in Figure 8. The top of the figure shows a hypothetical distribution of patients along Eysenck's introversion–extraversion dimension. Notice that the distribution is bimodal (i.e., it has two "humps"). This suggests that there are two groups of patients shown on this dimension, and that there is an area of relative discontinuity between these two groups. In other words, the distribution shown in the top half of Figure 8 would be consistent with the categorical model in which there are two types of patients: introverts and extraverts.

In contrast, the bottom half of Figure 8 shows a distribution of patients along the introversion–extraversion dimension that would favor the dimensional model. As shown by the latter distribution, there is only one mode, and no areas of relative discontinuity occur. If the data on psychiatric patients appear as in the bottom of Figure 8, to use a categorical model is superfluous. In other words, there is no information to be gained by dividing the patients into extraverts and introverts. In fact, subdividing the data shown in the bottom half of

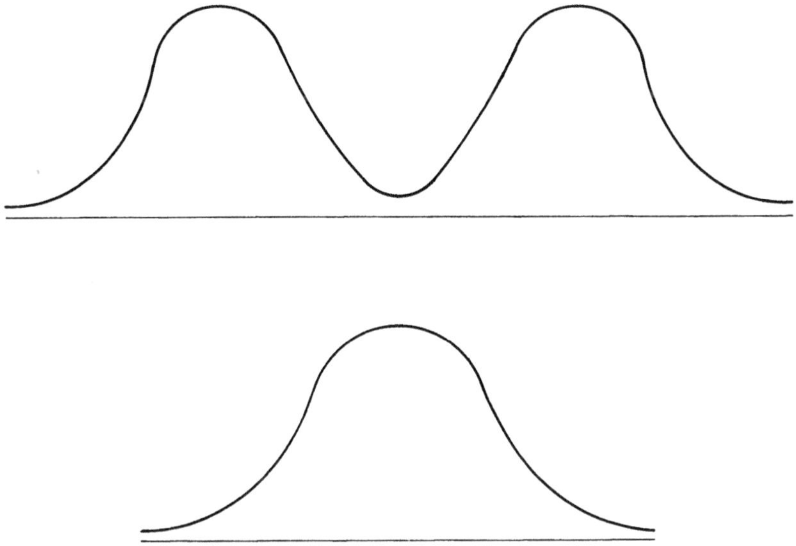

FIGURE 8. Comparison of bimodal (top) and unimodal (bottom) distributions.

Figure 8 into two categories would be misleading because there is no evidence that the categories are separate.

Although the above example is presented in terms of only a one-dimensional scale, the general issue of the use of discontinuity as a concept to decide which model is more appropriate can be generalized to a multidimensional case. For example, consider Figure 9. This figure shows a set of data on 90 patients plotted along two scales. Notice the clear "clusters" in this visual representation, and the areas of relative discontinuity appearing between the clusters. In contrast, Figure 10 also shows a visual representation of 90 patients along two scales. No clusters or areas of discontinuity appear. For Figure 10, the dimensional model would seem to make more sense. In contrast, for Figure 9, the categorical model would be the most appropriate because areas of discontinuity do appear.

To summarize, the concept of discontinuity has been presented as important to the debate about the dimensional model versus the categorical model because this concept could be used as a criterion to decide which model is more appropriate for a given data set. The concept of discontinuity has been used in this manner in the literature comparing one-depression versus two-depression classificatory systems (see Chapter 7 for more details) (Kendell, 1968).

Before leaving the concept of discontinuity, the reader should be

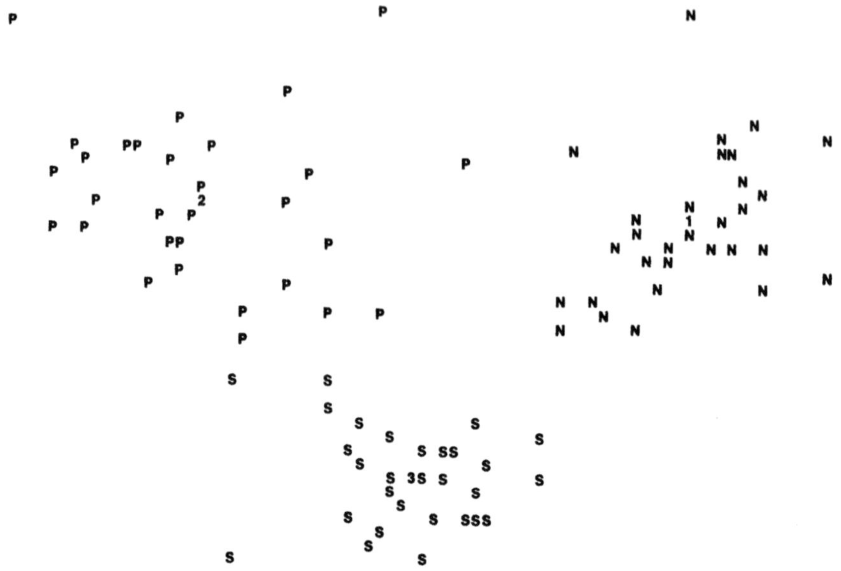

FIGURE 9. Scatter plot demonstrating a categorical model.

aware that although the use of this concept as a criterion to differentiate the two models is plausible, the fact is that applying this concept as a criterion may be further confounded by statistical aberrations. For instance, reconsider Figures 9 and 10. The data shown in these figures are the same! Only the dimensions by which the data are visually presented are different. When the data are presented in the form used in Figure 9, they appear to fit one model, but when the same data are plotted by means of different dimensions, a different model appears more reasonable. Knowing that Figures 9 and 10 are different plots of the same data should lead the reader to question the use of the concept of discontinuity as a criterion.

At this point, the reader may feel confused. A plausible solution to the debate between the two models was presented. Then a counterexample was given that showed that the plausible solution did not always work. Moreover, no details on the counterexample were presented; therefore, it was unclear why the use of the concept of discontinuity as a criterion had failed as a solution to the debate on the dimensional versus the categorical model. Unfortunately, to understand why the same data points appear so differently in Figures 9 and 10 requires some understanding of both factor analysis and discriminant analysis. Hence, further discussion of these figures will be postponed until later in the chapter.

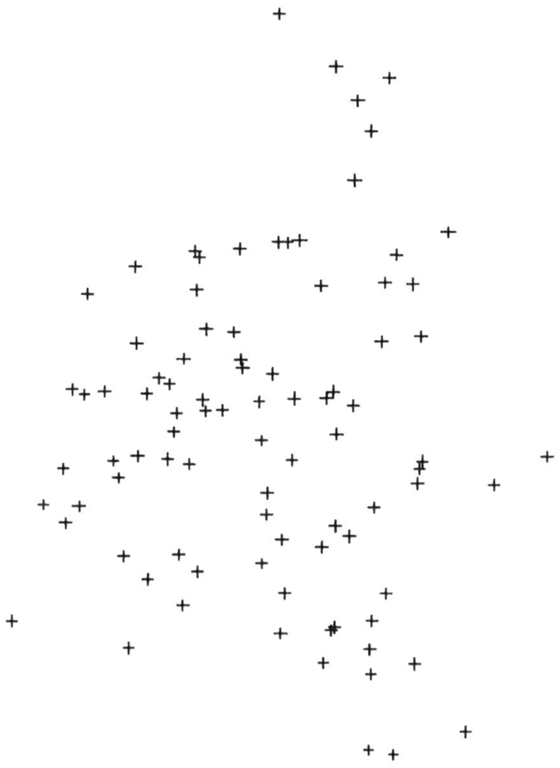

FIGURE 10. Scatter plot demonstrating a dimensional model.

The Statistics of the Methods Associated with the Models

The statistical method most commonly associated with the dimensional model is factor analysis, whereas cluster analysis is tied to the categorical model. Discussions of the models often make reference to these two methods, and the debate over the superiority of the models frequently requires comprehension of the statistical theory behind factor analysis and cluster analysis.

For instance, Eysenck (1960) has been a major advocate of the dimensional model. One of his arguments in favor of this model is that studies using factor analysis have noted that the distribution of factor scores shows no evidence of zones of discontinuity. In fact, plots of factor scores that are created from analyses on the data from psychiatric patients show the distribution to be unimodal (like the distribution shown at the bottom of Figure 8).

However, if one understands the statistical theory of factor analysis, Eysenck's argument can quickly be shown to be problematic. The statistical theory of factor analysis assumes that the data have a multivariate normal distribution. In simpler terms, factor analysis assumes that data will be unimodal. Because of this assumption, the factor scores created during a factor analysis will always tend to be unimodal. Therefore, learning that factor scores from the analysis of a particular data set are unimodal is not good evidence for the superiority of the dimensional model. To argue from the unimodal distribution of factor scores to the superiority of the dimensional model is rather like wearing red-colored sunglasses to prove that all colors are hues of red.

More generally, the debates between advocates of the two models have contained a number of references to technical points about the statistics of factor analysis and cluster analysis. These technical comments confuse general readers and often are misleading. As the reader will learn in Chapters 7 and 8, the statistical theory associated with factor analysis and cluster analysis is not nearly as clear as many authors seem to believe. In the debate between the models, when the reader notes an author making a technical reference to statistical issues, the reader should be suspicious of the author's arguments. The debate between these two models will not be resolved by reference to statistical theory.

Structure of the Categorical Model

In comparisons of the dimensional and the categorical model, the categorical model is often a "straw man." That is, the categorical model is associated with traditional psychiatric classification, and every complaint aimed at traditional psychiatric classification is assumed to be a viable argument against the categorical model. However, surprisingly little literature exists that has attempted to explicate the categorical model.

The categorical model has a number of other names. Some authors describe it as a *typological model;* others call it a *classification model;* occasionally, this model is described as a *cluster model.* The key concepts associated with the categorical model are *category, set, class, cluster,* and *type.* Although these terms are often used as if they are interchangeable, the concepts can have quite different meanings. The confusion about the structure of the categorical model can be illustrated by contrasting the meanings of these concepts.

The first of these concepts is the concept of *category.* This is the most neutral of the concepts and simply refers to any division in a classification. Because *category* has a relatively neutral meaning, it has been used as a generic term in this chapter to include the general shared meaning of all the concepts listed above.

A *set* refers to any collection of entities being classified. In psychiatric classification, a diagnostic category can be a set because the category can be conceptualized as a collection of patients who have the same diagnosis. Sets have members, that is, the entities that belong in the sets. The structural relationship among categories can be discussed through set theory (Gregg, 1954; Buck & Hull, 1966). In discussions of psychopathology, a basic confusion exists regarding who are the members of the diagnostic categories. Many authors assume that the members of diagnostic sets are patients. However, the authors of the DSM-III have argued that their system is a classification of disorders, not of patients; that is, mental disorders are the members of the diagnostic sets in the DSM-III. However, the characteristics used to define the diagnostic sets in the DSM-III are the characteristics of patients, not of disorders. In its organization, the DSM-III is a classification of patients. Nonetheless, because the principles of set theory have never been applied rigorously to the structure of psychiatric classification, a minor controversy has developed over the issue of whether patients or disorders are being classified (Panzetta, 1974; Spitzer *et al.*, 1977).

The next concept is that of *class*. As the term is used in classification theory, a class is a special type of set in which the members of the set have some basis for being similar. For example, Albert Einstein, Thomas Jefferson, Woody Allen, and Emil Kraepelin could be members of the same set. These men do not have any obvious characteristics in common, yet they can belong to a common set if they are defined as doing so.

In contrast, a class is a special type of set in which the members have at least some salient characteristics in common. All American presidents could comprise a class, for instance. This class contains Thomas Jefferson because he was elected to this position, but Albert Einstein does not belong to this class.

As standardly used in classification, the concept of *class* has an even more precise meaning. Generally, *class* refers to a set in which all members are reasonably similar to each other and are dissimilar from members from other classes. In biology, for example, *Canis familiaris* is the technical name for the class of domesticated dogs. All animals that fit in this class are reasonably similar to each other and are dissimilar from animals falling into other classes, such as *Canis lupis* (wolves).

The next concept is that of a *cluster*. The recent popularity of this concept can be attributed to the increased use of cluster analysis to generate classification systems. The concept of *cluster*, like the concept of *class*, refers to a set of similar entities. However, when the term *cluster* is used, similarity is being defined statistically. That is, a cluster of patients is a set of patients who can be shown statistically to have reasonably similar symptom patterns.

The concept of *cluster* can have a variety of possible specific meanings. For instance, a cluster can be a set of entities in which each entity is highly similar to at least one other member of the set; or a cluster can be a set of entities in which a member is highly similar to all other members of the same set; or a cluster can be a set in which a member has a high average similarity to the members of the same set. These three definitions of *cluster* are different and can lead to the creation of different classifications.

Type is another concept often used in discussion of the categorical model. The term *type* has a long history in personality theory, but as Levy (1970) pointed out, the exact meaning of this term has been a source of major confusion. Most commonly, the concept of *types* is assumed to refer to "Platonic ideals." In his philosophical system, Plato argued for the existence of conceptual ideals that could be used to explain the real world. For instance, there is a conceptual ideal that we have learned to call a *circle*. Circles, as perfect ideals, do not exist in the real world. In group therapy, we say the patients sit in a circle, but in fact, no group forms a geometrically exact circle. Nonetheless, the ideal concept of the circle helps us to desribe the important feature of the physical arrangement of patients in a group. In the same way, *introvert* is a conceptual ideal. That is, *introvert* refers to a specific "type" of person. In the real world, no pure introverts actually exist. Nonetheless, the concept of the introvert as a type of personality can be used to describe certain persons who approximate this conceptual ideal.

Recently, a literature has developed in cognitive psychology around the concept of *prototype* (Rosch, 1978). A prototype is a learned example of a category that one can use to decide whether new entities belong to the category or not. For example, when children learn the concept of *bird*, they do not learn it by memorizing a set of characteristics that can be used to differentiate birds from other animals. They learn about birds by having a parent point at a real bird, saying, "Bird." Thus, children learn categorical concepts by example. The prototype is the particularly salient example of the category that contains most of the important defining features of the category. For instance, a robin is a good prototype for the category of *bird.* A chicken is not. Nor is a penguin.

The concept of *prototype* has recently been applied to the classification of psychopathology. Cantor, Smith, French, and Mezzich (1980) performed a study that showed the superiority of a prototype perspective when compared with a traditional view of psychiatric classification. Unfortunately, many of the arguments in this article assume an unrealistic view of traditional psychiatric classification and would be irrelevant against a sophisticated analysis of the categorical model as applied to traditional psychiatric classification.

From the preceding discussion of key concepts associated with the categorical model, it should be evident that the structure of the categorical model is not precise. In fact, *categorical model* is a term that represents a number of possible structural variations, many of which are significantly different from the straw man that proponents of the dimensional model have been wont to attack. An important theoretical advance in classification would be the explication of the categorical model. Certainly, without a better understanding of this model, arguments that attempt to resolve whether this model is or is not superior to the dimensional model are futile.

Hierarchical Model

Now that the issues associated with the debate on the categorical versus the dimensional model have been discussed, the three remaining models of psychopathology can be introduced. The *hierarchical model* is a specific type of a categorical model in which there is a hierarchical relationship among the categories. An example of a hierarchical model is shown in Figure 11. This figure represents the classification of affective disorders in the DSM-III. Hierarchical models have two important characteristics. First, the diagnostic process is usually conceptualized in terms of a hierarchical tree. Suppose, for instance, that one wants to diagnose a patient with what appears to be a depressed mood. In the hierarchy from Figure 11, the first diagnostic question that the clinician should address is, Does the patient have an affective disorder? If the answer is yes, then the clinician should consider whether the patient has a bipolar or a unipolar disorder. If unipolar, the depression can be classified as "major depressive disorder, unipolar" or as "dysthymic disorder."

The second important feature of a hierarchical model is that descriptive specificity increases as one moves further down the hier-

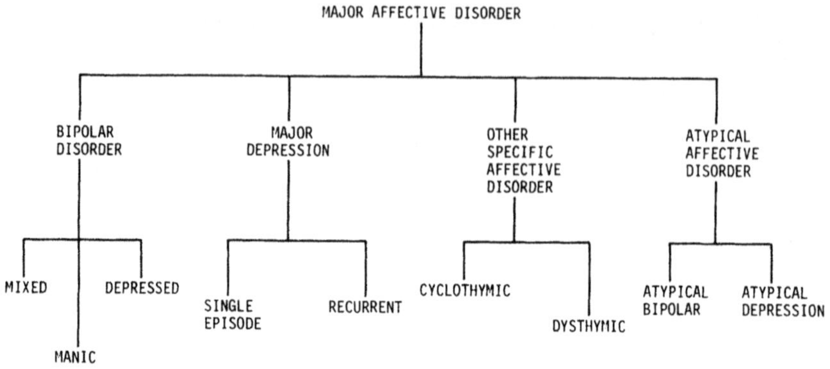

FIGURE 11. Hierarchical representation of DSM-III subcategories of "major affective disorder."

archy. For instance, two patients with a dysthymic disorder should be more similar than two randomly chosen patients who have an affective disorder.

Hierarchical models have been traditionally used in psychiatric classification. This model is inherent in all editions of the DSM, for instance. Among multivariate statistical procedures, there is a particular family of cluster-analytic techniques, called *hierarchical agglomerative procedures,* that generate classifications that fit the hierarchical model (Everitt, 1980; Sneath & Sokol, 1973).

Three major varieties of hierarchical models exist. The first of these is the hierarchical model inherent in biological classification. Biologists have been very concerned about the nature of the hierarchical relationships among categories because hierarchies represent the evolutionary relationships among different species. Readers interested in learning about alternative views of the hierarchical structure of biological classification should consult Simpson (1961), Sneath and Sokol (1973), and Nelson and Platnick (1981).

The second major variety of hierarchical models comes from the study of ethnobiology by cultural anthropologists. These researchers have analyzed how non-Western Cultures have organized their classifications of living organisms. Berlin (1978) has theorized about the regularities in these studies. He argued that all classifications of living organisms are hierarchical; that they contain at least four ranks; that the hierarchical nesting of categories in natural languages is not perfect; and that there are linguistic, psychological, and functional regularities to the hierarchies found in non-Western classifications of living organisms.

The final major variety of hierarchical models is the type of model

CLASS 4	DELUSIONS OF DISINTEGRATION	DELUSIONS OF DISINTEGRATION
CLASS 3	INTEGRATED DELUSIONS	DELUSIONS OF GRANDEUR DELUSIONS OF PERSECUTION DELUSIONS OF CONTRITION
CLASS 2	NEUROTIC SYMPTOMS	CONVERSION DISSOCIATION PHOBIA COMPULSION RUMINATION
CLASS 1	DYSTHYMIC STATES	ANXIETY DEPRESSION ELATION

FIGURE 12. Foulds and Bedford's hierarchical classification system.

proposed by Foulds and Bedford (1975). They have suggested a hierarchical arrangement among categories in terms of diagnostic priority, as shown in Figure 12. The details of the Foulds and Bedford classificatory system were reviewed in the "Suggested Readings" section at the end of Chapter 5. Foulds and Bedford have argued that the diagnostic process should occur in descending order among the classes in their hierarchical system. That is, the clinician should rule out the presence of integrated delusions (Class 3) before attempting to diagnose a dysthymic state (Class 1). Also, Foulds and Bedford have argued that there should be a perfect nesting of symptoms throughout the hierarchy. Thus, a patient who is diagnosed in Class 4 (Delusions of Disintegration) can have symptoms associated with any other class, whereas a patient diagnosed in Class 2 (Neurotic Symptoms) cannot have any of the defining symptoms of Classes 3 and 4.

If the Foulds and Bedford model were applied to traditional psychiatric classification, the principles of diagnostic priority and symptom nesting would suggest a different view of the hierarchical arrangement among diagnostic categories from the one that is often assumed. For instance, organic brain syndromes would have the highest diagnostic priority; schizophrenia would be next, followed by the affective disorders and finally the neuroses. Because of the principle of symptom nesting, a patient with an organic brain syndrome could have both neurotic symptoms and schizophrenic symptoms, but a neurotic patient could not have the defining symptoms of schizophrenia or of an organic brain syndrome.

Circumplex Model

The fourth model, the *circumplex*, is the least frequently used model in psychopathology. The circumplex model is a special case of a dimensional model in which the dimension wraps around on itself to form a circular arrangement. Figure 13 shows a circumplex model proposed by Timothy Leary for understanding interpersonal behavior (Wiggins, 1979). Notice that the 16 segments in Leary's model are represented in the form of a circle. There are two dimensions in this model that are usually denoted by the concepts of *affiliation* and *dominance.* Interpersonal behaviors that have similar meaning are next to each other (e.g., "love" and "tenderness"). Opposite behaviors appear at the opposite sides of the circle ("love" and "hostility"). Leary's representation of interpersonal behavior through a circumplex model has been a provocative stimulus for other research and the development of related uses of the circumplex model (Lorr & McNair, 1965; Benjamin, 1979; McLemore & Benjamin, 1979; Wiggins, 1979).

Wiggins (1982) has reviewed circumplex models as applied to psychopathology. His review suggested that this model has decided potential for the classification of psychopathology. In particular, Wiggins noted that the circumplex model could be profitably applied to the area of personality disorders. In fact, he suggested that Millon's classification (1982) of personality disorders, a classification that was the basis of the DSM-III categories on Axis II, was similar to the circumplex model of personality as proposed by Leary (1957).

Hybrid Model

The fifth type of model for classification is what Skinner (1981) called a *hybrid.* As the name implies, this model represents combinations of the other four models. For instance, Figure 14 represents a combination of the dimensional and the categorical models in what Degerman (1971) described as a *radex.* This particular radex is extrapo-

POWER

SUCCESS NARCISSISM

GENEROSITY EXPLOITATION

TENDERNESS PUNISHMENT

LOVE HOSTILITY

COLLABORATION REBELLION

TRUST DISTRUST

CONFORMITY MASOCHISM

WEAKNESS

FIGURE 13. Leary's circumplex system of interpersonal behavior.

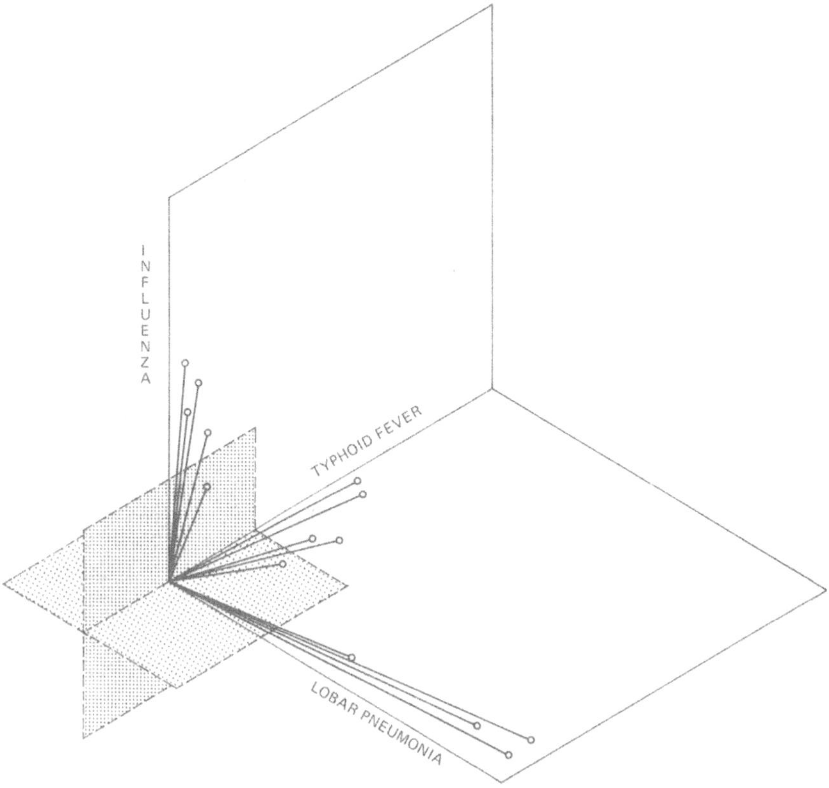

FIGURE 14. Radex contrasting type of medical disorder and severity. (From "A Vector Model of Disease for Teaching and Diagnosis" by P. H. E. Sneath. *Medical Hypotheses*, 1975, *1*, 12–22. Copyright 1975 by Churchill Livingstone. Reprinted by permission.)

lated from Sneath (1975), who contrasted the type of a medical disorder (categorical information) with its severity (dimensional information). In this three-dimensional figure, each disorder is represented by a cluster of patients along a dimension. Thus, patients with influenza all cluster along the vertical dimension. The severity of the disease is shown by the distance of each patient from the center (origin) of the space. One patient, for instance, has a relatively severe case of typhoid fever, whereas another patient, who also has this disease, does not fall beyond the normal limits and may not require certain aspects of treatment, such as hospitalization.

MULTIVARIATE STATISTICAL METHODS

The five models described by Skinner (1981) all represent different perspectives that can be used in forming a classification. A model is an abstract system. A *method* is a particular realization of a model. It is a procedure that can be applied to a data set so that one can view the data through the perspective of a model. For example, *factor analysis* is the name of a family of methods that presuppose a dimensional model. After a factor analysis has been applied to a data set, the data can be represented through a multidimensional presentation of the data. Strauss, Gabriel, Kokes, Ritzler, VanOrd, and Tarana (1979) had 20 clinicians rate five diagnostic categories using 25 symptoms. The result was 100 "archetypal patients" who represent the symptom patterns that clinicians associate with these five diagnoses. The results of a factor analysis allowed the patients and the various symptoms to be plotted in terms of a two-dimensional factor space (see Figure 15). The symptoms are shown in this plot as arrows. The archetypal patients are shown as dots. The figure shows that the patients fall into "clusters" and that swarms of symptoms are associated with each cluster. For instance, the paranoid schizophrenic archetypes are all collected in the lower right-hand corner of the plot, and the arrows represent the related symptoms, such as flat affect, delusions of passivity, delusions of reference, auditory hallucinations, and so on.

The association of particular methods and models is not always via one-to-one relationships. In other words, most models can be associated with more than one statistical method. Also, some methods can be used as a realization of more than one model. For instance, factor analysis is frequently used to form a dimensional representation of a data set. However, factor analysis is not the only method associated with the dimensional model; discriminant analysis, canonical analysis and multidimensional scaling are other methods that can be used to create a dimensional representation of a data set. Also, it is worthwhile noting that factor analysis can be used to create a categorical model of data. This latter use of factor analysis is not frequent, but it is possible.

This section of the chapter introduces a number of multivariate statistical methods that have been applied to the topic of classification. In a gross way, these methods can be subdivided into two major types: *exploratory* methods, which can be used to create classification systems, and *confirmatory* methods, which can be used to evaluate and test existing classificatory systems. The most commonly used exploratory methods are factor analysis, cluster analysis, and multidimensional scaling. Discriminant analysis, multivariate analysis of variance, canonical analysis, and multiple regression are well-known confirmatory methods.

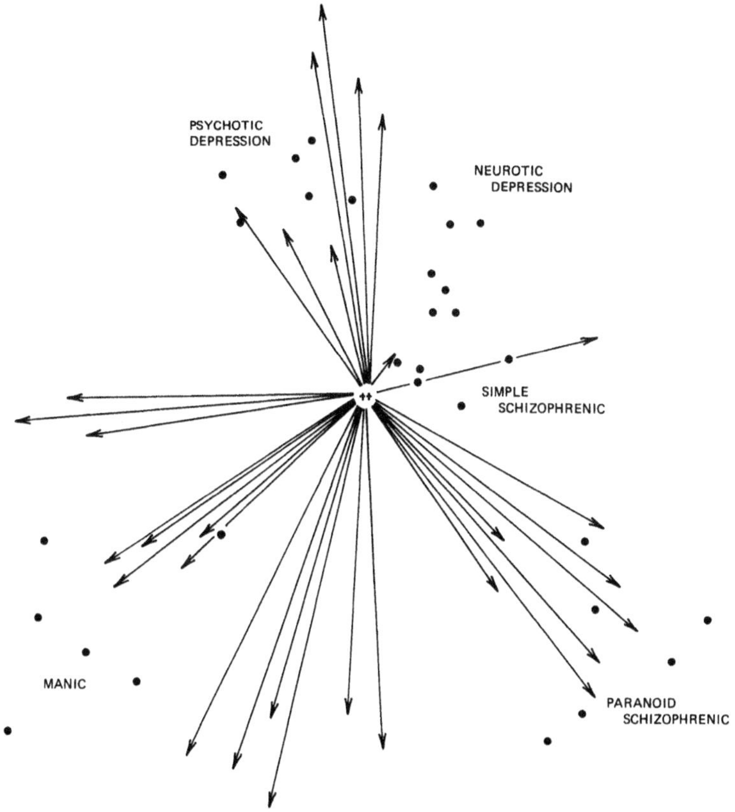

FIGURE 15. Factor analysis of Strauss *et al.* (1979) data plotted in 2-dimensional factor space. (From "Do Psychiatric Patients Fit Their Diagnoses?" by J. S. Strauss, K. R. Gabriel, R. F. Kokes, B. A. Ritzler, A. VanOrd, and E. Tarana. *Journal of Nervous and Mental Disease*, 1979, *167*, 105–113. Copyright 1979 by The Williams & Wilkins Co., Baltimore. Reprinted by permission.)

Exploratory Methods

Factor analysis is the oldest of the exploratory multivariate statistical methods. Generally, factor analysis is concerned with creating dimensions that can be used to represent the data. The dimensions (called *factors*) that are found by factor analysis are linear composites of the variables, with the constraint that these factors must account for as much variance in the data as possible.* Factor analysis has two

* A reasonably detailed introduction to factor analysis is presented in Chapter 7.

general uses: (1) to form a small set of new variables (i.e., factors) that account for most of the variance that was being represented by the larger number of original variables; and (2) to discover the "underlying factors" that explain the interrelationships observed among the original variables.

The second exploratory multivariate method is *multidimensional scaling* (MDS). Like factor analysis, multidimensional scaling forms dimensions that can be used to represent the data. However, the statistical requirements and the mathematical procedures of MDS are quite different from the procedures used in factor analysis to create the dimensions. One way to state the difference between MDS and factor analysis is to say that MDS is a "nonlinear" form of factor analysis. In other words, multidimensional scaling does not create dimensions that are linear composites of the original variables. Another way of making the distinction between the two methods is to state that factor analysis requires at least *interval scaling* among the original variables, whereas MDS requires only *ordinal scaling.*

To understand the distinction between ordinal scaling and interval scaling, consider the following two scales. The first is a scale for velocity as measured by a car speedometer. This scale would be called an interval scale because equal-sized intervals on the scale have the same meaning. That is, the difference of 10 mph between a velocity of 50 mph and 60 mph has the same meaning as the the 10-mph interval between 16 mph and 26 mph. On the other hand, consider a scale formed by asking a patient to rate his depression from 0 (extremely depressed, the pits) to 10 (not at all depressed). On this scale, if the patient says that his mood was a 7 on one day and a 5 on another, then the clinician could conclude that the patient was more depressed on the second day. The depression rating could transmit ordinal information (i.e., Was the patient more or less depressed relative to some other time point?), but the clinician could not claim that the difference between ratings of 5 and 7 was the same as that between ratings of 0 and 2 (i.e., this subjective rating of depression does not fit an interval scale). By only requiring ordinal scaling among the variables, MDS has a broader range of applicability than factor analysis. However, factor analysis has a greater history of use in the study of psychopathology because it is better suited to the task of test development.

Cluster analysis is the third exploratory method. Unlike the first two methods, cluster analysis is associated with the categorical model. Also, some methods of cluster analysis are quite appropriate to the hierarchical model. Cluster analysis is a relatively young multivariate statistical method and has become a generic term used to refer to a broad range of specific methods. The major feature that these methods have in common is that they attempt to partition a data set in such a way that the entities being clustered form homogeneous classes. Because of the wide range of methods associated with the term *cluster analysis,* further discussions of clustering will be postponed until Chapter 8.

Factor analysis has the longest history of use of the exploratory methods. In the last two decades, the most common use for factor analysis has been in the creation of new scales for the assessment of psychopathology. Two examples of this use will be mentioned briefly. Lorr, Klett, and McNair (1963) created the Inpatient Multidimensional Psychiatric Scale (IMPS). This assessment device contains 90 items that rate the behaviors and the self-reports of psychiatric patients during intake interviews. Factor analysis organizes these 90 items into 10 scales. Scoring sheets for the IMPS contain norms for male and female inpatients on these 10 scales. A related use of factor analysis was performed by Spitzer, Fleiss, Endicott, and Cohen (1967). They created a 248-item structured psychiatric interview. A factor analysis of the data on these 248 items led to the formation of 13 scales that summarize how patients respond to this collection of interview questions. The final version of the structured interview, called the Mental Status Schedule, contains scoring keys for the 13 factor-derived scales.

Cluster analysis has been applied primarily to research on psychopathology in order to form categorical classificatory systems. That is, cluster analysis has been used to organize the patients on whom data have been gathered into relatively homogeneous sets. For instance, Overall and his colleagues have used cluster analysis to study a number of data sets. A review of these studies (Overall, Pull, Carranza, & Cassano, 1977) suggests that there are eight major categories of psychopathology:

1. Florid thinking disorder
2. Withdrawn-disorganized thinking disturbance
3. Hostile-suspiciousness syndrome
4. Anxious depression
5. Agitated depression
6. Hostile depression
7. Withdrawn-retarded depression
8. Agitation-excitement syndrome

Cluster analysis has been used with increasing frequency over the last two decades. Topics such as depression, alcoholism, childhood disorders, learning disabilities, and delinquency have been the major

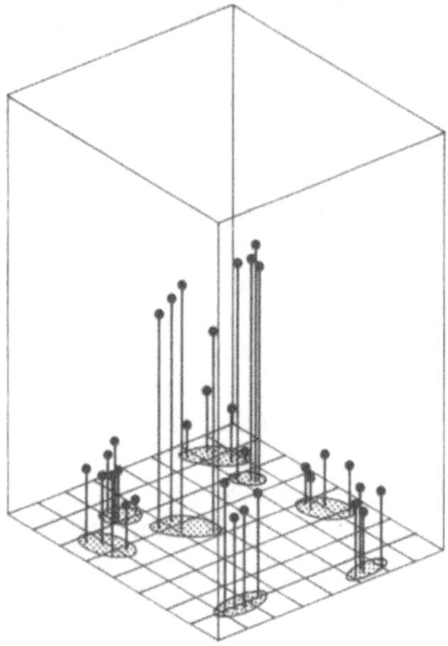

FIGURE 16. Three-dimensional solution from multidimensional scaling of Overall *et al.* (1977) data. (From "Phenomenological Classification of Psychiatric Patients" by J. E. Overall, C. Pull, J. Carranza, and G. Cassano. *Journal of Psychiatric Research*, 1977, *13*, 225–236. Copyright 1977 by Pergamon Press. Reprinted by permission.)

focuses of the clustering research. In Chapter 8, brief reviews are presented of the application of cluster analysis to depression and to alcoholism.

In contrast to factor analysis, multidimensional scaling has rarely been used to form scales for new assessment systems. Instead, MDS has typically created visual "maps" of the data that help a researcher to "see" the underlying structure of the data. An example of this use of MDS is contained in the previously cited article by Overall *et al.* (1977). They had psychiatrists from France, Italy, Mexico, and the United States rate the eight diagnostic categories found in their clustering research along 18 symptom scales. A three-dimensional solution from multidimensional scaling is shown in Figure 16. Each point on this "map" shows the location of the mean ratings for a category by the psychiatrists in a country. This MDS map shows that the psychiatrists in all four countries basically agreed on the descriptive meanings of the eight cluster-derived categories. Also, the location of the categories in this multidimensional space reflects information about the relationships among the categories. For instance, two categories (anxious depression and agitated depression) were rated as being fairly similar, and probably could be relatively easily confused. In contrast, another category (withdrawn-retarded depression) appeared to be a separate form of depression; that is, it was located in an isolated part of the three-dimensional space.

Confirmatory Methods

There are a number of multivariate statistical methods that can be considered confirmatory methods. Only four of these methods are described below: multiple regression, canonical analysis, multivariate analysis of variance, and discriminant analysis.

Multiple regression has been applied primarily in situations in which a researcher wishes to develop a mathematical system for making predictions. Suppose, for example, that a researcher wants to predict the scores of patients on some variable. In research on psychopathology, this variable might be a measure of response to drug therapy. In order to make a prediction of patients' scores on this variable, the researcher gathers a great deal of information on the patients' symptoms before the start of therapy. This symptom information can be summarized in terms of a set of scales. Multiple regression would be the statistical technique used to form an equation from the symptom scales that would most accurately predict the outcome variable of response to therapy. After an equation is developed, it can be applied clinically to predict which patients will show the most favorable response to a particular drug therapy. To summarize, multiple regression is a multivariate statistical method that can be used to form an equation to predict an outcome variable.

Canonical analysis is an extension of multiple regression. As described above, multiple regression is used when the researcher has a number of predictor variables and one variable to be predicted. However, in some situations, a researcher may decide that more than one variable needs to be predicted. For instance, in drug therapy research, the researcher may decide that a single measure of response to therapy is insufficient and that multiple estimates of the response to therapy would be better. In this condition, canonical analysis would be appropriate because it can be used to form the equations that best relate one set of variables to another set of variables. In other words, canonical analysis can form equations that allow the prediction of scores on a set of outcome measures from a set of scores on the predictor variables.

Another commonly discussed multivariate statistical procedure is *multivariate analysis of variance.* As its name implies, this method is a generalization of the analysis of variance. Specifically, multivariate analysis of variance uses the statistical procedures of hypothesis testing to decide whether various classes of patients appear the same or different across a set of variables. For instance, the Inpatient Multidimensional Psychiatric Scale was described earlier as having 10 dimensions derived from factor analysis. A multivariate analysis of variance could be applied to test the hypothesis that various diagnostic categories have different profiles on these 10 dimensions.*

The last confirmatory method to be discussed in this chapter is *discriminant analysis.* Discriminant analysis is like multivariate analysis of variance in that it is concerned with testing whether various categories are different across a set of variables. That is, one question posed by discriminant analysis is, Do these variables discriminate among the different categories? In addition, discriminant analysis also forms dimensions that are designed to differentiate optimally among the categories. That is, discriminant analysis is like a multiple regression in which a set of variables is used to predict a particular dependent measure, where the dependent measure is category membership.

Discriminant analysis has developed into a commonly used multivariate technique in classification research because it can be used to help one decide on the descriptive validity of diagnostic categories. If discriminant analysis fails to differentiate a set of categories along variables that should be relevant to these categories, then the validity of the categories would be in question.

Discriminant analysis is often confused with both factor analysis and cluster analysis. The confusion between discriminant analysis and factor analysis generally occurs because both will create dimensions that represent relations among the variables. However, the dimensions created by factor analysis attempt to account for as much variance as possible among the data. In contrast, the dimensions formed by discriminant analysis are designed to optimally predict group membership in the categories.

To show how different are the dimensions generated by these two methods, reconsider Figures 9 and 10. As noted earlier, these figures show two different representations of a common data set. Figure 10 was formed from the initial two dimensions of a factor analysis of the data. The other figure, Figure 9, was formed by discriminant analysis when the membership of the 90 patients in three categories was spec-

* Readers who are familiar with statistical principles will note that this statement is *not* an accurate representation of the "null hypothesis," which is tested when performing a multivariate analysis of variance. However, this statement does represent accurately the type of hypothesis that most researchers want to test when applying multivariate analysis of variance to research about the classification of psychopathology.

ified *a priori*. Compared with Figure 10, discriminant analysis separated the categories relatively clearly, as is shown in Figure 9. This occurred because discriminant analysis already knew the category membership of the patients and was concerned only with forming dimensions that would optimally separate the known categories. In contrast, Figure 10 suggests that there are no clusters in the data. This occurred because factor analysis assumes that the data are unimodal, and that the results of factoring a data set will lead to the formation of factors that have unimodal distributions even if the original data did not.

Also confused with discriminant analysis is cluster analysis. Both methods partition the entities in a data set into classes. However, in cluster analysis, the number and the composition of the classes are not known in advance. In fact, the whole point of using cluster analysis is to form classes. In contrast, discriminant analysis requires that the categories be known *a priori*. Discriminant analysis does not discover classes; it assumes their existence. In other words, the distinction between discriminant analysis and cluster analysis parallels the distinction that biologists make between *identification* and *classification*. Classification is the act of forming groups (classes, or categories). Identification is the process by which a new entity is assigned to a class that is already known. In terms of these definitions, cluster analysis is a classification procedure, whereas discriminant analysis is an identification procedure.*

An example may help the reader understand how multivariate analysis of variance and discriminant analysis can be applied to research on psychopathology. Lasky, Klett, Caffey, Bennett, Rosenblum, and Hollister (1962) performed one of the early studies on the effectiveness of psychotropic medications in the treatment of schizophrenia. Their sample consisted of 512 schizophrenic men with acute psychotic symptoms who had been admitted to VA hospitals. These schizophrenic patients were randomly assigned to drug therapy with one of six drugs:

Chlorpromazine (Thorazine)
Chlorprothixene (Taractan)
Fluphenazine (Prolixin)
Reserpine (Serpasil)

* Statisticians have not made the preceding distinction between classification and identification. Hence, in statistical textbooks, discriminant analysis and cluster analysis are often both described as classification procedures, a terminology that has promoted the confusion of these two very different statistical procedures.

Thioridazine (Mellaril)
Triflupromazine (Vesprin)

Patients were continued on these drugs for 24 weeks, or until the patient was discharged. Each patient was interviewed at the beginning of the study, after 8 weeks, and after 24 weeks. The Inpatient Multidimensional Psychiatric Scale was completed after each interview.

A multivariate analysis of variance was used to compare the six groups of patients after 8 weeks of treatment. This statistical method tested whether the different drug groups had the same IMPS profiles before and after treatment. The results of the multivariate analysis of variance showed chlorpromazine (Thorazine) and thioridazine (Mellaril) to be significantly better than reserpine (i.e., the patients receiving either chlorpromazine or thioridazine exhibited less psychotic behavior than patients receiving reserpine). The difference on the multivariate analysis of variance between fluphenazine (Prolixin) and reserpine (Serpasil) also approached significance.

Lasky *et al.* also performed a discriminant analysis on the before and after data and found one dimension that managed to separate all six drug groups. This discriminant dimension is shown in Figure 17. Notice how the groups are rank-ordered in terms of drug response from chlorpromazine (Thorazine) to reserpine (Serpasil). These findings are interesting historically because, at the present time, only three of the six drugs that Lasky *et al.* studied are still commonly used as antipsychotic medications: chlorpromazine, thioridazine, and fluphenazine. These are the three drugs that the discriminant analysis on the 1962 data showed to lead to the most favorable therapeutic response. The drug that Lasky *et al.* found to perform least successfully was reserpine, a common antipsychotic in the 1940s and the early 1950s that disappeared from standard clinical use with the appearance of the phenothiazines.

FINAL REMARKS

This chapter has been concerned with the introduction of the models and methods associated with the application of multivariate statistical analysis to classification problems. The chapter has focused primarily on five major models and the issues that have been raised in a debate between the proponents of two of the models. These five structural models for the classification of psychopathology are important, but despite their importance, a careful enunciation of these mod-

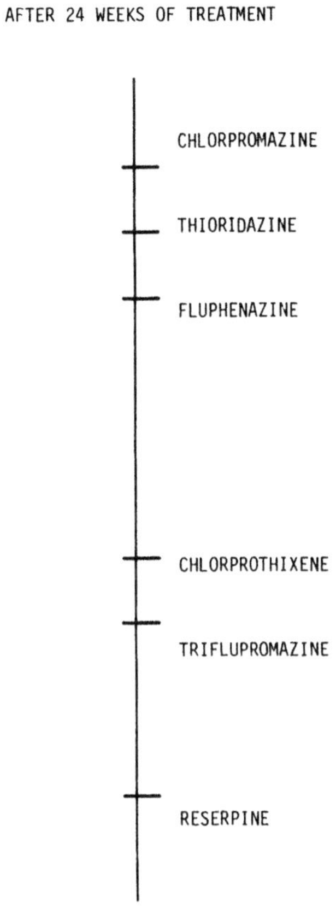

FIGURE 17. Discriminant vector showing relative status of drug groups. (From "Drug Treatment of Schizophrenic Patients" by J. J. Lasky, C. J. Klett, E. M. Caffey, J. L. Bennett, M. P. Rosenblum, and L. E. Hollister. *Diseases of the Nervous System,* 1962, *23,* 698–706. Copyright 1962 by Physicians Postgraduate Press. Reprinted by permission.)

els and their implications has been lacking. As the following two chapters of the book should demonstrate, this failure to enunciate classificatory models is symptomatic of a more general failure in the quantitative approach: the failure to develop clear and well-elaborated theories of psychopathology.

SUGGESTED READINGS

Sage University Paper series on Quantitative Applications in the Social Sciences. Beverly Hills, Calif.: Sage Publications, 1976–1983.

The Sage Publications series on quantitative procedures is one of the best sources currently available for introductory presentations of multivariate statistical methods. Currently, this series contains 36 papers published independently in the form of reasonably brief and inexpensive monographs that are intended to introduce novices to specific statistical methods. Particularly relevant to clinicians and researchers concerned with classification are the monographs by Kruskal and Wish (1978) on multidimensional scaling, Kim and Mueller (1978) on factor analysis, Carmines and Zeller (1980) on reliability, and Klecka (1980) on discriminant analysis. Other topics addressed in these papers are causal modeling, applied regression analysis, exploratory data analysis, time series analysis, canonical analysis, and the analysis of nominal data.

Tatsuoka, M. M. *Multivariate analysis: Techniques for educational and psychological research.* New York: Wiley, 1971.

For readers in the social sciences who have a solid background in analysis of variance and have had some introduction to correlational procedures, the Tatsuoka book is a solid and well-developed presentation of multivariate statistics. Tatsuoka started the book by reviewing some of the basic issues in univariate statistics and then generalized the univariate procedures to multivariate statistics. Like most texts on multivariate statistics, this book will require that the reader develop some familiarity with matrix algebra. Contained in this book are excellent presentations on multivariate analysis of variance, discriminant analysis, and canonical analysis. The exploratory methods are handled poorly. Factor analysis is not well described. The suggested readings at the end of Chapter 7 are much better resources for the last topic. Tatsuoka did not address multidimensional scaling and cluster analysis.

Other textbooks on multivariate statistics that are of value (listed in increasing order of difficulty) are by Thorndike (1978), Harris (1975), Overall and Klett (1972), Morrison (1967), and Anderson (1958).

Skinner, H. A. Towards the integration of classification theory and methods. *Journal of Abnormal Psychology,* 1981, *90,* 68–87.

Skinner is one of the bright young psychologists who is interested in quantitative approaches to classification. This recent paper contains a clear statement of Skinner's approach to the classification of psychopathology. Within this paper, he has presented a three-stage outline by which research on classification systems can be organized, and he has used this outline to critically discuss the advances made by the DSM-III.

Like most persons concerned with the quantitative approach to classification, Skinner has addressed the issue of the categorical versus the dimensional model. He has rejected the simplistic notion that either the dimensional or the categorical model must be superior; instead, he has advocated the application of hybrid models. Representative discussions of his ideas can be found in Skinner (1977, 1979a,b).

Chapter 7

Factor Analysis and Psychopathology

Knowledge comes from noticing resemblances and recurrences in the events that happen around us.

Author unknown

A major contributor to medical classification was the seventeenth-century British physician Thomas Sydenham, who promoted the concept of a *syndrome.* Sydenham argued that a careful observer of patients could note that certain sets of symptoms tended to co-occur. If these co-occurring sets of symptoms were seen repeatedly across a number of patients, this observance would suggest that the syndrome may represent more than a chance collection of symptoms. Instead, the consistent appearance of a syndrome would suggest a disease with a common etiology and a common treatment.

Factor analysis, the multivariate statistical technique to be discussed in this chapter, can be thought of as a statistical tool used to isolate syndromes of co-occurring symptoms. For example, consider the fellowing set of symptoms that occur among adolescent patients (Jenkins & Glickman, 1946):

S1—pupils fixed to light
S2—pupils irregular, unequal or sluggish
S3—spinal Wassermann or Kahn test positive
S4—irritability

S5—nervousness
S6—emotional instability
S7—temper tantrums
S8—exclusion, expulsion or suspension from school
S9—deterioration

As shown in Table 16, these symptoms all tend to covary. That is, if one of these eight symptoms is noted, then most of the other symptoms will also be observed. This particular set of eight symptoms constitutes a clinical syndrome known as *juvenile paresis*. The syndrome was first noted clinically during the 1880s (before factor analysis) and is now known to be caused by a syphilitic infection of the central nervous system that is acquired by the fetus from the mother during pregnancy, but that does not become clinically manifest until adolescence.

Factor analysis, in effect, attempts to identify a collection of co-occurring symptoms such as the collection shown in Table 16. In addition, factor analysis provides a statistical estimate of the "underlying factor" that explains the association among a collection of related symptoms. From the correlations presented in Table 16, a factor analysis was performed. The column furthest to the right in this table displays the correlation of each symptom with the underlying paretic factor. In the jargon of factor analysis, these correlations of the symptoms with the factor are called *loadings*.

Historically, factor analysis was developed by psychologists to study intelligence. During the early twentieth century, a popular area of psychological research was the study of mental abilities. A British psychologist, Charles Spearman, noted that the scores on all ability

TABLE 16. Correlation Matrix of Eight Symptoms Constituting Juvenile Paresis Syndrome

	S1	S2	S3	S4	S5	S6	S7	S8	Paresis factor
S1		.68	.53	.30	.33	.56	.28	.45	.80
S2	.68		.14	.19	.27	.25	.45	.11	.60
S3	.53	.14		.14	.51	.13	.05	.37	.52
S4	.30	.19	.14		.40	.55	.68	.11	.65
S5	.33	.27	.51	.40		.44	.22	.38	.67
S6	.56	.25	.13	.55	.44		.35	.35	.72
S7	.28	.45	.05	.68	.22	.35		.43	.66
S8	.45	.11	.37	.11	.38	.35	.43		.60

Note. From Jenkins and Glickman (1946).

tests were positively correlated. In other words, if a person performed well on one ability test, it was likely that she or he would perform well on another test. From this observation, Spearman postulated the existence of a "general factor" (called *g*) that would account for these positive correlations. Factor analysis was initially developed by Spearman as a procedure for isolating this general factor. The g factor is now considered highly similar to the concept of intelligence.

Initially, as proposed by Spearman, factor analysis was primarily a heuristic procedure used to analyze correlation matrices. However, the statistical theory of factor analysis has been advanced. In particular, the recent contributions by Joreskog (1969) have had a major influence on the development of a rigorous statistical theory for factor analysis.

Before discussing the literature on factor analysis and its application to clinical data, a general introduction to factor analysis is presented. This introduction is not very technical, and it focuses on an example from a clinical data set in order to help the reader understand how factor analysis proceeds. After this introduction to factoring, there is a discussion of the use of factor analysis to study psychotic behavior.

AN INTRODUCTION TO FACTOR ANALYSIS

The earliest clinical research using factor analysis was performed by Thomas V. Moore. He gathered descriptive data on 367 psychotic patients and used factor analysis to isolate empirical syndromes among these patients. Moore published two articles about this research: a journal article in 1930 and a monograph in 1933. Because the monograph contained most of the details of his research, the following discussion emphasizes information obtained from the monograph.

Step I: Obtaining Descriptive Data

Moore started his research by proposing a list of 40 symptoms that he believed were clinically important to the study of psychotic patients:

Insane relatives	Mutism
Alcoholism in parents	Negativism
Number of previous attacks	Refusal of food
Disorientation in time	Stereotypism of attitudes
Disorientation in space	Stereotypism of actions
Rational delusions	Stereotypism of words

Bizarre delusions	Giggling
Hypochondriacal delusions	Destructiveness
Auditory hallucinations	Talking to voices
Visual hallucinations	Irritability
Tactual hallucinations	Tantrums
Other hallucinations	Homicidal tendency
Reasoning	Excitement
Logical fallacies	Euphoria
Perception	Depression
Total memory	Anxiety
Memory ratio	Tearfulness
Absence of insight	Retardation
Shut in	Neurasthenia
Loss of finer sensibilities	Suicidal tendency

Moore provided definitions for all the symptoms in this list. These definitions were subdivided into three types: (1) symptoms with prose definitions; (2) symptoms defined by the patient's performance on a test; and (3) symptoms defined via scores on behavioral rating scales. The definition of *stereotypism of attitudes* is a good example of a definition by prose.

> Stereotypism of attitudes: This symptom was considered positive if during the course of the mental disorder the patient had manifested a phase or period in which he maintained himself for long periods in fixed postures, e.g., standing on one leg, maintaining a peculiar attitude and staring into space, etc. (Moore, 1933, p. 23)

The second type of symptom was defined in terms of test performance. For instance, patients were shown different sets of common objects. After each set was removed, the patient was asked to remember the objects. The number of items remembered determined the patient's "total memory" score. Finally, the last type of symptom was defined by a score on a set of behavioral ratings. *Excitement* was defined in terms of ratings on the patient's motor movements, rate of talking, loudness, and frequency of excited spells as these were observed in a clinical interview and on the ward.

After defining his symptoms, Moore proceeded to gather data on the 367 psychotic patients. Most of these patients were in one of two mental institutions in the Washington, D.C., area.

Critical comments can be directed at the procedures that Moore used to gather data on the psychotic patients. For instance, Moore did not provide any indication of how these patients were sampled, nor did he describe the types of clinicians who were used to rate the symptoms of these patients. Also, no data were provided on the reliability of the symptoms, an important step, given the frequently voiced concern about the consistency of clinical judgments. Finally, it is apparent from Moore's reported results that there were a fair number of missing data; only 221 of the 367 patients took the memory test, for example. Moore did not describe his procedures for handling missing data.

All four of the above criticisms are serious, and their resolutions could directly affect the results. However, because Moore's study was the earliest research of this type, judging its methodological adequacy by contemporary standards seems harsh.

Step II: Forming a Correlation Matrix

The second major step in the process of performing a factor analysis is to correlate all symptoms with each other. The result from Moore's data was a 40 × 40 correlation matrix (which had a total of 780 unique correlation values). Because these correlations had to be calculated by hand in the 1930s, the formation of the correlation matrix was not a trivial step. The entire 40 × 40 matrix was published in Moore's monograph (1933).

Table 17 presents part of Moore's correlation matrix. Only part is shown because of the size of a 40 × 40 matrix. Table 17 contains 45 unique values, about 1/17 the size of the entire matrix. In visually analyzing this matrix, the reader will note that some of the 10 symptoms do tend to covary (e.g., "excitement" and "euphoria" correlate at $r = .880$), whereas other symptoms have little or no relationship with each other ("mutism" and "sucidal tendency" correlate at $r = -.041$).

One way of conceptualizing factor analysis is as a procedure for reorganizing the correlation matrix in order to find a syndrome of covarying symptoms. If one reads the list of the 10 symptoms in Table 17, one will notice that 4 of the symptoms are related to the clinical syndrome called *mania* ("excitement," "euphoria," "tantrums," and "irritability"), and another 4 are related to a syndrome we could call *catatonia* ("negativism," "stereotypism of attitudes," "refusal of food," and "mutism"). Table 18 displays the same correlations as are shown in Table 17, except that the symptoms are organized into these two groups.

Notice that the intercorrelations among the symptoms *within* a group are relatively high, whereas the correlations among symptoms from different groups are relatively small. When the symptoms are reorganized in this manner, the appearance of the clinical syndromes in the correlation matrix can easily be seen.

TABLE 17. Correlations among 10 Symptoms Reported by Moore (1933)

	Excitement	Negativism	Alcoholism	Suicidal	Stereotypism	Tantrums	Irritability	Refusal	Euphoria	Mutism
Excitement		.22	.17	.01	.07	.41	.42	.05	.88	.04
Negativism	.22		−.18	−.02	.61	.32	.44	.63	−.09	.77
Alcoholism in parents	.17	−.18		−.02	−.02	−.02	−.09	−.24	.22	−.12
Suicidal tendency	.01	−.02	−.02		−.05	.00	−.26	.19	−.06	−.04
Stereotypism of attitudes	.07	.61	−.02	−.05		.15	.19	.56	−.23	.60
Tantrums	.41	.32	−.02	.00	.15		.59	.24	.39	.20
Irritability	.42	.44	−.09	−.26	.19	.59		.32	.47	.36
Refusal of food	.05	.63	−.24	.19	.56	.24	.32		−.07	.66
Euphoria	.88	−.09	.22	−.06	−.23	.39	.47	−.07		−.05
Mutism	.04	.77	−.12	−.04	.60	.20	.36	.66	−.05	

TABLE 18. Reorganized Correlations among 10 Symptoms

	Mutism	Negativism	Refusal	Stereotypism	Excitement	Euphoria	Tantrums	Irritability	Alcoholism	Suicidal
Mutism		.77	.66	.60	.04	−.05	.20	.36	−.12	−.04
Negativism	.77		.63	.61	.22	−.09	.32	.44	−.18	−.02
Refusal of food	.66	.63		.56	.05	−.07	.24	.32	−.24	.19
Stereotypism of attitudes	.60	.61	.56		.07	−.23	.15	.19	−.02	−.05
Excitement	.04	.22	.05	.07		.88	.41	.42	.17	.01
Euphoria	−.05	−.09	−.07	−.23	.88		.39	.47	.22	−.06
Tantrums	.20	.32	.24	.15	.41	.39		.59	−.02	.00
Irritability	.36	.44	.32	.19	.42	.47	.59		−.09	−.26
Alcoholism in parents	−.12	−.18	−.24	−.02	.17	.22	−.02	−.09		−.02
Suicidal tendency	−.04	−.02	.19	−.05	.01	−.06	.00	−.26	−.02	

Reorganizing a 40 × 40 correlation matrix in order to search for a set of correlated symptoms is not a trivial exercise. Part of what factor analysis does is to perform this reorganization.

Step III: Factoring the Correlation Matrix

The next step is the most technical aspect of factor analysis. The explanation that follows will not provide sufficient information for an understanding of the complete details of how factor analysis methods actually work. Readers who are interested in learning more about factoring should read Gorsuch (1974) or Harman (1976).

Factor analysis assumes that there are two types of variables that are of concern to a research project: (1) a set of observed variables and (2) a set of unobserved variables or "factors" that are the underlying dimensions represented by the observed variables. In Moore's study, the observed variables were the 40 symptoms on which he obtained present/absent ratings. The unobserved variables or factors were estimated by the factoring procedures.

Factor analysis assumes a linear model. That is,

$$F_1 = a_1 V_1 + a_2 V_2 + \ldots + a_{10} V_{10}$$

where F_1 designates the first factor; V_1 is Variable 1; and a_1 is the "loading" that determines how important Variable 1 is when estimating Factor 1.

When creating the first factor, the factoring method forms a linear combination of the observed variables that accounts for as much variance as possible in the correlation matrix. This means that the first factor is chosen so that the loadings (a_1, a_2, a_3 . . . a_{10}) are as large as possible. This step is performed by using iterative trial-and-error procedures to determine the optimal values of a_1 to a_{10}. At the end of the iterations, the correlations of the observed variables with the first factor are as large as possible. The correlations of all variables with the first factor are shown in Table 19. Notice that the absolute values of these correlations exceed $r = .35$ for 7 of the 10 variables.

After creating the first factor, the variance associated with this factor is removed from the correlation matrix. This revised correlation matrix is shown in Table 20. Notice that the values of the correlations shown in this table are smaller than the values presented in Table 18. In general, as factors are removed, the correlations remaining in the residual matrix will decrease in size. Then, a new set of iterations is performed to estimate the a_1 to a_{10} values for the second factor. The

TABLE 19. Correlations of Variables with First Factor (Unrotated)

	Factor I
Mutism	.80
Negativism	.86
Refusal of food	.75
Stereotypism of attitudes	.67
Excitement	.42
Euphoria	.23
Tantrums	.57
Irritability	.69
Alcoholism in parents	−.16
Suicidal tendency	−.05

correlations of each variable with the second factor are shown in Table 21.

The amount of variance associated with each factor is indicated by the *eigenvalue* of the factor. The eigenvalue is simply a linear composite of the values of the constants a_1 to a_{10}. For instance, the eigenvalue for the first factor of the 10 symptoms is 3.43, which means that that factor accounts for almost 35% of the variance in the data.

This process of estimating factors and then forming a revised correlation matrix can be repeated as often as the researcher desires. Because there is reason to believe that two factors are represented among these 10 variables, factoring is stopped after two factors are formed. These two factors account for 59% of the variance in the correlation matrix (Factor I accounts for 35% of the variance, and Factor II accounts for 24%).

The particular factoring procedure that forms Table 19 is called *principal components analysis.* Other factoring methods do exist, such as the *centroid method, principal axis factoring,* and *maximum likelihood factoring.* The *centroid method* was promoted during the 1930s and 1940s by L. L. Thurstone. The reason that Thurstone emphasized centroid factoring was that this method is relatively simple to use when performing computations. Because all computations during this era were performed by hand, simplicity was a powerful reason to use centroid factoring. Even so, the performance of a factor analysis study was very expensive in terms of time. For instance, Gulliksen (1974) remembered supervising 20 clerks in Thurstone's laboratory during the extraction of a seven-factor solution from a correlation matrix of 50

TABLE 20. Residual Correlation Matrix after First Factor Removed Variable Numbers

	Mutism	Negativism	Refusal	Stereotypism	Excitement	Euphoria	Tantrums	Irritability	Alcoholism	Suicidal
Mutism	.36	.08	.36	.06	−.29	−.23	−.25	−.19	.02	.00
Negativism	.08	.26	.31	.03	−.14	−.28	−.17	−.16	−.03	.03
Refusal of food	.36	.31	.86	.31	−.11	−.15	.03	.06	−.18	.21
Stereotypism of attitudes	.06	.03	.31	.55	−.21	−.38	−.23	−.27	.09	−.01
Excitement	−.29	−.14	−.11	−.21	.83	.79	.18	.13	.24	.03
Euphoria	−.23	−.28	−.15	−.38	.79	.95	.26	.31	.26	−.05
Tantrums	−.25	−.17	.03	−.23	.18	.26	.67	.29	.77	.03
Irritability	−.19	−.16	.06	−.27	.13	.31	.29	.52	.03	−.22
Alcoholism in parents	.02	−.03	−.18	.09	.24	.26	.77	.03	.97	−.03
Suicidal tendency	.00	.03	.21	−.01	.03	−.05	.03	−.22	−.03	.99

TABLE 21. Correlations of Variables with Second Factor (Unrotated)

	Factor II
Mutism	−.34
Negativism	−.26
Refusal of food	−.35
Stereotypism of attitudes	−.40
Excitement	.78
Euphoria	.92
Tantrums	.44
Irritability	.42
Alcoholism in parents	.33
Suicidal tendency	−.14

ability tests. The total computation required almost one year to complete!

The next method of factoring is called *principal components analysis*. In theory, this method was recognized in the 1930s but did not become feasible until high-speed computers were available to perform the calculations. A modern computer such as an IBM 370 can perform a principal components analysis on a 50 × 50 correlation matrix in about three seconds. The advantage of principal components analysis is that it extracts factors that account for as much variance in the observed correlation matrix as possible. The statistical model for principal components analysis is given below:

$$V_1 = b_{11} F_1 + b_{12} F_2 + b_{13} F_3 + \ldots + b_{1k} F_k + E$$

where there are k factors in the solution; V_1 refers to Variable 1; F_1 refers to Factor 1; E refers to random error; and b_{12} is the weight of Factor 1 when estimating Variable 2. In effect, the model associated with principal components analysis assumes that all systematic variance among the variables can be accounted for by the factors.

The third method is called *principal axis factoring*. Principal axis factoring and principal components analysis differ only in terms of the precise statistical model that is being used. Principal axis factoring assumes a slightly different model than does principal components analysis:

$$V_1 = b_{11} F_1 + b_{12} F_2 + b_{13} F_3 + \ldots + b_{1k} F_k + U_1 + E$$

where all notation has the same significance as above and U_1 refers to the unique variance associated with Variable 1. In other words, the principal axis model does not assume that all systematic variance for a variable is associated with the extracted factor. Instead, some variance (perhaps a great deal for certain variables) is associated with unique aspects of that variable that are independent of the factors.

In practical terms, principal components analysis and principal axis factoring yield very similar solutions. The only difference is that, with principal axis factoring, the researcher must provide *communality estimates* for the variables. The communality estimate of a variable is simply some *a priori* guess (or an empirical approximation) by the researcher of the multiple correlation of the variable with all the factors to be extracted. These communality estimates are placed in the diagonal of a correlation matrix prior to factoring. In contrast, principal components analysis has values of 1.0 in all diagonal cells of a correlation matrix (because, according to its statistical model, the factors should account for all variance in the variables; hence, the multiple correlations between the variables and the factors for a complete solution should be 1.0).

Maximum likelihood factor analysis is the most recent method of factoring to gain general popularity and use. This method is discussed later in the chapter.

Step IV: Number of Factors

The fourth major step when performing a factor analysis is to determine how many factors are present in the data. Factor analysis by itself does *not* determine the number of factors. Any factor analysis method will generate as many factors as there are variables in the data. However, it is very rare that anyone would interpret all of these factors.

There are two major goals when estimating the number of factors. The first goal is to select a small number of factors in order to make interpretation of the factors relatively easy. As the number of factors increases, interpretation usually becomes more difficult. The second goal is to account for as much variance in the original data as possible. Some early factor analysts suggested choosing the number of factors so that at least 80% of the variance is accounted for.

Unfortunately, these two goals usually conflict. In order to account for a large proportion of the variance, the researcher often must select a large number of factors. For instance, if one decided to select the number of factors that would account for at least 80% of the total variance, one would need to settle on a four-factor solution to the

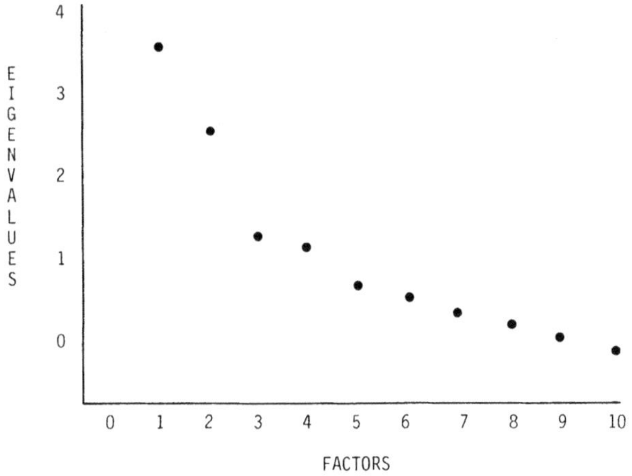

FIGURE 18. Plot of eigenvalues for 10 factors.

correlations in Table 18. On the other hand, two factors give a more parsimonious interpretation, and as shown in Table 18, the clinical meaningfulness of a solution with two factors is good. How is the conflict between the goal of a small number of factors and the goal of accounting for as much variance as possible usually resolved?

One commonly used rule in selecting the number of factors is to choose all factors that have an eigenvalue greater than 1.0. The logic behind this rule is that, when performing a principal components analysis, the mean value for all eigenvalues is 1.0. (The mean value is not 1.0 when principal axis factoring is the method being used.) Thus, by selecting all factors with eigenvalues greater than 1.0, the researcher is choosing factors whose eigenvalues are greater than the average. This rule for selecting the number of factors is a default in many of the commonly used statistical packages that contain factor analysis procedures, such as SPSS, SAS, and BMDP. However, this rule has the well-known property of generally overestimating the number of factors (Rummel, 1970). For instance, when factor analysis is applied to a data set of random numbers, the "eigenvalue greater than 1.0" rule usually estimates the presence of two or more factors in the data even though no factors are actually present.

The second commonly used procedure is called the *scree test* (Cattell, 1966). In this procedure, a plot is made of the numbers of factors versus the eigenvalues. This plot for the 10-symptom data is shown in Figure 18. Notice that in this figure the sizes of the eigenvalues are large initially and then drop rapidly. Cattel, in his scree test, suggested drawing a straight line from the smallest eigenvalue along the nearby values. When an eigenvalue is markedly above the straight line, the number of factors should be estimated at that value. Notice that in Figure 18, the first point that is clearly off the straight line occurs at two factors. Thus, two factors appear to give the best estimate of the number of factors for the 10-symptom data.

A third method for selecting the number of factors is to examine the residual correlation matrix after the variance associated with the factors has been removed. In the example, when the residual correlation matrix is examined after two factors have been removed, the correlation values are quite small. Removing an additional third factor does not reduce the residual correlations very much. Thus, two factors appear to give a good estimate of the systematic variation in the correlation matrix. It should be noted that the examination of residual correlations is always a good step when performing a factor analysis, but most standard statistical packages do not present these correlations (even as options).

Step V: Rotation of the Factors

After the factor analysis method is completed and the number of factors is selected, the researcher obtains a listing of the correlation of each variable with each factor. This listing is contained in a *factor-loading* matrix.* This matrix is shown in Tables 19 and 21 for two factors from the 10-symptom data. In these tables, note that the first factor that accounts for 35% of the variance is moderately to highly correlated with most of the variables (except "suicidal tendency," "alcoholism of parents," and "euphoria"). This first factor is a "general" factor that correlates with most variables. The second factor is somewhat more specific. It is highly correlated with two variables ("euphoria" and "excitement"), and it also has moderate correlations with four variables ("tantrums," "irritability," "mutism," and "alcoholism of parents"). This factor appears to be similar to the mania factor postulated from Table 18 after the reorganization of the correlation matrix. But the presence of correlations with the variables of "mutism" and "alcoholism of parents" in this factor does not make much sense. Also, in the visual reorganization of Table 18, a "catatonia" factor had been suggested because "mutism," "negativism," and so on showed moder-

* This definition of a factor-loading matrix is accurate only for orthogonal rotations. In oblique rotations, these correlations are contained in the *factor structure* matrix. Consult Harman (1967, Section 13.3) for more details.

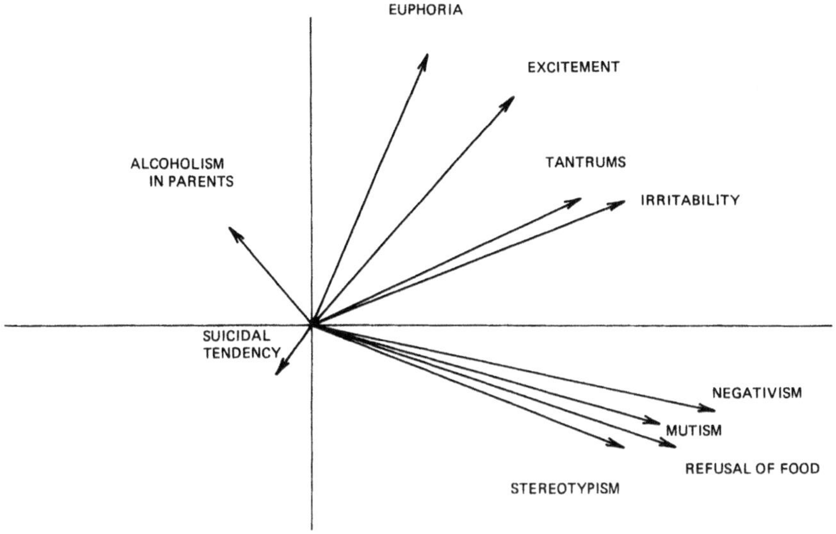

FIGURE 19. Two-dimensional space defined by first two factors with 10 symptoms represented by vectors.

ate to high intercorrelations. However, no catatonia factor was found in Tables 19 or 21.

In short, the interpretation of the factor-loading matrix in Tables 19 and 21 is disappointing. Instead of finding two clear factors, mania and catatonia, as is suggested by the visual analysis of the correlation matrix, one general factor and a somewhat messy mania factor are noted. The clinical meaningfulness of the solution is not very great.

Factor rotation is a procedure that is commonly used in order to increase the meaningfulness of a factor solution. In order to understand factor rotation, one needs to learn something about the graphic representation of factors. A factor can be considered a basic dimension associated with a given set of observed variables. These dimensions can be conceptualized as axes in a multidimensional space. In this space, the variables can be represented as vectors.

Figure 19 shows the two-dimensional space defined by the first two factors. The 10 symptoms are presented as vectors in the space. The coordinates for the point of each vector are the correlations of the symptoms with the factors. The length of the vector is proportional to the amount of variance for a given symptom that is explained by the factor solution.

Notice in Figure 19 that one can see two clusters of symptoms: a tight cluster containing the catatonic symptoms and a somewhat

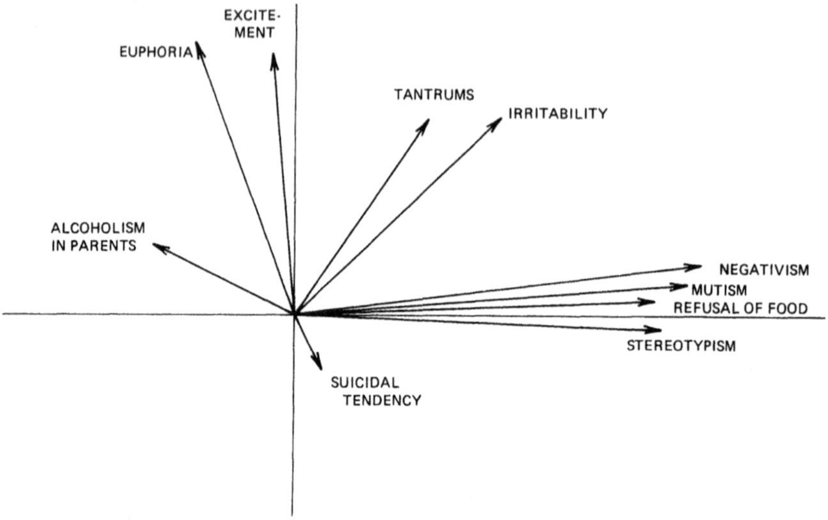

FIGURE 20. Rotation of the first two factors.

looser cluster of the manic symptoms. The factors do not directly point at either cluster of variables. In particular, the manic cluster seems to fall partway between the two factors. If the factors were rotated somewhat (as shown in Figure 20), the factors could be oriented so that they come relatively close to pointing at the cluster of variables. The factor-loading matrix associated with the rotation is presented in Table 22. Notice in Table 22 that the manic factor and the catatonic factor are clearly defined. On Factor I (shown by the horizontal axis on Figure 20), four variables have high loadings: "negativism," "mutism," "refusal of food," and "stereotypism of attitudes," whereas "excitement," "euphoria," "tantrums," and "irritability" load highly on Factor II (shown by the vertical axis on Figure 20).

In other words, factor rotation is used to increase the meaningfulness of a factor solution. Experience with factor analysis has shown that most applications of this method do not yield meaningful solutions unless followed by a rotation. Graphically, factor rotation can be conceptualized as a process that points the factors at clusters of variables.

The method of factor rotation used to change from Figure 19/Tables 19 and 21 to Figure 30/Table 22 is called a *varimax rotation.* This rotational procedure is commonly used because it is relatively likely to point at clusters of variables. More technically, a varimax rotation is likely, but is not guaranteed, to meet Thurstone's criteria of *simple structure.* Thurstone proposed simple structure criteria to evaluate the psychological meaningfulness of a factor solution. Without providing technical definitions of these criteria, briefly the criteria can be explained as ensuring that the factors will point at "clusters" of similar variables that are reasonably exclusive and that each factor will be defined by a sufficient number of variables to make its interpretation meaningful.

A number of methods for rotating a factor solution are in existence. Generally, these rotational methods are organized into two classes: *orthogonal* methods and *oblique* methods. Orthogonal methods keep the factors (dimensions) at right angles to each other, thereby ensuring that factors will be independent and uncorrelated, whereas oblique rotations permit the formation of correlated factors. Varimax is the frequently used orthogonal rotational method. Quartimax and equimax are other orthogonal rotational procedures (see Harman, 1976). Promax is an oblique method that has been developing increasing popularity and may be the best of the oblique techniques (Gorsuch, 1974).

Some researchers working with clinical data have attacked the implicit assumption when using orthogonal rotations that factors must be uncorrelated. These researchers have argued that syndromes of psychopathology could be correlated. For instance, mania and catatonia are syndromes that are not necessarily independent. Catatonic excitement is a well-known clinical phenomenon, which can be confused with the behavior seen in some manics. Thus, mania and catatonia need not be independent concepts.

TABLE 22. Rotated (Varimax) Factor Loadings

	Factor I	Factor II
Mutism	.86	.08
Negativism	.88	.18
Refusal of food	.83	.05
Stereotypism of attitudes	.78	−.04
Excitement	.00	.88
Euphoria	−.24	.91
Tantrums	.29	.66
Irritability	.41	.70
Alcoholism in parents	−.30	.21
Suicidal tendency	.02	−.14

FACTOR ANALYSIS OF THE COMPLETE MOORE DATA

Now that the reader has been introduced to the five major steps of factor analysis, a factor solution to all the 40-variable Moore data will be presented.

Step I, obtaining the descriptive data on the patients, has already been described for the Moore data, as has Step II, the formation of the correlation matrix. It should be noted, regarding Step II, that because Moore had both binary and continuous data he chose the tetrachoric correlation coefficient to form his 40 × 40 table. For Step III, a principal components analysis was performed on the 40 × 40 correlation using the SPSS statistical package. This factoring method was selected because the correlation matrix was singular (presumably because of missing data), and yet this method can still provide some estimate of the factors.*

Step IV is the selection of the number of factors. The "eigenvalue greater than 1.0" rule would suggest the presence of 13 factors. This is a large number of factors—probably too large, as the factor-loading matrix for 13 factors shows a number of factors that are correlated with only one or two variables. A scree test suggests five factors. Five factors are also plausible because this was the number of factors that Moore (1933) used as his estimate for the number of factors. Thus, the final number of factors is estimated at five. These factors account for a total of 57.5% of the variance.

The decision at Step V concerning rotation of the factors is to use a varimax orthogonal rotation. This rotation yields a reasonably good simple structure whose clinical interpretation makes sense. The factor-loading matrix after rotation is shown in Table 23. In order to help the reader analyze this table, only correlations that are greater than .30 are shown.

The first factor is the *catatonia* factor. Its highest loadings are with the variables "stereotyped motor behavior" and "negativism." A total of 11 of the 40 variables is associated with this factor.

The second factor was called a "deluded-hallucinatory" syndrome by Moore (1933). This factor is correlated with all symptoms involving delusions and hallucinations and is correlated just above .40 with two other symptoms ("stereotypism of words" and "talking to voices").

* An exact definition of a "singular" correlation matrix will not be presented here. Mathematically, a factor solution to a singular matrix can be hard to estimate, although many of the popularly available methods of factoring in contemporary statistical packages, such as SPSS and BMDP, do provide good estimation routines.

TABLE 23. Varimax Rotated Factor Matrix for Moore Data

	Factors				
	1	2	3	4	5
Insane relatives					
Alcoholism in parents					
Number of previous attacks				.337	
Disorientation in time	.302		.724		
Disorientation in space			.681		
Rational delusions		.583			
Bizarre delusions		.698			−.383
Hypochondriacal delusions		.486			
Auditory hallucinations	.304	.830			
Visual hallucinations		.835			
Tactual hallucinations		.858			
Other hallucinations		.823			
Reasoning			.847		
Logical fallacies			.776		
Perception			.746		
Total memory			.543		
Memory ratio			.472		
Absence of insight					−.406
Shut in	.770		.354		
Loss of finer sensibilities	.647				.361
Mutism	.793				
Negativism	.703				
Refusal of food	.795				
Stereotypism of attitudes	.718				
Stereotypism of actions	.655		.301		
Stereotypism of words	.513	.402	.309		
Giggling	.654				−.311
Destructiveness	.560			.634	
Talking to voices	.429	.421			
Irritability	.329			.677	
Tantrums				.636	
Homicidal tendency	.311				
Excitement				.858	
Euphoria	−.385			.890	
Depression				−.345	.795
Anxiety					.800
Tearfulness					.627
Retardation				−.340	.660
Neurasthenia					.466
Suicidal tendency					.387

This syndrome is a collection of symptoms that is frequently seen among state hospital patients and that is often associated with a diagnosis of schizophrenia (Moore, 1933).

Moore (1933) labeled the third factor a "cognitive syndrome" because it was associated with poor performance on intellectual tasks ("reasoning," "logical fallacies," "memory ratio," and so on). In addition, the factor is correlated with the two disorientation variables. This suggests that the syndrome associated with these variables could be a general *organic brain syndrome* because disorientation, memory disturbance, and impaired intellectual functioning are all clinically associated with OBS.

The fourth factor is a *mania* syndrome. It is most correlated with five symptoms that are often associated with mania: euphoria, excitement, tantrums, homicidal tendency, and destructiveness. This factor was labeled mania by Moore (1933).

The last factor is a *depression* factor that has high correlations with anxiety, depression, tearfulness, retardation, and neurasthenia. This factor also has a moderate negative correlation with "absence of insight" (i.e., patients who were depressed knew they were depressed).

To summarize, the factor analysis of all of the 40 × 40 Moore data yielded a solution with five factors. These syndromes could be named:

Catatonia
Schizophrenia
Organic brain syndrome
Mania
Depression

All five of these syndromes are well-known to clinicians who work with severely disturbed persons. The clinical meaningfulness of this solution is high and suggests a useful classification of major syndromes among psychiatric inpatients.

REVIEW OF FACTOR-ANALYTIC STUDIES OF PSYCHOTICS

The next section of this chapter focuses on a review of the factor-analytic studies of psychotic and severely distrubed patients. No attempt is made to perform a comprehensive review; instead, by focusing on the research of four influential persons (Moore, Wittenborn, Lorr, and Overall), the evolution of the use of factor analysis in psychopathology research is traced, and comments are made about the strengths and weaknesses of this research.

Moore

The seminal work by Moore has already been described, and a modern factor analysis of his data was presented above. The earliest report by Moore was published in 1930 in the *American Journal of Psychiatry.* This report contained relatively little information about the methodological details of his study. For instance, Moore did not mention how many patients were examined as the basis for this study. Moore's writing style in the journal article was also somewhat confusing. For instance, Moore discussed an *eight*-factor solution in the summary section of his paper; however, seven factors were shown in the tables that purported to show his results. Not surprisingly, different reviewers have disagreed about the number of factors from the Moore (1930) study. The eight syndromes discussed by Moore in his study were:

Cognitive defect
Catatonic syndrome
Uninhibited or kinetic syndrome
Noneuphoric manic syndrome
Euphoric manic syndrome
Delusional hallucinatory syndrome
Syndrome of constitutional hereditary depression
Syndrome of retarded depression

The second paper that Moore published was in a monograph series titled *Studies in Psychology and Psychiatry* from the Catholic University of America (Moore, 1933). This paper was much longer and contained details about the methodological and clinical aspects of his study. Moore stated that the results published in the 1933 monograph were based on a somewhat larger patient sample and a recomputation of the correlation matrix. (Moore had used 41 symptoms in his 1930 article but dropped 1 symptom for the 1933 monograph. Also Moore was using tetrachoric correlation coefficient, and in 1933, he had changed the computational procedures for estimating this correlation.) Much of the 1933 monograph was devoted to describing the clinical picture associated with each syndrome and to presenting representative case histories of patients who were uniquely associated with each syndrome.

In performing his factor-analytic studies, Moore had been in contact with a major figure in the history of factor analysis, L. L. Thurstone. Thurstone was the proponent of multiple-group factor analysis and the centroid method as a practical solution to the computational aspects of factoring. Thurstone also proposed the "simple structure" criteria, which could be used to determine if a factor solution was meaningful. Thurstone's ideas on factor analysis dominated the topic from the 1930s through the 1950s.

In 1934, Thurstone published an important theoretical article in *Psychological Review* called "Vectors of the Mind." In that paper, Thurstone proposed his new approach to factor analysis, which differed significantly from the position advocated by previous researchers. Also, Thurstone published a table in this paper that contained his own factor solution to Moore's data. Thurstone found *five* factors. These five factors were very similar to the five factors found in the principal components solution noted above. In his theoretical article, Thurstone did not describe the details of his factor procedure, nor did he provide tables that listed the eigenvalues or the rotated factor loadings.

One additional reanalysis of the Moore data has been published in a doctoral dissertation by a student named J. W. Degan who had studied with Thurstone. Degan published his dissertation in monograph form in 1952. In his reanalysis of the Moore data, Degan decided to drop the eight cognitive tests because about one-third of the patients had missing data on these tests. Also, Degan decided to recompute the correlation matrix because better hand computational algorithms were available for estimating the tetrachoric correlation coefficient than had been used by Moore. Degan factored the resulting 32 × 32 matrix with the centroid method, and an oblique rotation followed. Degan found nine factors in his factoring of the Moore data, which he called:

Hyperprojection
Depression
Manic hyperexcitability
Catatonia
Schizophrenic dissociation
Traumatic hysteria
Hyperirritability
Deterioration
Neurasthenia

The names of these nine factors were different from the five factors isolated by Moore and by Thurstone. Because Degan had performed an oblique rotation, his nine factors were correlated, and a "second-order" factor analysis could be performed on the 9 × 9 correlation matrix among the first-order factors. Degan interpreted a four-factor solution. These second-order factors were:

Mania
Hebephrenic schizophrenia
Depression versus paranoia
Catatonic schizophrenia

This review of the various solutions to the Moore data should point out two major difficulties that are present in factor-analytic studies of psychopathology. The first difficulty concerns the methodological difference in how the factor analyses are performed. The five different analyses (Moore, 1930, 1933; Thurstone, 1934; Degan, 1952; and this chapter) involved the use of somewhat different sets of variables: Moore (1930) used 41; Moore (1933), Thurstone (1934), and this chapter analyzed 40 symptoms; Degan(1952) studied only 32. Also the studies differ in terms of the factoring method. Moore (1930, 1933) used a very early procedure called *Spearman's method of tetrads,* which really did little more than form clusters of correlated symptoms. No factor estimates nor factor-loading matrices were generated. Thurstone and Degan used a centroid factoring method, which was the standard method of factor analysis prior to high-speed computers. This chapter contains a principal components solution. Also, the different solutions vary in the number of factors obtained. The range in the number of factors varies from five (Thurstone's 1934 solution) to nine (Degan's solution).

A second major difficulty in factor analysis studies that can lead to interpretive variability is the naming of the factors. To understand the factor naming difficulty, consider the following set of symptoms:

Bizarre delusions
Auditory hallucinations
Other hallucinations
Visual hallucinations

In most solutions to the Moore data, these four symptoms appear together in a common factor. However, the exact symptom composition of this common factor has shown some change. Moore (1930) noted the four symptoms of auditory hallucinations, bizarre delusions, disorientation, and stereotypism of words as belonging to one factor. He titled this factor *delusions–hallucinations.* In 1933, Moore found essentially the same syndrome but gave it the clinical name of *paranoid dementia praecox.* Thurstone (1934) had a group of four symptoms to which he gave the name of *hallucinatory syndrome.* These symptoms were bizarre delusions, auditory hallucinations, other hallucinations, and speaking to voices. Degan (1952) noted a factor that had nine symptoms with significant factor loadings. He called this factor *hyperpro-*

jection because the symptoms reflected an overuse of the psychological defense of projection. The symptoms and their loadings were:

Tactual hallucinations	.73
Other hallucinations	.73
Visual hallucinations	.68
Auditory hallucinations	.65
Bizarre delusions	.55
Hypochondriacal delusions	.47
Rational delusions	.45
Suicidal tendency	.26
Absence of insight	.19

In this chapter, a factor containing hallucinatory and delusional symptoms was noted. It was titled a *schizophrenic* factor because these symptoms are commonly associated with schizophrenia among inpatients and because Moore had mentioned these symptoms as indicators of schizophrenia in the clinical section of his monograph. Notice how different these five names are for this one factor, which is basically the same across the five studies. This variance in naming factors has been a common problem in the clinical application of factor analysis to the problems of classification. Meehl, Lykken, Schofield, and Tellegen (1971) provided an interesting suggestion on how this problem can be avoided. This suggestion involves the use of different experts in naming the factors and a procedure to resolve any differences of opinion.

From the interpretation of this factor, one can begin to sense the variability that can occur in factor analysis studies. All five solutions started with basically the same data. Also all five resulted in a similar factor. However, the exact set of symptoms occurring in this factor does show variation from one study to another. Variability also occurred in the names of factors. This variability in factor analysis studies has been a major limitation on its acceptance and certainly suggests the dangers inherent in reifying any factor solution.

Wittenborn

Wittenborn is a clinical psychologist who received his Ph.D. from the University of Illinois in 1942. From 1942 until 1954, he was on the faculty at Yale University. Since then, he has been a professor at Rutgers.

Wittenborn's major research on psychopathology and factor analysis occurred in the late 1940s and the early 1950s. The goal of Wittenborn's research was the development of a quantified method for multiple psychiatric diagnoses (Wittenborn, 1951, p. 290). Wittenborn devised an instrument containing 55 symptom items. He called these items "rating scales." According to Wittenborn, the following were the guidelines he used when creating and selecting the items (or rating scales):

> 1. The rating scales must sample adequately the important symptoms in mental hospital patients.
> 2. The rating scales must be restricted to currently discernible behavior so that their repeated use can reveal changes in the patient.
> 3. The scales must provide ratings which are relatively independent of the insights and sophistication of the rater.
> 4. The rating scales must provide ratings which are relatively independent of any bias or theoretical persuasion of the rater.
> 5. The rating scales must be of such a form that each rating scale can be checked for every patient.
> 6. The rating scales must be feasible insofar as they draw upon behavior features which are likely to be known or may be readily ascertained.
> 7. The rating scales must be simple in form, and economical and convenient to use.
> 8. The rating scales must be in a form which does not bias the result of any analysis based upon them.
> 9. The rating scales must be reliable. (Wittenborn, 1951, pp. 290–291)

Notice that the guidelines proposed by Wittenborn are all aimed at reducing irrelevant sources of variance in the ratings and yet keeping maximal coverage of the symptoms that inpatients can present.

The set of 55 items constitutes the Psychiatric Rating Scales (Wittenborn, 1955). After creating the PRS, Wittenborn had it administered to 140 psychotic inpatients in a Veterans Administration hospital. The results were factor-analyzed with the use of the centroid method, and a visual orthogonal rotation was performed to yield the final factor loadings.

Wittenborn decided that seven factors accounted for the results. The seven factors are listed below as Wittenborn named them. Also listed are the four symptom items with the highest correlations (loadings) with each factor:

Factor I: Hebephrenic or Deteriorated Schizophrenia
- Characteristically oppositional (.625)
- Lies or steals (.561)
- Unaware of the feelings of others (.518)
- Incontinent because of own negligence (.474)

Factor II: Manic State vs. Depressed State
- Temper tantrums (.564)
- Opinions exceptional to physical laws (.558)
- Dramatically attention-demanding (.545)

vs.

- All overt activity is at a minimum (−.737)
- Cannot make decisions (−.684)
- Failure of affective response (−.611)

Factor III: Conversion Hysteria
- Use made of physical disease symptoms (.727)
- No organic basis for complaints (.631)
- No concern over physical handicaps (.621)
- Organic pathology with emotional basis (.610)

Factor IV: Catatonic Excitement
- Ideas change with spontaneous rapidity (.683)
- Great variation occurs in rate of speech (.679)
- Memory faults (.674)
- Repudiates earlier insights (.502)

Factor V: Anxiety
- Desperately distressed by his anxiety (.781)
- Delusional belief that he is evil (.731)
- Feelings of impending doom (.693)
- Shows failure and blocking (.685)

Factor VI: Paranoid Schizophrenia
- Believes others influence him (.643)
- Patient's thinking clearly delusional (.620)
- Feels systematically persecuted (.356)
- Hallucinations (.335)

Factor VII: Paranoid Conditions
- Cannot resist compulsive acts (−.538)
- Compulsive acts continuous (−.467)
- Dramatically attention-demanding (.464)
- Feels systematically persecuted (.437)
- Grandiose notions (.419)

As in the case of the analysis of the Moore data, it again can be noted that independent interpretations of the factors might yield different names, thereby suggesting different conditions. Consider Factor VII, for instance, which Wittenborn named "Paranoid Conditions." Wittenborn believed that Factors VI and VII were fairly similar, except that Factor VI contained items referring to delusions and hallucinations. However, when analyzing the list of the five symptoms with the highest loading on Factor VII, the name of *paranoid conditions* is not immediately obvious. This factor, like Factor II, is bipolar because some items have a negative correlation, and others are positive. The negatively correlated items refer to compulsive behaviors. The three items with positive correlations have a mixture of hysterical, paranoid, and manic connotations. To this author, the clinical meaning of Factor VII is not apparent.

Wittenborn believed that it was important to test the generality of his factor by analyzing a solution using another sample of patients. In other words, a meaningful factor solution should be replicated. Consequently, Wittenborn and Holzberg (1951) used the 55-item PRS to rate patients in a state hospital in Connecticut. All ratings were made by the staff during an intake conference. Patients who had organic brain syndromes, alcoholism, or sociopathy or who were over 60 years old were excluded. The results were factor-analyzed by means of a centroid method followed by an orthogonal rotation. Again, seven factors were isolated. Wittenborn and Holzberg (1951) interpreted the results as supporting the generality of six of the seven original factors. The only factor that Wittenborn and Holzberg believed was not replicated was the "Hebephrenic (Deteriorated) Schizophrenia" factor. They believed that this factor was not present in the state hopsital sample because the selection procedure at the state hospital would have favored selecting acutely disturbed patients, not chronic patients who showed deterioration.

Again, an independent analysis of the factor loadings from Wittenborn and Holzberg (1951) leads to questions about their interpretation. For example, one factor from the state hospital patients contained the same four items as did the hebephrenic schizophrenia factor from the VA sample. Hence, the hebephrenic factor did seem to be replicated, despite Wittenborn and Holzberg's claim to the contrary. On the other hand, a factor that Wittenborn and Holzberg regarded as having been replicated was the "paranoid condition" factor; however, a comparison of the two factors labeled "Paranoid Condition" appeared quite different across both samples. Among the state hospital patients, the paranoid condition factor correlated with the following five items:

Factor VI: Paranoid Conditions (State Hospital patients)
- Engrossed in plans (.404)
- Exaggerations of ability and well-being (.358)
- Unable to stick to or carry out any plan (.351)
- Speech is stilted (.316)
- Grandiose notions (.316)

If the list of symptoms is compared with the list of symptoms under Factor VII ("Paranoid Conditions") for the VA patients (see page 194), the reader will note that only one item is common to both lists: "grandiose notions." The other four items in each list are different and suggest meaningful differences among the factors. The Wittenborn and Holzberg interpretation that the paranoid conditions factor was replicated does not seem appropriate, given these results.

To further study the PRS, Wittenborn, Herz, Kurtz, Mandell, and Tatz (1952) performed two factor analyses on ratings that differed only in terms of the two psychiatrists who were making the ratings. They interpreted the factors as consistent and inferred that the factors were stable across raters.

Wittenborn and his associates also studied three patient groups using the factors from their earlier studies. The patient groups were involutional psychotics (Wittenborn & Bailey, 1952); organic psychotics (Wittenborn, 1952); and manics (Wittenborn & Weiss, 1952). In all three studies, the authors argued that the use of the factor analysis showed that the clinical diagnostic categories were too heterogeneous, and that further subdivisions of these diagnoses were needed.

Wittenborn's initial research with the PRS was received favorably. Eysenck (1952), when reviewing factor analysis studies of clinical relevance, commented that the Wittenborn studies were relatively well done for this research area. Lorr's review of the same literature also had a generally positive appreciation of Wittenborn's accomplishments: "The early Wittenborn studies were exceptionally well planned and meticulously carried out" (Lorr *et al.*, 1963, p. 49).

Despite this positive reception, some methodological criticisms were made of Wittenborn's use of factor analysis. Lorr (1957) and Eysenck (1952) believed that Wittenborn should have performed an oblique rotation, not an orthogonal rotation. Lorr (1957) reanalyzed the Wittenborn and Holzberg (1951) data on state hospital patients. He found seven oblique factors, which he named:

- Paranoid projection
- Melancholy agitation
- Conversion hysteria

Phobic-compulsive reaction
Excitement with hostile belligerence
Excitement with expansiveness
Thinking disorganization

The first three factors were very similar to factors noted by both Wittenborn (1951) and Wittenborn and Holzberg (1951). The fourth factor was the same as a factor found only in the Wittenborn and Holzberg study. The last three oblique factors were new.

Wittenborn's factor-analytic research on psychopathology did not end with this collection of studies. For instance, for an article titled "The Dimensions of Psychosis" (1962), Wittenborn factor-analyzed data from 150 male VA inpatients who were rated on 98 items. Ten meaningful factors were found, which he interpreted as having similar meanings as factors found by Moore (1933), Degan (1952), and his earlier research (Wittenborn, 1951; Wittenborn & Holzberg, 1951).

To conclude, Wittenborn's research in the early 1950s helped to stimulate the use of factor analysis to study psychiatric data. His methodology and interpretation of results appear sloppy by current standards. Nonetheless, his guiding principles show a concern about carefully gathering descriptive data in order to improve the classification of psychopathology.

Lorr

The third researcher who has performed important factor-analytic studies of adult patients is Maurice Lorr. Lorr completed a master's degree at the University of Illinois in 1935. Then, he went to work for the Illinois Institute in Juvenile Research in Chicago. At that time, an important descriptive study on the disturbed behaviors of children was being performed by Luton Ackerson (1931, 1942). In this research, the records on over 3,000 children were analyzed for the presence of 100 different traits. The Ackerson research was the stimulus for the development of factor-analytic studies of childhood psychopathology (see Achenbach & Edelbrock, 1978).

While in Chicago, Lorr enrolled at the University of Chicago, where he studied with Thurstone. After completing his Ph.D., Lorr was employed in a research section of the Veterans Administration. Lorr's factor-analytic research on psychopathology began in the 1950s. In 1953, Lorr and Jenkins published a factor analysis of the Ackerson data. Richard Jenkins, the coauthor of this paper, was a psychiatrist who had also worked at the Illinois Institute of Juvenile Research and had been promoting the use of factor analysis to study the abnormal behavior of children. Jenkins's research studies (including the Lorr and Jenkins paper) were the basis for the classificatory categories in the DSM-II concerning children.

About the same time, Lorr and Jenkins also became interested in using factor analysis to study adult psychotics. Using Thurstone's analysis of the Moore data, they created an instrument called the Northport Record to assess 12 hypothesized syndromes. There were 92 items on the scale, half of which were to be rated from observations made during an interview and the other half from ward behavior. They factor-analyzed the 92 items and found a total of 11 factors—a solution somewhat different from what had been hypothesized (Lorr, Jenkins, & O'Connor, 1955).

The Northport Record was significantly revised, and Lorr developed a new instrument called the Multi-dimensional Scale for Rating Psychiatric Patients (MSRPP) (Lorr, 1953). Lorr, O'Connor, and Stafford (1957) factored the MSRPP plus Wittenborn's PRS. They found a total of eight confirmed syndromes that they felt could be generalized across these two measurement instruments.

In the early 1960s, Lorr and his colleagues developed a new instrument called the Inpatient Multidimensional Psychiatric Scale (IMPS). The IMPS and its development were summarized in a book by Lorr *et al.* (1963). The formation of the IMPS represents the major achievement of Lorr's research career.

The IMPS consists of items that are to be rated by two clinicians after a 30- to 45-minute interview of a patient. The IMPS was designed to be used with functionally psychotic and severely maladaptive neurotic patients. Originally, the IMPS contained 90 items that were designed to assess nine basic syndromes. Lorr *et al.* (1963) stated that these nine hypothesized factors were derived from their reviews of Degan (1952), Wittenborn (1951), Guertin (1952), and Lorr (1954).

By means of this early form of the IMPS, data were gathered by interviewing 566 patients from 44 different inpatient settings. The factor studies of the data led to the discovery of a new tenth factor and to the elimination of 15 of the 90 items. The initially published version of the IMPS consisted of 75 items (subsequent research led to the addition of 8 more items) and 10 basic scales. The names of these scales are:

Excitement (EXC)
Hostile belligerence (HOS)
Paranoid projection (PAR)

Grandiose expansiveness (GRN)
Perceptual Distortion (PCP)
Anxious Intropunitiveness (INP)
Retardation and Apathy (RTD)
Disorientation (DIS)
Motor Disturbance (MTR)
Conceptual Disorganization (CNP)

Lorr *et al.* (1963) reviewed 13 earlier studies in which factor analysis had been used on data gathered from the observation of psychotic patients. They argued that the 10 factors in the IMPS were very similar to the scales noted in earlier research. For instance, according to Lorr *et al.*, the motor distrubance (MTR) factor is similar to the catatonic factor found in the Moore data. In addition, the paranoid projection factor from the IMPS is very similar to what Wittenborn described as a paranoid schizophrenic factor.

A second-order factor analysis was performed on the scales derived from the 10 first-order factors. These second-order factors were called:

Excitement versus retardation
Schizophrenic disorganization
Paranoid process

Notice that these three second-order factors correspond to the major categories of functional psychosis as recognized in the DSM systems: schizophrenia, paranoid disorders, and affective disorders (i.e., manic-depressive psychosis).

In addition to the descriptive factoring studies, Lorr *et al.* (1963) reported reliability and validity data on the 10 IMPS scales. The internal consistency estimates of reliability when two raters were used ranged from .75 to .92. These reliability values are very high especially, when performing ratings on psychiatric patients, whose behavior can be quite extreme and inconsistent. The validity data for the IMPS were not as impressive. These investigators did show that patients on open as opposed to closed wards and with different psychotic diagnoses could be differentiated on the IMPS scales. The correlations of IMPS scales and scales from ratings of ward behavior (Lorr & O'Connor, 1962) were lower than might be derived to demonstrate the generality of the IMPS syndromes across the types of information. Lorr *et al.* suggested that the IMPS might be useful in treatment outcome studies but admitted that data showing the utility of the IMPS for this type of research were not available then.

The IMPS became popular fairly quickly within the research literature about psychopathology. An investigation of citations made in the psychological literature during 1967 showed 23 references to the IMPS. However, the number of citations per year has shown a steady decrease since that time. In 1973, there were 17 citations; 9 in 1976; and 5 in 1979.

The most frequent use of the IMPS in early research was as a behavioral measure in drug therapy studies. For instance, Hollister, Overall, Shelton, Pennington, Kimbell, and Johnson (1967) used the IMPS plus another measure to compare the efficacy of two types of drugs (plus a placebo) when treating depressed inpatients. Drug treatment studies in which the IMPS was used were also performed by Goldberg, Mattison, Cole, and Klerman (1967); Simpson, Angus, and Edwards (1967a,b); Smith, Surphlis, Gynther, and Shimkunas, (1967); and Spiegel and Keith-Spiegel (1967).

Reviews of the IMPS are not easy to find in the literature. The critical commentary about the IMPS has generally been positive. Verma and Eysenck (1973) employed the IMPS in a factorial study of psychoticism. They used the IMPS, they said, because it "has certain advantages of theoretical construction, factorial stability, and validity over all other rating scales, and appears to have been carefully prepared" (p. 575). Coming from Eysenck, this is high praise. Pauker (1972) reviewed the IMPS for the seventh edition of Buros's *Mental Measurements Yearbook:* "The reviewer was impressed with the amount, quality, and control of work which has gone into the development and research of the IMPS" (p. 89). Pauker agreed that the reliability and the validity of the IMPS had been shown to be good. Mariotto and Paul (1974) commented favorably on the available reliability and validity data about the IMPS. They cited studies by Klett and McNair (1966), Lorr and Klett (1965), Lorr *et al.* (1962), Klett and Moseley (1965), Lorr and Cave (1966), Lorr and Hamlin (1971), and Mefferd (1968) as supporting the favorable impressions concerning reliability and validity. Given all of these favorable comments, what problems could be associated with the IMPS that have led to its steady decline in use since the mid-1960s?

One major difficulty with the IMPS is its limitations of scope in terms of the patient population. First, the IMPS was designed for use only with inpatients; thus, it does not cover the types of psychopathology seen in outpatients. Second, among inpatients, the IMPS most heavily emphasizes the symptomatology seen among schizophrenics. Although schizophrenics are commonly seen within inpatient settings, other types of patients are also common, such as patients with organic brain syndromes, patients recovering from substance abuse, patients with relatively maladaptive character disorders, and neurotic patients. More important, the IMPS is not a particularly good instrument when used with depressed patients. Raskin and Crook (1976) argued from their data on depressed patients who were treated with antidepressants that the IMPS was not a very sensitive measure of depression. The major scale concerned with depression in the IMPS is the Anxious Intropunitiveness scale (INP). Derogatis, Bonato, and Yang (1968) showed that the INP scale is one of the least powerful scales in the IMPS. Mariotto and Paul (1974) also found the INP scale to be a poor scale in terms of its psychometric properties when compared with the other, much stronger IMPS scales. The INP scale assesses only cognitive aspects of depression and does not have items concerning the physiological features (sleep difficulties, appetite loss, and so on). The Retardation and Apathy (RTD) scale does assess some of the motor behavior features of depression but also contains items about the slowed motor behavior that is likely to be seen among patients who have been chronically hospitalized.

A second difficulty with the IMPS is related to clinical meaningfulness. Scales such as "Anxious Intropunitiveness" and "Perceptual Distortion" do not have names that make their meaning immediately obvious, especially to clinicians who have never used the scales. Names such as "Depression" and "Hallucinations" would convey more direct meaning about the clinical concepts represented in the scales. The issue of clinical meaningfulness becomes apparent when one looks at other interview-based rating instruments that were developed during the late 1960s and the early 1970s. For instance, the Mental Status Schedule is a 342-item instrument that is designed for rating the symptoms of psychiatric patients after a structured intake interview. This instrument was created by Spitzer and his colleagues (1967) and was designed to assess the range of symptoms on which clinicians are likely to focus during an initial interview of a patient. It was designed to be used by practicing clinicians, and the description of the symptoms is very similar to the way in which most clinicians think of symptoms. These symptom items from the MSS have been factor-analyzed, and the resulting scales were given such clinical names as "Depression," "Anxiety," "Thought Disorder," and "Delusions/Hallucinations."

To conclude, Lorr has used factor analysis to create at least 14 different instruments for assessing psychopathology. The most carefully researched of these is the IMPS. The factorial structure of this behavior-rating system has been applauded uniformly by reviewers. Nonetheless, the IMPS can be faulted for the fairly narrow range of patients for which it is appropriate; in addition, the names of the 10 factor scales do not add to the clinical meaningfulness of the IMPS.

Overall

The last researcher to be discussed is John Overall. Overall attended graduate school at the University of Texas and completed his Ph.D. in 1958. He was employed for two years in conducting research within the Veterans Administration, during which time he became acquainted with the factor studies performed by Lorr. In 1963, Overall took a position with the University of Texas Medical Branch in Galveston, where he remained until 1978.

Overall has been the most prolific of the researchers discussed in this chapter, averaging five publications per year since completing his Ph.D. Among the topics that he has addressed in his research have been factor analysis, classification, drug research, analysis of variance, multivariate statistics, and breast cancer. Besides his journal articles, Overall has published one book (with C. James Klett), entitled *Applied Multivariate Statistics* (Overall & Klett, 1972).

The most important development by Overall during his productive career has been the creation of the Brief Psychiatric Rating Scales (BPRS) (Overall & Gorham, 1962). There are 18 separate scales on the BPRS. Overall and Gorham acknowledged that the major stimulus for the first 14 scales was the factors found by Lorr in his studies of the MSRPP and the IMPS. The remaining 4 scales were added at the recommendation of clinicians in order to increase the clinical coverage of the BPRS. (Historical note: The earliest version of the BPRS [Gorham & Overall, 1960] had 14 scales, all suggested by Lorr's factor analysis studies. Subsequent descriptive articles in the early 1960s about the BPRS listed 16 scales [Overall & Gorham, 1962]. The last 2 scales were added in 1966 [Hedlund & Vieweg, 1980].) The 18 scales of the BPRS are listed below:

Somatic concerns	Hostility
Anxiety	Suspiciousness
Emotional withdrawal	Hallucinatory behavior
Conceptual disorganization	Motor retardation
Guilt feelings	Uncooperativeness
Tension	Unusual thought content
Mannerisms and posturing	Blunted affect

Grandiosity	Excitement
Depressive mood	Disorientation

Each of the 18 scales is to be rated along 7 points, ranging from "not present" to "moderate" to "extremely severe." Overall and Gorham (1962) recommended that these ratings be completed after a brief 15 to 20-minute interview of the patient. Each of the BPRS scales is estimated directly by clinicians. These procedures differ from the IMPS in that clinicians perform ratings on the 78 individual symptoms of the IMPS and then the scores are summed to create the 10 final IMPS scales. In effect, scale scores on the IMPS are created in a two-step process, but the BPRS involves a single, direct step.

Besides the difference in the scoring procedure, the BPRS has another major difference from the earlier factor-derived instruments: the clinical meaningfulness of the scales. The names of all scales in the BPRS refer to concepts that have been traditionally used when describing psychopathology (e.g., depression, somatic concerns, and anxiety). As noted earlier in this chapter, other instruments, such as the IMPS, have scales whose names do not have an obvious meaning to clinicians (e. g., anxious intropunitiveness).

Overall and Gorham (1962) reported little information about the psychometric properties of the BPRS. They noted reliability estimates from the ratings of two clinicians that ranged from .56 ("Tension") to .87 ("Guilt feelings"). No specific validity data were mentioned.

Hedlund and Vieweg (1980) recently completed a thorough review of studies involving the BPRS. Eight research papers were found that listed reliability estimates on the individual scales. The median reliability of the scales was .75 (range of .63 for "Emotional withdrawal" to .88 for "Hallucinatory behavior"). Hedlund and Vieweg suggested that the scales with the highest reliability were those that used patient verbalizations in order to be scored. These reliability estimates for the BPRS are not outstanding, but they certainly are satisfactory for 7-point rating scales that measure fairly broadly defined clinical concepts.

The major validity evidence about the BPRS comes from psychopharmacology research. The original purpose for the development of the BPRS was to create an easily administered instrument that could be used in drug studies of psychiatric patients. Hedlund and Vieweg (1980) found over 150 studies of psychopharmacology research in which the BPRS had been used in conjunction with other clinical measures. They concluded, "the results of these studies is that, with only rare exception, BPRS change scores have consistently reflected treatment changes that are corroborated and supported by other clinical ratings" (p. 52).

In addition to reliability and validity studies, Overall and his colleagues have performed a number of factor analysis studies of the BPRS (Gorham & Overall, 1960; Overall, Hollister, & Pichot, 1967: Overall & Porterfield, 1963), as have other investigators (e.g., Guy, 1976). The results have repeatedly shown four factors:

Factor I—Thinking Disorder
- Conceptual disorganization
- Hallucinatory behavior
- Unusual thought content

Factor II—Anxious Depression
- Anxiety
- Guilt feelings
- Depressive mood

Factor III—Hostile Suspiciousness
- Hostility
- Suspiciousness
- Uncooperativeness

Factor IV—Withdrawal/Retardation
- Emotional withdrawal
- Motor retardation
- Blunted affect

Since its publication, the BPRS has had reasonably frequent use in studies of psychiatric patients, especially in psychopharmacology research. For instance, in 1967 there were 23 journal articles which cited the basic article describing the BPRS (i.e., Overall & Gorham, 1962). Interestingly, in the same year the IMPS had the same number of citations. However, by 1979 the citations to the BPRS had shown a clear growth with 72 citations but the IMPS was used only infrequently (i.e., 5 citations). Why the difference in the use of these two instruments, especially as the IMPS was so carefully derived and as the BPRS was based on IMPS factors?

Three answers are suggested by reviewing these instruments. One advantage of the BPRS is its ease of administration. Less time is required to complete and score the BPRS than to complete and score the IMPS. A second answer is that, although the BPRS was designed to focus on inpatients, experience has shown it to be useful with a wider range of patients (Hedlund & Vieweg, 1980). In contrast, very few symptoms on the IMPS concern characteristics of patients who are likely to be seen in outpatient settings, on alcoholism treatment units, and so on. The third and final reason for the relatively good acceptance of the BPRS is the clinical meaningfulness of the scales in this instrument. As a quick reading of the scale names on pages 202–203 will show, the BPRS scales all refer to clinical concepts that every mental health professional will recognize. This claim cannot be made for the IMPS scales.

One final comment needs to be made about the BPRS and Overall's research. This discussion of factor-analytic studies of psychotic patients has shown a historical progression from Moore to Wittenborn to Lorr to Overall. The BPRS appears to be the culmination of the work performed by these researchers. When viewed in this manner, it is particularly important to consider the four general factors that have been consistently found in the BPRS: thinking disorder, anxious depression, withdrawal/retardation, and hostile suspiciousness. These four factors represent four major areas of symptom description that are important for the understanding of psychopathology. A thinking disorder, for instance, concerns the cognitive aberration that is often associated with schizophrenia (at least, as schizophrenia is defined by American clinicians), but it can also occur in patients with brain damage, mania, obsessive styles, and so on. A great deal of psychological research during the mid-twentieth century was concerned with attempts to clarify the concept of *thought disorder* and the psychological mechanisms that underlie disturbed cognitive processes (Chapman & Chapman, 1973). The second factor from the BPRS concerns disorders of mood (anxiety, depression). These were particularly popular topics during the 1960s and 1970s (Akiskal & McKinney, 1975). In particular, a great deal of the recent biochemical and psychopharmacological research has focused on this area. The third factor, hostile suspiciousness, refers to the clinical constellation of anger, hostility, and paranoia. Research in this area is relatively sparse. Patients who are associated with this conceptual group are likely to be diagnosed as paranoid or as having character disorders, neither of which are categories with a careful research literature. The final factor of withdrawal/retardation concerns a syndrome that is often noted in chronically hospitalized patients but that also is seen in severe depressions, advanced dementias, and so on. Again, little research has been performed in this area, and the mechanisms by which such symptoms develop are not well understood despite their frequent appearance in clinical settings.

FACTOR STUDIES OF DEPRESSION

So far in this chapter, the focus has been on the study of inpatient populations by four important researchers who used factor analysis.

The reason for this specific focus was to provide the reader with an understanding of how the use of factor analysis has evolved and also of what the strengths and limitations of factor studies are. Obviously, however, a great deal of the factor analysis literature concerning psychopathology has not been covered here. A good example of a content area in which factor analysis had been hoped to resolve an important classificatory issue is depression. A brief discussion of this issue and the factor studies follows.

During the early-to-mid-twentieth century, two competing perspectives of the classification of depression were developed. One perspective arose from the alternative views of Kraepelin and Freud and advocated the presence of two major forms of depression. The first form, called *psychotic depression,* was assumed to have an endogenous etiology (often assumed to be hereditary), had the most severe clinical picture, and was believed to be the form most likely to respond to electroconvulsive therapy. *Neurotic* depression was distinctly milder, was generally considered a reaction to the loss of an important love object, and required psychotherapy for adequate treatment. An important reason for the neurotic versus psychotic distinction was Freud's belief that only neurotics could benefit from the transference events stimulated by therapy.

Other professionals, most notably a number of British psychiatrists, favored the notion that depression was a unitary concept. These professionals suggested that the clinical picture of depression could vary in degree of severity. They did not believe, however, that the clinical picture of different depths of depression differed sufficiently to warrant the inference of two kinds of depression. Also, these British investigators disputed the etiological differences assumed for the endogenous (psychotic) and the reactive (neurotic) forms. In particular, they found that neurotic depressives generally had experienced no more stressful losses before becoming depressed than had patients clinically described as having a psychotic depression.

In his excellent monograph, *The Classification of Depressive Illnesses* (1968), Kendell provided a well-written history of this controversy. He also documented how this controversy had led to competing schools of thought about classification within Great Britain.

In the early 1960s, the technique of factor analysis began to be applied to depressed patients in order to resolve this controversy. Factor analysis, it was believed, would be able to empirically resolve the one-depression versus two-depression argument. An early factor study by Hamilton and White (1959) in Great Britain is indicative of the results found in these factor analysis studies. Hamilton (1960) gathered ratings on 49 depressed male inpatients. A total of 17 symptoms was used as input into the factor analysis. Hamilton found that his first factor was a bipolar factor, in which retarded symptoms (slowed motor behavior, excessive guilt, and suicide attempts) had positive loadings and anxiety symptoms had negative loadings. Hamilton and White (1959) showed that those patients who scored high on the retardation pole of the factor were independently rated as endogenous depressives.

By the late 1960s, a number of factor-analytic studies of depression had been performed. Eleven of these were reviewed by Mendels and Cochrane (1968). These authors noted that all seven studies that reported *unrotated* factor loadings had found an "endogenous" factor of depression, whereas the four studies that reported only rotated loadings had not. Among the seven studies with an "endogenous," unrotated factor, Mendels and Cochrane showed that six studies found this to be a bipolar factor (i.e., some symptoms had high positive correlations and others had high negative correlations). However, the studies were in little consistent agreement regarding which symptoms were associated with the negative, or reactive, pole of the factors. In contrast, the endogenous pole had all positive loadings and a consistent collection of symptoms, including:

Motor retardation
Deeply depressed
No reactivity to environmental change
Loss of interest in life
Somatic complaints

Mendels and Cochrane concluded that these studies had all demonstrated the validity of psychotic/endogenous depression, but the status of neurotic depression as a classificatory concept was uncertain.

Two years later, Costello (1970) also reviewed the factor-analytic literature on depression. He disagreed sharply with the conclusions reported by Mendels and Cochrane (1968). Costello stated that Mendels and Cochrane had dropped a number of symptoms from the list of symptoms used by more than one study when comparing the endogenous factors. Also Costello intercorrelated the endogenous factors noted across different studies. These correlations were all positive but were moderately low (two-thirds of the correlations were $r = .56$ or less). Costello noted that most factor analysis studies of depression did not locate just one bipolar factor or two independent factors but found a number of factors, many of which did not appear in other studies. Costello concluded that "the question of the endogenous-reactive distinction remains an open one" (p. 21).

At this point, it is useful to interject a methodological issue that was being discussed among British researchers. Factor analysis is concerned with the identification of abstract *dimensions* that account for a large proportion of variance among symptoms. Factor analysis *per se* does not identify *categories* of patients. Because psychotic depression and neurotic depression are separate categories, at least in the theories of many clinicians, the discovery of a bipolar neurotic versus psychotic (or reactive-endogenous) factor does *not* directly support those clinicians who believe in the two-category classification system. In order to support the two-category classification, a researcher would need to show that there is a bimodal distribution of patients along the psychotic-neurotic continuum. That is, there would be some patients who fit in the psychotic portion of the continuum; other patients would fall in the neurotic portion; and, most important, there would be relatively few patients who fall midway between. In contrast, however, if the distribution of patients on the continuum is unimodal, then the evidence would be supportive of a one-dimensional classification of depression that would vary along a severity continuum from psychotic to neurotic to everyday depressive episodes.

Kendell (1968) was a clear proponent of the unimodal versus bimodal distribution test as a way of differentiating the two classificatory theories. In his 1968 monograph, Kendell studied 696 patients with a clinical diagnosis of depression. A principal components analysis was performed, followed by an oblique rotation. Twelve factors were found from the 42 symptoms, none of which appeared to be an endogenous/psychotic factor or a reactive/neurotic factor. The first two factors from a second-order factor analysis, however, were labeled *neurotic* and *psychotic,* respectively. (Note: A second-order factor analysis is performed on the intercorrelation matrix among the oblique first-order factors.) The plots of patient scores on these second-order factors had unimodal distributions. Kendell also used a discriminant analysis to separate patients with a neurotic diagnosis from patients with a psychotic diagnosis. Again, a unimodal distribution was found. Kendell concluded that "depressive illness is best regarded as a single continuum extending between the traditional neurotic and psychotic stereotypes" (p. 83).

Not all studies have found a unimodal distribution of scores. Carney, Roth, and Garside (1965) found a bimodal distribution among scores on the discriminant function derived in their study. Another study by these investigators also showed a bimodal distribution (Gurney, Roth, Garside, Kerr, & Schapira, 1972). In contrast, Kendell's additional research found only unimodal distributions (Kendell, 1969; Kendell & Gourlay, 1970).

What should be concluded from these factor studies of depression? Given the apparent variability in results, firm conclusions are difficult. However, a few general ideas do emerge. One important point is that cluster analysis is a more appropriate statistical tool for analyzing a categorical classification of depression. Cluster analysis studies of depression will be mentioned in the next chapter. Another important point, granted by Kendell (1976), is that the concept of endogenous/psychotic depression has relatively sound data to support it but that evidence supporting the concept of reactive/neurotic depression is lacking.

In his 1976 review of this controversy, Kendell made cogent comments on the ability of factor analysis to resolve the issues in the classification of depression:

> In the last twenty years countless attempts have been made to improve on the intuitive classifications of the past by subjecting clinical ratings, systematically collected from large populations of patients, to various forms of multivariate analysis. The result, as we have seen, has been disappointing. . . . The attempt to resolve our classification problems by the statistical analysis of clinical data has failed up to now and may well continue to do so. (p. 26)

METHODOLOGICAL AND THEORETICAL COMMENTS

Before concluding this chapter, final comments need to be made about the methodological and theoretical aspects of factor analysis. These comments focus on factor analysis as an *exploratory* procedure that is used to discover new syndromes of psychopathology.

In an exploratory factor-analysis study, four steps are usually followed. First, the researcher gathers data on a set of patients using a collection of symptoms that seem relevant to those patients. Next, the symptoms are correlated. Third, the correlation matrix is factor-analyzed. Finally, the resulting set of factors is interpreted. These four steps appear to provide a good outline of the procedure followed by Moore (1930, 1933), for instance. Major concerns about this approach are discussed under three headings: (1) sampling; (2) factor methodology; and (3) reification.

Sampling

The first concern in exploratory factor analysis is sampling. The initial step of any exploratory study is to select a set of variables (symptoms) and to select the patients. In effect, the researcher must sample from the domain of relevant variables and also sample from the patient population of interest.

The sampling of the symptoms has a major effect on what factors can appear in a solution. Notice, for example, that in the Moore study no alcoholism factor appeared. The lack of an alcoholism factor occurred despite the fact that alcoholism has been a prevalent psychiatric problem in the United States. Why was no alcoholism factor found? The most obvious answer is that the data were insufficient to document a possible history of alcohol abuse. That is, the patients were not asked questions about the frequency and regularity of their alcohol use; nor were the patients asked if alcohol use had had any effect on their psychosocial functioning (e.g., at home or on the job); nor were the patients' records searched for an indication that DTs had occurred after admission. In short, an alcoholism factor could not appear in an exploratory factor analysis of the Moore data because the variables used in this study precluded the appearance of an alcoholism factor. The selection of symptoms, by and large, determines the compostion of the factors that are found. Factor analysis cannot create factors *de novo*.

Sampling of the patients can also influence a factor solution. Suppose, for instance, that all patients with a suspected affective disorder were deleted from the Moore study. The remaining patients would have uniformly low scores on the symptoms associated with depression and mania. Thus, little variance would occur on the affective symptoms, and the correlation of these symptoms with other symptoms would be very poorly estimated. As a result, it would be unlikely that the presence of depression and manic factors would be discovered.

To summarize, any exploratory factor-analysis study is clearly limited by the sampling of variables and of patients used in the study. The sampling procedures limit the factors that can be discovered. Thus, when selecting symptoms and patients, researchers must be careful to adequately represent the clinical problem to which factor analysis is being applied.

Factor Methodology

The second area of concern is the variety of methodological choices associated with factor analysis. Variability in solutions as a function of methodology was shown in the comparison of four separate analyses of the Moore data.

The major methodological choices associated with most factor-analysis studies include the following:

1. Type of similarity measure
2. Factor extraction method
3. Estimation of communalities
4. Determination of the number of factors
5. Method of rotation
6. Estimation of factor scores

Some of these choices, such as the determination of the number of the factors, were briefly discussed when the steps of factor analysis were introduced. Other choices were not adequately discussed (e.g., estimation of communalities). Understanding which methodological decisions are appropriate for a given factor study is difficult. Researchers who intelligently use factor analysis must be sophisticated in matrix algebra and multivariate statistics if they are to understand the various options that can be selected at each choice point. To the novice, the array of possible factor-analysis options usually appears overwhelming, especially given the amount of associated jargon. Even more disconcerting is the realization that authorities in the field offer different advice about which options are to be selected.

In light of the variety in factor procedures and their effects on the resulting solutions, Armstrong and Soelberg (1968) suggested three general steps that should be associated with any exploratory factor-analysis study: (1) the use of split samples; (2) Monte Carlo simulation; and (3) confirmatory follow-up studies. The first suggested step, the use of split samples, permits a replication of any factor solution. Replication is a basic procedural step in all sciences. If an empirical result is to be given credence, it should be possible to show that the result can be replicated under similar circumstances. If a result fails the replication, then it becomes suspect.

The second step suggested by Armstrong and Soelberg (1968) is the use of Monte Carlo simulation. Armstrong and Soelberg provided an example of factor analysis being applied to random numbers in which no factors existed. However, the factor results did suggest the presence of an apparently meaningful nine-factor solution. Because factor analysis methods almost always "find" a factor solution even though none may exist, Armstrong and Soelberg suggested that researchers also factor random data to determine if the solutions to real data and to random data are different. If the solutions are nearly the same for both real and random data sets, again there is reason to be suspicious of the factor solution to the real data.

Armstrong and Soelberg also suggested that a third step after any exploratory factor analysis would be the performance of a follow-up factoring study. The initial exploratory study should suggest what major factors are present, what symptoms might need to be added if additional factors are to be isolated, and what symptoms can be deleted as not yielding useful estimates of the existing factors. A follow-up study would be essentially confirmatory in its intent and would permit a more precise specification of the expected factor solution. If this solution is not confirmed, further factoring studies are probably needed to illuminate the factor structure of these data.

Reification

The third concern in exploratory factor analysis is the ease with which a factor solution can be reified. Now that factor analysis is available in most statistical packages (such as SPSS, SAS, and BMDP), researchers find it relatively easy to use this tool even if they do not understand exactly what factor analysis does. It is easy for researchers to gather a large mass of descriptive data, submit it to a factor-analysis computer program, and assume that the obtained factors are meaningful. Tom Swift would have loved to have a magic factor-analysis machine in order to explain his data to him.

Overstated as they are above, the dangers of the naive use of factor analysis and the subsequent reification of a factor solution should be obvious. Factor analysis is not magic; it is not a procedure that always helps a researcher to understand his or her data better; it does not always estimate the underlying syndromes in a data set. Factor analysis, instead, greatly removes the researcher from his or her data; factoring involves many decisions, a number of which require the subjective judgment of the researcher. The results of factoring *per se* do not tell the researcher whether the solutions are meaningful or not. In other words, a factor solution is best conceptualized as suggesting hypotheses about the information contained in a data set. These hypotheses need independent testing and exploration if they are to clarify the scientific issues being studied by a researcher.

Confirmatory Factor Analysis

A major theoretical revolution has occurred in factor analysis over the last decade. This revolution has focused on the development of maximum-likelihood factor analysis and its use in confirmatory studies. In confirmatory factor analysis, the researcher specifies the expected factor solution *before* performing the factor analysis. Confirmatory factor analysis involves a statistical test of whether the expected factor solution is obtained from the data. Maximum-likelihood factor analysis is the method that is used in confirmatory factoring because statistical tests of fit have been derived. This method used mathematically complex procedures to estimate the population factor structure from a data sample.

Confirmatory factor analysis is very useful when developing new clinical tests, scoring systems for structured interviews, behavioral rating systems, and so on. The researcher can specify, for instance, how many factors she or he expects from a rating-scale system and which items will load highly on which factors. Confirmatory factor analysis then permits a test of this expected factor solution. If the solution fails, the researcher can examine the results to decide what aspects of the expected solution need to be revised. For instance, the researcher may have underestimated the number of factors required to account for the systematic variance of the data; or the particular item composition of a given factor may not have occurred as expected.

Confirmatory factor analysis is not limited just to the study of the dimensional structure of clinical measurement instruments. Joreskog (1970) and Joreskog and Sorbon (1978) proposed a general model for the analysis of covariance structures that has potentially wide application to the study of longitudinal data, to categorical systems, to traditional factor-analysis problems, and so on. Unfortunately the mathematics, thus the computer programs, needed to perform studies using maximum likelihood procedures have proved to be very complicated and to require a great deal of technical sophistication. Hence, the frequency of confirmatory studies is still fairly low despite the promise of these techniques. Interested readers should consult a paper by Meredith, Frederiksen, and McLaughlin (1974) that contains a relatively clear discussion of the statistical models associated with confirmatory factor analysis and the study of covariance structures. Muliak's textbook (1972) on factor analysis introduces the mathematics for understanding maximum likelihood methods of factoring. Finally, a recent paper by Joreskog (1978) summarized many of his important ideas about the application of maximum-likelihood factor analysis. Joreskog (1978) proposed a general model that includes exploratory factor analysis, confirmatory factor analysis, regression, and path analysis within one conceptual system.

SUGGESTED READINGS

Gorsuch, R. L. *Factor analysis.* Philadelphia: Saunders, 1974.

A number of different textbooks introduce factor analysis. The Gorsuch book is the best of the recent offerings on this topic. The book is reasonably detailed, covers the range of topics associated with modern factor analysis, and is clearly written. The latter point is important. Given the plethora of jargon and the potentially complex statistical theory associated with factor analysis, finding a book that is well written and that has a fair amount of statistical detail, which permits a detailed understanding of factor options, is not an easy task. For both coverage and clarity, the Gorsuch book is recommended as the best currently available text.

Among other factor analysis books, the classic contemporary text is by Harman (1976). However, it is quite difficult to read. Muliak (1972) is also difficult mathematically but has a thorough treatment of recent maximum-likelihood factoring methods. A book written for novices is Rummel (1970). This book has a practical focus, including, for example, a chapter on the information that should be submitted when publishing a factor analysis study. However, the Rummel book is poor in handling the more statistical aspects of factor analysis, such as how factor extraction methods work.

Brief, but readable, introductions to factor analysis are also available. One is a paper by Cattell (1965), in which he attempts to explain factor analysis to biologists who have not used this method. Another is the general description of factor analysis in the SPSS manual, which covers many of the major points (but ignores a number of important details and cautions).

Two historical sources on factor analysis are also of use. One is by Eysenck, titled *The Structure of Human Personality* (1953). This book provides an excellent rationale for the use of exploratory factor analysis. The other is Thurstone (1947). This book is a very clear, though out-of-date, treatment of the basic issues in factor analysis.

Comrey, A. L. Common methodological problems in factor analysis. *Journal of Consulting and Clinical Psychology,* 1978, *46,* 648–659.

The quality of the factor analysis studies that have appeared in the published journal literature is variable. Comrey's article provides a basic discussion of the most common issues surrounding the use of factor analysis to study psychopathology. For instance, Comrey emphasizes the dangers of naive exploratory studies and suggests that factor analysis may yield more useful results when the researcher has a set of hypotheses about what factors will appear. This article is "must" reading for any researcher who plans to use factor analysis to study psychopathology.

Jackson, D. N. The dynamics of structural personality tests: 1971. *Psychological Review,* 1971, *78,* 229–248.

A major use of factor analysis in the clinical literature is to develop new psychological tests. Here Jackson has offered an interesting commentary on how to involve factor analysis when empirically creating a test. In addition, he discussed how a feedback system between clinical theories and the use of multivariate procedures can lead to the development of clinically meaningful and psychometrically sound instruments. Jackson (1978) provided an example of his approach to test development when he discussed the creation of the Jackson Personality Inventory.

Chapter 8

Cluster Analysis and the Classification of Patients

THE LAW OF THE HAMMER

Give a small boy a hammer and suddenly he will discover everything needs pounding.

Abraham Kaplan, 1964

Fads occur in all sciences. Clinical psychology is no exception. Within clinical psychology, topics such as behavior modification, personality testing with the MMPI, client-centered therapy, and research on hypnosis were at one time appropriately considered fads. Nonetheless, these areas have managed to sustain interest beyond faddism and have introduced a corpus of knowledge that has had a substantial effect on the field of clinical psychology.

Cluster analysis became a small, but nontrivial, fad during the 1970s. Clustering was popular because cluster-analytic methods were relatively easy to understand, because a number of researchers became convinced that generating new and better classification systems would be useful, and because cluster-analytic methods were recognized as objective, empirical procedures that could efficiently form new classification systems.

However, there are a number of problems with cluster-analytic methods. Many of these problems are associated with the relative infancy of these methods and the lack of a systematic understanding of their statistical properties. This chapter attempts to introduce the reader to some of the methods of cluster analysis, to discuss the ways in which clustering has been used to form classifications of psychopathology, and to discuss critically many of the methodological and theoretical issues surrounding this topic.

INTRODUCTION TO CLUSTER-ANALYTIC METHODS

Cluster analysis is best characterized as a loosely organized collection of quasi-statistical procedures that are used to create classificatory systems. A large number of specific methods are included under the generic term of *cluster analysis.* Currently, the most popular two families of clustering methods are known as *hierarchical agglomerative methods* and *iterative partitioning methods.* In addition, *inverse factor analysis* has been frequently used in studies of psychopathology. At the present time, there are at least 150 different methods that are identified as cluster analysis techniques. Bailey (1974) has suggested 12 ways of describing important differences among these various clustering techniques: (1) agglomerative versus divisive; (2) monothetic versus polythetic; (3) natural versus artificial; (4) number of clusters predetermined or not; (5) single level versus hierarchical level; (6) overlapping versus nonoverlapping; (7) outliers permitted or not; (8) form of linkage; (9) combinatorial versus noncombinatorial; (10) compatible versus incompatible; (11) objective versus subjective similarity level; and (12) iterative versus noniterative.

The most popular cluster-analytic methods are the *hierarchical agglomerative methods.* These methods require the formation of a similarity–dissimilarity matrix, each cell of which describes the degree of similarity between two entities. From this matrix, agglomerative methods gradually build clusters by putting similar entities into the same cluster. The four most popular methods of agglomerative hierarchical cluster analysis are single linkage, complete linkage, average linkage, and Ward's method (Anderberg, 1973; Everitt, 1980). To help the reader understand how these methods create classificatory systems, various clustering techniques are demonstrated through the use of a data set of MMPI profiles on 10 patients.

Table 24 gives the MMPI profiles for 10 hypothetical patients (Lanyon, 1968). For those readers who are not familiar with the MMPI (the Minnesota Multiphasic Personality Inventory), this self-report inventory is a psychological measurement device commonly used to assess psychopathology. Over 4,000 studies have been published in

TABLE 24. MMPI Data on 10 Patients

	MMPI scales										
	L	F	K	Hs	D	Hy	Pd	Pa	Pt	Sc	Ma
Patient 1	53	58	57	58	64	60	(72)	61	63	61	55
Patient 2	49	59	54	53	57	53	(70)	58	60	64	64
Patient 3	56	61	51	53	59	55	(71)	61	63	64	63
Patient 4	48	68	46	57	55	55	(70)	63	65	(72)	68
Patient 5	50	54	54	(76)	(70)	(73)	59	55	62	60	58
Patient 6	52	54	56	(79)	(71)	(72)	59	57	66	62	56
Patient 7	52	58	52	(75)	(72)	(77)	64	63	67	(70)	55
Patient 8	54	54	56	(70)	62	(71)	57	50	54	54	57
Patient 9	56	60	57	(72)	(83)	(72)	65	57	62	(72)	52
Patient 10	51	61	48	(82)	(85)	(74)	67	62	(78)	(74)	61

Note. Circled scores are above 70 and standardly are considered significantly deviant.

which the MMPI was used. The MMPI consists of 566 true-false items that are stated in the first person. The task of the person who is completing the MMPI is to answer each item true or false as the item applies to himself or herself. Traditionally, the 566 items are scored along 13 scales. Of these 13 scales, only 11 are listed in Table 24*:

L—Lie scale
F—"Fake bad" scale
K—"Correction" for denial of psychopathology
Hs—Hypochondriasis
D—Depression
Hy—Hysteria
Pd—Psychopathic deviate
Pa—Paranoia
Pt—Psychasthenia
Sc—Schizophrenia
Ma—Hypomania

In the interpretation of the MMPI profiles, it is important to know that the average score of a "normal" person is 50. Any score above 70 (circled on Table 24) is considered a clinically significant indicator of deviance in that area of psychopathology. The careful reader will observe that the first four profiles in Table 24 are all fairly similar. All have elevated profiles (scores above 70) on the Pd scale. These Pd

* The MF (Male–Female) and Si (Social Introversion) scales are the remaining two that are commonly scored on the MMPI, but these two scales are not emphasized in most clinical interpretations of the MMPI; hence, they are not listed here.

profiles (named *Pd* because the psychopathic deviate scale is the only clinical scale that is significantly elevated) are commonly associated with the type of psychopathology frequently designated as "passive aggressive personality disorder" or "emotionally unstable personality."

The next four profiles also are fairly similar. All four showed elevations on Hs, D, and Hy, with the depression scale (D) having the consistently lowest score of the three. Thus, all four profiles contain the "conversion V" pattern. The reason for the appellation *conversion V* is that these profiles all have a *V* pattern among the first three clinical scales (D less than Hs and Hy), and the term *conversion* is used because this MMPI profile is associated with a "conversion reaction" diagnosis. Notice that these second four patients would all be expected to have neurotic diagnoses, whereas the first four would be expected to have diagnoses in the character disorder spectrum.

The last two profiles are elevated in the neurotic triad of the Hs, D, and Hy (without the *V* pattern) and also are elevated on Pt and Sc. These profiles are often seen in patients with some type of affective disorder. The last two profiles will be called *miscellaneous profiles* because, despite their general similarity to the group of "neurotic" profiles, these profiles represent patients whose psychopathology cannot be clearly classified in either of the first two groups.

From the visual examination of these 10 MMPIs, one would expect the first 4 patients to form a cluster and the next four to form a second cluster. The last two patients would either join to make a small third cluster or merge with the second cluster. The clustering methods to be introduced are all discussed in terms of how they analyze these data.

The first clustering method to be discussed is *single linkage,* which is the simplest of the cluster-analytic methods to understand. In fact, the first major book on cluster analysis, by Sokol and Sneath (1963), described single linkage as "elementary cluster analysis." The first step when performing single linkage is to create a pairwise similarity matrix among the patients. A similarity matrix is shown in Table 25. In this matrix, the similarity measure being used is Euclidean distance.*

* Euclidean distance is defined here as

$$d = \frac{1}{p} \sum_{j} (x_{a,j} - x_{b,j})^2$$

where p is the number of variables, $x_{a,j}$ is the datum for Patient a on variable j, and $x_{b,j}$

TABLE 25. Matrix of Euclidean Distances among 10 MMPI Patients

	Patient									
Patient	1	2	3	4	5	6	7	8	9	10
1		23.7	18.5	59.6	70,8	77.1	75.7	69.7	82.7	160.3
2	23.7		8.2	27.3	119.1	136.1	138.7	100.9	154.1	233.5
3	18.5	8.2		24.4	117.3	129.2	121.1	106.0	136.5	205.3
4	59.6	27.3	24.4		146.5	158.5	132.9	155.5	170.3	198.2
5	70.8	119.1	117.3	146.5		4.3	24.4	23.1	44.3	83.8
6	77.1	136.1	129.2	158.5	4.3		18.3	39.0	37.6	65.9
7	75.7	138.7	121.1	132.9	24.4	18.3		76.7	25.0	39.6
8	69.7	100.9	106.0	155.5	23.1	39.0	76.7		92.1	185.5
9	82.7	154.1	136.5	170.3	44.3	37.6	25.0	92.1		53.2
10	160.3	233.5	205.3	198.2	83.8	65.9	39.6	185.5	53.2	

Euclidean distance, for reasons to be discussed later, is the most popular of the similarity (dissimilarity) measures for clustering. When this measure is used, the most similar patients are the ones with the *smallest* pairwise value; that is, these patients are the "least distant." In contrast, the most dissimilar patients are those with the largest value ("most distant"). From the matrix of similarity values, it is easy to note that Patients 5 and 6 are the most similar (distance of only 4.3), and that Patients 2 and 3 are also quite similar (8.2). In contrast, Patients 2 and 10 are very dissimilar (with a much larger distance of 223.5).

After forming the similarity matrix, the next step is to merge individuals in order to form clusters. As noted above, single linkage is a *hierarchical agglomerative* method. It starts by combining Patients 5 and 6 to form the first cluster because these are the most similar patients. The next two most similar patients are 2 and 3, who are merged at a Euclidean distance of 8.2. The similarity matrix then is searched for the next smallest distance value, which is 18.3. At this value, Patients 6 and 7 are merged. Because Patients 5 and 6 have already been merged, Patients 5, 6, and 7 all belong to the same cluster at this point. Next, Patient 1 joins with 2 and 3. Then, Patient 8 and,

is the datum for patient b on variable j. This mathematical definition of Euclidean distance is *not* the standard definition, but it is used here because it is the definition used in the most popular program for performing cluster analysis, called *CLUSTAN*, which generated the results and figures for this chapter.

later, Patients 9 and 10 join with Patients 5, 6, and 7. Patient 4 also merges with 1, 2, and 3. In the last step, 1, 2, 3, and 4 are combined with 5, 6, 7, 8, 9, and 10 to make one all-encompassing cluster. The entire clustering process using single linkage on these 10 MMPI profiles is summarized in Table 26.

An important point to note about single linkage is its simplicity. A complete description of the clustering process for 10 patients was accomplished in one paragraph. Although single linkage is one of the simplest of the clustering algorithms, most other methods of cluster analysis are also relatively easy to understand (compared with factor analysis, for example). Part of the reason for the simplicity of many cluster-analytic methods is that these methods are *heuristics*. That is, these methods are "rules-of-thumb" that seem to be plausible procedures as methods for creating clusters. Very few cluster-analytic methods have been derived from a well-enunciated statistical theory (Bailey, 1974).

The clustering process that was just described verbally can also be described visually. Figure 21 presents a *dendrogram* (or tree diagram) that shows the clustering process as the mergers occur from left to right. After examining Figure 21, the reader can understand why single linkage is considered a "hierarchical agglomerative" method. It is a *hierarchical* method because the entire clustering process can be represented in the form of a hierarchical tree. Single linkage is an *agglomerative* method because the clusters are gradually built by small groups of patients merging into successively larger groups. A few other

TABLE 26. Agglomerative Steps Used by Single Linkage on 10 MMPI Patients

Patients	Merge at a Euclidean distance of
5 and 6	4.3
2 and 3	8.2
5, 6, and 7	18.3
1, 2, and 3	18.5
5, 6, 7, and 8	23.1
1, 2, 3, and 4	24.4
5, 6, 7, 8, and 9	25.0
5, 6, 7, 8, and 10	39.6

Note. All patients merge at a Euclidean distance of 69.7.

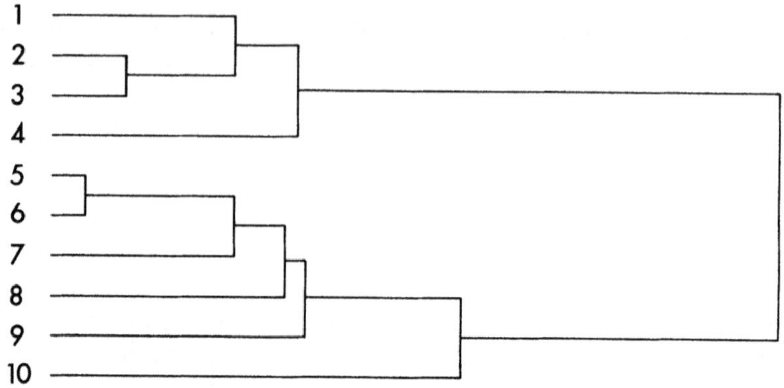

FIGURE 21. Dendrogram, or tree, using single linkage.

characteristics of hierarchical agglomerative methods can be noted. In general, hierarchical agglomerative methods always require $N - 1$ steps to complete the clustering process, where N denotes the number of entities being clustered. Also hierarchical agglomerative methods do not tell the researcher how many clusters exist in a data set. These methods always start with N clusters, in which each entity is the only member of its own cluster. If the researcher wants to find a classification in which the number of categories is neither 1 nor N (as is usually the case), the researcher must analyze the dendrogram and choose his or her own estimate of the numbers of cluster. For instance, a visual analysis of Figure 21 suggests the presence of two clusters because the spread between the first two "branches" of the tree is relatively large. The first cluster is the "personality disorder" group, and the second cluster contains the four conversion reaction profiles plus the two miscellaneous profiles.

Single linkage is not the only hierarchical agglomerative method. Other methods in this family of cluster analysis are complete linkage, average linkage, median method, centroid method, Ward's method, flexible beta method, and so on. Three of these methods—complete linkage, average linkage, and Ward's method—have been used frequently. Each of these is briefly discussed here.

Complete linkage (Sorenson, 1948) begins the clustering process in the same way that single linkage does—by forming a similarity matrix among the patients. Thus, Table 25 can also be used as the basis for the clustering by complete linkage. The first step is the merging of the two most similar individuals. These are Patients 5 and 6. Next, Patients 2 and 3 are joined.

For the first two steps, both single linkage and complete linkage form the same groups. It is in the third step that a slight change occurs. At Step 3, single linkage merges Patient 7 with the cluster containing Patients 5 and 6. Instead, complete linkage adds Patient 1 to the cluster of Patients 2 and 3; immediately thereafter, it merges Patient 7 with the dyad of Patients 5 and 6. The reason for the difference is that complete linkage requires "complete links" among all members of any groups that are to be merged. Thus, Patient 7 cannot join the cluster of Patients 5 and 6 at a Euclidean distance of 18.3, as he does in single linkage because the distance between Patients 5 and 7 is 24.4. In the same way, Patient 1 cannot be added to the cluster of Patients 2 and 3 at a Euclidean distance of 18.5, which is the distance between Patients 1 and 3, because the distance between Patients 1 and 2 is 23.7. However, at a Euclidean distance of 23.7, the latter merger can occur because, at this value, the requirement of complete links among all members is satisfied. In the same way, Patient 7 merges with Patients 5 and 6 at a value of 24.4.

This difference between the mergers of single linkage and complete linkage at Steps 3 and 4 is relatively minor. It is at a later stage of the clustering that a more serious difference occurs. This difference can be noted by analyzing Table 27, which contains a description of the agglomerative steps performed by complete linkage, and also by examining the dendrogram in Figure 22. As shown in both the table and the figure, one of the last steps of the clustering process is for Patient 8 to be merged with the cluster of Patients 1, 2, 3, and 4 at a relatively high value (a Euclidean distance of 155.5). This inclusion of Patient 8 in the group of "personality disorder" patients is surprising. However, if one

TABLE 27. Agglomerative Steps Used by Complete Linkage on 10 MMPI Patients

Patients	Merge at a Euclidean distance of
5 and 6	4.3
2 and 3	8.2
1, 2, and 3	23.7
5, 6, and 7	24.4
5, 6, 7, and 9	44.3
1, 2, 3, and 4	59.5
5, 6, 7, 9, and 10	83.8
1, 2, 3, 4, and 8	155.5

Note. All patients merge at a Euclidean distance of 233.5.

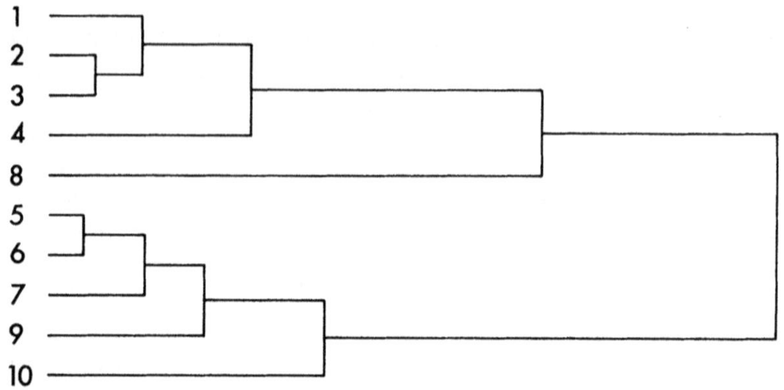

FIGURE 22. Dendrogram, or tree, using complete linkage.

examines Table 25, one will note the reason for the difference. Patients 8 and 9 are quite dissimilar. In fact, the Euclidean distance between them is 185.5. This dissimilarity occurs even though Patient 8 is reasonably similar to Patients 5, 6, and 7, as well as to Patient 10. However, because of the requirement of complete linkage before any merger can occur, Patient 8 cannot join the cluster of Patients 5, 6, 7, 9, and 10. Instead, he merges with patients 1, 2, 3, and 4.

As a result, the final clustering solution found by complete linkage also suggests the presence of two clusters: one cluster that contains the MMPI profiles suggesting some form of "personality disorder" and a second cluster containing the "neurotic" profiles. However, complete linkage has difficulty deciding where Patient 8 should be located, and the dendrogram shown in Figure 22 would suggest that Patient 8 might be considered an *outlier,* who does not belong to either of the two main clusters.

Average linkage, more precisely known by the unwieldy name *unweighted pairwise group mean averaging* (UPGMA) (Sneath & Sokol, 1973), has been suggested as a "compromise" between single linkage and complete linkage. Single linkage is considered too liberal because it joins groups at any excuse (i.e., with a single link); in contrast, complete linkage is too conservative because it requires complete links among all members of both groups before a merger takes place. Average linkage is a compromise because it joins groups based on the value of the average similarity among the groups.

The dendrogram showing the results of average linkage is presented in Figure 23. Notice that this figure implies two clusters that have the same members as the two-cluster solution from single linkage. However, Patient 10 is the last to join the "neurotic" cluster when

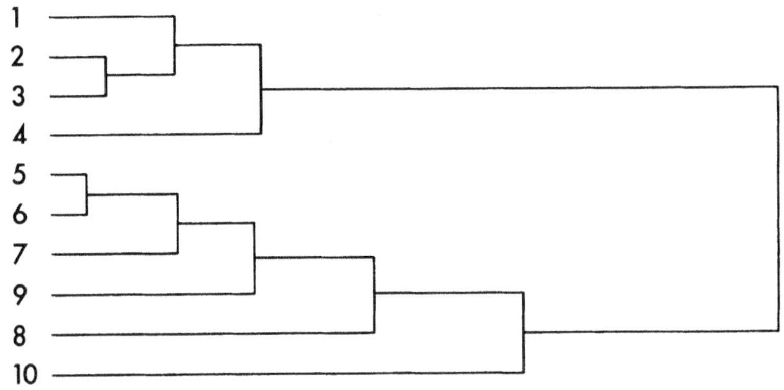

FIGURE 23. Dendrogram, or tree, using average linkage.

using average linkage, thereby suggesting that this patient, more than Patient 8, is a relative outlier.

The fourth method, Ward's method (Ward, 1963), is unique among the hierarchical agglomerative methods because it attempts to optimize a statistical criterion: the error sum of squares within the clusters. Thus, two groups are merged only if that is the step that will keep the variance within the clusters at a minimum.

The dendrogram for Ward's method occurs in the next figure, Figure 24. As in average linkage and single linkage, the presence of two clusters is also implied by this method. However, Ward's method does do something different with the "neurotic" cluster: it subdivides the cluster into parts. One subgroup contains Patients 5, 6, and 8, and the other subgroup contains Patients 7, 9, and 10.

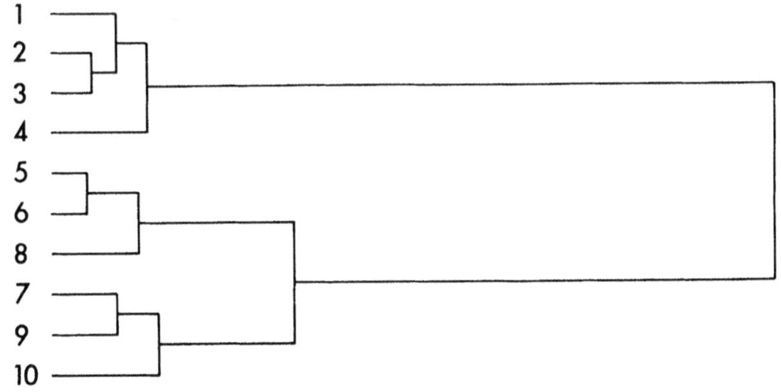

FIGURE 24. Dendrogram, or tree, using Ward's method.

To summarize the results from the four hierarchical methods, these methods generally agree that there are two clusters in the data. Only complete linkage disagrees on the actual membership of the clusters. However, all four methods found somewhat different solutions to this small data set.

The finding that different clustering methods generate different solutions to a common data set has been frequently noted in the literature. For example, Bartko, Strauss, and Carpenter (1971) asked 20 clinicians to rate what they considered the "typical" symptom patterns for five different diagnostic categories. Thus, 100 "archetypal patients" ($20 \times 5 = 100$) were created who represented five ideal clusters, each cluster being associated with a diagnostic category. Bartko *et al.* used 13 clustering methods in order to classify the archetypal patients. They found 13 different solutions to the data set, and only one solution exactly matched the classification into the five diagnostic categories that had been used by the clinicians to create the data! The implications of this finding that different cluster-analytic methods can yield different solutions to the same data is discussed further on in the chapter.

The preceding material described four hierarchical methods of cluster analysis. This presentation did not exhaustively cover the range of clustering methods. For instance, two popular cluster-analytic methods in studies of psychopathology are inverse factor analysis and the k-means iterative partitioning method. *Inverse factor analysis* uses exactly the same methods that are involved in standard exploratory factor analysis, except that the matrix being factored represents the similarity relationships among the *patients* instead of the relationships among the *variables* (Stephenson, 1953). In other words, for the MMPI example data set, inverse factor analysis would start by forming a 10×10 correlation matrix that would show the correlations between pairs of patient profiles. This approach would contrast with standard factor analysis, in which a 11×11 correlation matrix would be formed among the variables. After the patient $\times$ patient correlation matrix is formed, inverse factor analysis proceeds in the same way that standard factor analysis would. The results of inverse factor analysis on the 10 MMPI profiles is shown in Table 28. In this solution, there are only two eigenvalues greater than 1, and the two-factor solution accounts for 87% of the variance. The varimax rotation of the factor-loading matrix as represented in Table 28 shows that Factor I is highly correlated with the profiles of Patients 5–10, and that Factor II contains Patients 1–4. Notice that the inverse factor-analysis solution suggests that Patient 3 is the best examplar of the "personality disorder" clus-

TABLE 28. Inverse Factor Analysis Results on 10 MMPI Patients

	Factor loadings	
Patient	Factor I	Factor II
1	.36	.75
2	−.09	.95
3	−.14	.96
4	−.01	.89
5	.97	−.09
6	.97	−.13
7	.95	.16
8	.81	−.37
9	.88	.08
10	.91	.29

Note. Factoring method was principal components analysis followed by a varimax rotation of two factors. The eigenvalues for the first three factors were 5.21, 3.46, and 0.56.

ter, and that Patients 5 and 6 are the best exemplars of the "neurotic" cluster.

The other approach to cluster analysis that has been showing increasing popularity is the family of *iterative partitioning methods.* The best known of these methods is called the *k-means* method, as originally proposed by MacQueen (1967). The *k*-means method begins by choosing *seed points,* which are estimates of the initial centroids of the clusters. For instance, the results from inverse factor analysis suggest that Patient 3 is the best exemplar of the "personality disorder" cluster. Thus, Patient 3 could be used as a seed point in starting a *k*-means method because this patient profile would be a good initial estimate of the "personality disorder" cluster. In the same way, Patients 5 and 6 could be used as seed points for the "neurotic" cluster. After the seed points are chosen, the *k*-means method computes the similarities between the seed points and the remaining patients. The left side of Table 29 shows the Euclidean distances between each patient and the seed points for each cluster. Next the *k*-means method assigns all patients to the cluster with which they are most similar. Thus, Patients 1, 2, and 4 are assigned to the "personality disorder" cluster, and Patients 7, 8, 9, and 10 are assigned to the "neurotic" cluster. However, it should be noted that patient 10 is reasonably dissimilar from both clusters, and that the *k*-means method does not assign him to either

TABLE 29. *k*-Means Iterative Partitioning Method Applied to 10 MMPI Patients

	Initial Euclidean distances from seed points		Distances of patients from final clusters	
Patients	Cluster I	Cluster II	Cluster I	Cluster II
1	18.5	72.9	27.3	59.8
2	8.2	126.5	8.4	114.4
3	0.0	122.2	4.7	106.6
4	24.4	151.4	31.5	137.3
5	117.3	0.0	103.3	6.0
6	129.2	0.0	115.1	6.9
7	121.1	20.3	107.0	21.0
8	106.0	30.0	97.9	48.1
9	136.5	39.9	125.8	38.1
10	205.3	73.8	189.2	70.2

cluster. After all the patients are assigned to the seed points, the centroids of the clusters are computed by calculating the means on all variables for all members of the cluster. Then, the similarities between the cluster centroids and all patients are again recomputed, and all patients are assigned to the most similar cluster. This process is *iteratively* repeated until no reassignments occur. For the MMPI example data set, the recomputed distances between the patients and the cluster centroids are shown on the right side of Table 29. Because the first assignment led to a stable solution, no iterations are necessary. At this point, the *k*-means method has *partitioned* the data set into two groups: the "personality disorder" cluster contains Patients 1–4, and the "neurotic" cluster has Patients 5–9.

Other families of clustering methods that have not been discussed in this chapter are *hierarchical divisive* methods, which are similar to hierarchical agglomerative methods except that the divisive methods start with one cluster and successively divide clusters into smaller and smaller subsets; *mode searching* methods, which attempt to find areas of relative density in the multidimensional space being defined by the variables; *graph theoretical* methods, which are based on the mathematics of graph theory; *k-linkage* methods, which have some of the mathematically positive features of single linkage but avoid some of its practical limitations; and *Lorr's* method, which forms clusters one at a time and has internal rules for estimating how many clusters exist in the data. In all, there are over 150 different methods of cluster analysis that have been proposed in various parts of the scientific liter-

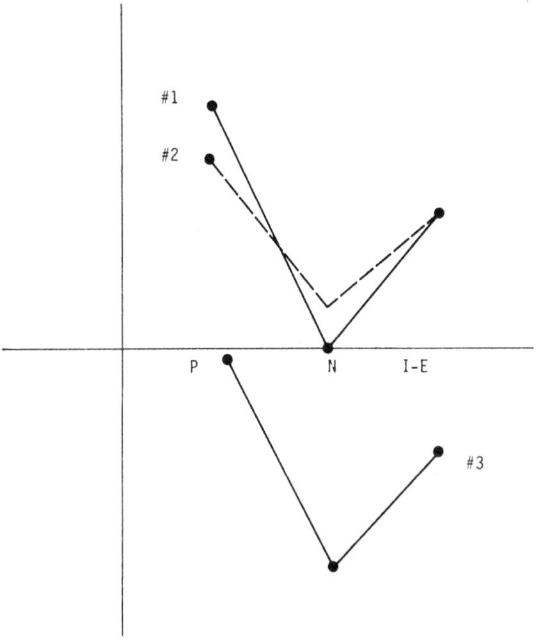

FIGURE 25. Profiles of three patients on Eysenck's three dimensions of personality (r_{12} = .99, r_{13} = 1.00; d_{12} = .56, d_{13} = 3.46).

ature (Anderberg, 1973; Sneath & Sokol, 1973; Bailey, 1974; Clifford & Stephenson, 1975; Everitt, 1980).

Another important issue regarding the use of a cluster-analytic method is the choice of the similarity measure. Clustering methods usually require the researcher to define what similarity measure should be used with a particular data set. The most commonly used similarity measures in research about psychopathology have been Euclidean distance and Pearson product-moment correlation. Psychologists are more familiar with correlation as a measure of similarity than of distance. Hence, casual users of cluster analysis often choose Pearson correlation as the similarity measure. It should also be noted that inverse factor analysis and Lorr's method were designed to be used with Pearson correlation.

However, experts concerned with clustering have generally preferred the use of Euclidean distance to the use of correlation. The reason for this preference can be gleaned from Figure 25. This figure shows the profiles for three patients—1, 2, and 3—on Eysenck's three dimensions of personality: psychoticism (P), neuroticism (N), and introversion/extraversion (I-E). Patients 1 and 2 have high scores on psychoticism and moderate scores for introversion on the I-E scale. The third patient is fairly different from the first two because the most deviate score for this patient is a particularly low score on the neuroticism scale. When the correlations among the three profiles are calculated, Patients 1 and 2 correlate at $r = .99$, and Patients 1 and 3 correlate at $r = 1.00$. This correlation of 1.00 occurs because the profiles for Patients 1 and 3 have exactly the same shape. As a measure of similarity, Pearson product-moment correlations are sensitive only to shape (Fleiss & Zubin, 1969). In contrast, Euclidean distance is sensitive to shape and elevation of profiles. The practical importance of this difference can be shown by noting that the Euclidean distance between Patients 1 and 2 is $d = .56$, whereas the distance between Patients 1 and 3 is 3.46. Because 3.46 is a much larger distance than .56, Euclidean distance shows that Patients 1 and 3 are distinctly less similar than Patients 1 and 2. Thus, Euclidean distance as a similarity measure does a better job than Pearson correlation of representing the way in which most clinicians would perceive the relationships among these three patients. Because Euclidean distance is sensitive to both elevation and shape, most writers discussing similarity measures have favored the use of Euclidean distance.

As with clustering methods, there are a large number of similarity measures that can be used in clustering. At least 50 different measures have been proposed (Hartigan, 1967; Sneath & Sokol, 1973). Many of these similarity measures were designed for use with binary data.

Now that a number of clustering methods and similarity measures have been introduced, some broad conclusions can be drawn about this generic family of statistical procedures. These conclusions concern four major methodological problems associated with the topic of cluster analysis. The reader will note that these problems are stated in fairly general and nonspecific terms, largely because statistical knowledge about cluster analysis is just developing and a detailed understanding of these procedures does not currently exist.

Different clustering methods can generate decidedly different solutions to the same data. This point was demonstrated above in the presentation of the Bartko *et al.* (1971) study. Other research studies that show a similar variability associated with the choice of the clustering method are by Everitt, Gourlay, and Kendell (1971); Byrne (1978); Mezzich (1978); and Blashfield and Morey (1980). The reason that this conclusion presents a serious problem is the large number of clustering methods that are available. New cluster methods are being proposed at the rate of about one per month (Milligan, 1981). In addition, there are in excess of 50 different similarity measures that can be used with these clustering methods (Sneath & Sokol, 1973). Because different similarity measures can be used with each clustering method, there may be as many as 750 different combinations (i.e., 150 methods × 50 similarity measures) from which a researcher can choose. This range of possibilities is truly staggering, especially if the methods in combination with various measures are likely to generate different solutions to a common data set.

No method of cluster analysis has gained clear ascendancy as the preferred method for future research applications. In the methodological literature about cluster analysis, a number of authors have attempted to contrast different clustering methods in order to decide which methods are better. Some authors have created analytical criteria and compared the methods in terms of these criteria (Jardine & Sibson, 1971; Fisher & Van Ness, 1971). The method that has appeared most desirable according to these criteria is single linkage. For instance, single linkage is the only hierarchical agglomerative method that is invariant to the order of the data. Other hierarchical methods can find different solutions simply if the order of the patients is scrambled. However, despite the desirability of single linkage according to mathematical criteria, most applied researchers have found single linkage to be unusable because its solutions often appear meaningless (Sneath & Sokol, 1973; Williams, Lance, Dale, & Clifford, 1971).

Another approach to comparing cluster-analytic methods is to choose data sets for which there is some reason to believe that the "true" classification of the data is known before clustering. For instance, a commonly used data set contains 150 flowers that represent three different species of iris (Fisher, 1936). Rice and Lorr (1969), Goldstein and Linden (1969a), Bartko *et al.* (1971), Cunningham and Ogilive (1972), Rogers and Linden (1973), Mezzich (1978), and Mezzich and Solomon (1960) have used the approach of clustering data with known solutions in order to compare methods. Unfortunately, these studies have not generated any consensus on which clustering method to use.

The final type of comparison study uses artificial data sets generated by "Monte Carlo" methods. These studies specify *a priori* the statistical characteristics of a set of clusters. Then by means of random-number generator routines, data are generated that fit the specifications of the different clusters. These data are mixed together. Because all data have been generated from known parent clusters, the true classification for each data set is precisely known. Milligan (1981) reviewed a dozen studies that have used this approach. Most of the early Monte Carlo studies (e.g., Kuiper & Fisher, 1975; Blashfield, 1976) favored Ward's method. However, the more recent studies have been less consistent and have led to a questioning of the early results (e.g., Edelbrock, 1979; Milligan, 1980).

The practical implication of the lack of a preferred method of cluster analysis is that no method is likely to become dominant in applied use. Hence, it will be difficult to decide whether the variability in future studies is due to variations in statistical procedures or to problems in the generality of a particular solution. In addition, until some consensus is achieved about which methods are to be preferred, many researchers will avoid using all cluster-analytic methods, as apparently arbitrary decisions about methodology can have a large effect on the meaning of a solution.

Given the problems associated with cluster analysis methods, researchers should carefully validate any clustering solution in order to test its validity. One important point about clustering methods is that most methods will form clusters even if none exist. For instance, if an iterative partitioning method has its control parameters set to generate four clusters, it will generate a four-cluster solution even if the data are completely random. Also, as noted earlier, cluster-analytic methods often give different solutions to the same data. Thus, a researcher should be cautious before accepting *prima facie* any classification formed with these methods.

Four different types of procedures exist for deciding whether a cluster solution is meaningful or not: (1) internal validity statistics; (2) replication procedures; (3) graphic methods; and (4) external validation.

Internal validity statistics are measures that attempt to assess how homogeneous the clusters are. These statistics attempt to give the researcher some idea of how "tight" each cluster is and whether the clusters represent relatively discrete groupings in the data. Examples of these statistics are the homogeneity statistic proposed by Tryon and Bailey (1970) and Wilks's lambda, as discussed by Friedman and Rubin (1967).

The second family of methods for testing a cluster solution is *replication procedures.* This procedure assumes that if a cluster solution is to have generality, the same solution should be found across different samples. This procedure was used by Skinner, Jackson, and Hoffman (1974), for instance, when they decided to accept only those inverse factors that were found in three subsets of a large data set on alcoholics.

The third family of methods is *graphic procedures,* which allow the researcher to visualize the clusters. As an example of a graphic procedure, Figure 26 contains a presentation of a three-cluster solution

FIGURE 26. Scatter plot of 90 artificial data points with known three-category structure.

of MMPI data in a two-dimensional space defined from a factor analysis of the variables (Blashfield & Morey, 1980). The three clusters are easily visualized in this figure.

The final testing approach is the most important: *external validation.* In this approach, the validity of the clusters is examined by comparing them in terms of variables not used to form the clusters. A good example of a study concerned with the external validity of a cluster solution was performed by Finney and Moos (1979). These researchers generated an eight-cluster solution, and they tested for cluster-by-treatment interactions using a number of treatment outcome measures. More studies should be performed with the thoroughness of the Finney and Moos article.

The literature about cluster analysis is diverse and is spread over a wide range of sciences. Classification is a fundamental topic in every science. Because cluster analysis has been touted as a statistical procedure for creating classifications, interest in cluster analysis has grown rapidly in a number of sciences (Scoltock, 1982). Blashfield and Aldenderfer (1978) attempted to read all the journal articles concerning cluster analysis that were published in 1973. To accomplish this, we looked at 287 articles in journals such as *IEEE Transactions on Computers, Journal of the American Statistical Association, Bulletin of the Japanese Society of Scientific Fisheries, Behavioral Science, Journal of Biochemistry, American Antiquity, Journal of Ecology,* and *Quarterly Journal of Medicine.* These 287 articles were spread across 162 journals. An obvious implication of this review was that staying abreast of the literature on cluster analysis is an enormous task. In addition, we noted that the diversity of this literature had led to the formation of apparent "cliques" of researchers who share a common jargon. For example, we found seven different names for the method most commonly described as *single linkage.* Unless a researcher was aware of the seven names, she or he might not realize that an ecologist who is using the "nearest neighbor" method is performing the same clustering technique as a cognitive psychologist who is using "Johnson's minimum method" or as a biologist who is using "single linkage."

ALCOHOLISM AND CLUSTER ANALYSIS

Cluster analysis has been used in approximately 200 studies concerning psychopathology. The majority of these studies have been performed since 1970. In fact, the growth of the cluster-analytic literature about mental disorders has had a steady increase over the last two decades; only a few scattered studies were published before 1965.

A bibliography of all cluster-analytic studies about psychopathology is listed in Appendix II. An analysis of this appendix will show that the studies cover a wide range of topics. However, a majority of the clustering articles focus on three topics: (1) alcoholism; (2) depression (including suicide); and (3) childhood disorders (including mental retardation).

In order to help the reader develop an understanding of the literature about cluster analysis and psychopathology, a brief review of the alcoholism literature is presented here. The intent is to give some idea of how cluster analysis has been applied as well as of the strengths and limitations of current uses of this methodology.

Alcoholism is a topic in psychopathology that, despite its obviously widespread prevalence, has not attracted a great deal of interest among clinical scientists. In general, the quality of research about psychopathology is mediocre when compared with the quality of research about pharmacology, experimental psychology, genetics, and other related sciences. More specifically, research about alcoholism is even poorer in terms of methodological quality than the general literature about psychopathology. Part of the reason is that no satisfactory theories of alcoholism have been proposed that discuss the psychological, medical, social, pathophysiological, and economic aspects of alcoholism. Another related reason is that no adequate classification of alcoholics has been developed. A number of classifications have been proposed (e.g., Bowman & Jellinek, 1941, reviewed 17 different classifications that had been suggested before World War II), but none has gained acceptance. Thus, alcoholism has been tacitly assumed to be a unitary mental disorder.

Researchers have attempted to correct this state of affairs by creating new classifications of alcoholics. Of the studies listed in Appendix II, 27 have used clustering to study alcoholism. All but 4 of these studies have incorporated self-report measures such as the MMPI, Cattell's 16PF, and the PRF (Personality Research Form). In order to keep this review manageable, only the studies concerned with the MMPI are discussed in detail.

The first study to stimulate much interest in the cluster analysis of alcoholics was performed by Goldstein and Linden (1969b). These investigators studied 513 alcoholics who were participating in a one-month treatment program at a state hospital. As part of their evaluation, all patients were required to complete the MMPI. Goldstein and Linden subdivided the patients into two groups and used Lorr's method (Lorr & Radhakrishnan, 1967) to cluster both subsamples. Four major clusters were noted across the two subsamples (only scales with significantly deviant scores—i.e., scores above 70—are listed):

Type I	Pd
Type II	D-Pt-Sc-Pd-Hs-Hy
Type III	No scales above 70
Type IV	Pd-Ma

Goldstein and Linden suggested the following diagnostic descriptions for each cluster:

Type I	Emotionally unstable personality
Type II	Neurosis with anxiety reaction or depression
Type III	Alcoholism in a psychopathic personality
Type IV	Alcoholism with other drug addictions and paranoid features

Goldstein and Linden noted that the number of patients in the fourth cluster was so small that they questioned its generality. No validation research was presented to support the existence of these

TABLE 30. Correlations among Cluster Centroids from Two Studies of Alcoholics

	Goldstein & Linden (1969b)				Whitelock *et al.* (1971)			
	I	II	III	IV	A	B	C	D
I	1.00			.42	.77			
II		1.00				.95	.80	.88
III			1.00	.41	.57	.47	.63	.53
IV	.42		.41	1.00	.40	.42	.50	○
A	.77		.57	.40	1.00		.48	
B		.95	.47	.42		1.00	.80	.89
C		.80	.63	.50		.80	1.00	.77
D		.88	.53	○		.89	.77	1.00

Note. Only correlations greater than +.40 are shown. The diagonal values that are circled should be high if Whitelock *et al.* replicated the Goldstein and Linden clusters. All other off-diagonal values should be small.

clusters. However, the clinical descriptions of the clusters were well written and contributed to the subsequent popularity of the article.

Another study in which clustering was used on MMPI data was published by Whitelock, Overall, and Patrick (1971). The second author of this trio is John Overall, whose work on the BPRS was discussed in the preceding chapter. Whitelock *et al.* studied MMPIs gathered on 136 male alcoholic inpatients in a state hospital. They clustered half the data using a version of inverse factor analysis developed by Overall (1968), and they replicated the clusters on the second subsample of the data. As in the Goldstein and Linden study, four clusters were discovered. The high scales for each cluster were:

Type A Pd
Type B D-Sc-Hs-Pd-Pt
Type C Pd-D-Pt
Type D D

The conclusion of Whitelock *et al.* was that they had discovered "mean cluster profiles [which were] very similar to the MMPI profiles Goldstein and Linden found in their psychiatric populations" (p. 16). In point of fact, this conclusion by Whitelock *et al.* was not accurate. That is, the cluster solution they found was not very similar to the cluster solution noted by Goldstein and Linden. To understand this point, examine Table 30. This table shows the correlations among the mean MMPI profiles for the four Goldstein and Linden clusters and for the four Whitelock *et al.* clusters (only correlations above .40 are shown). If Whitelock *et al.* were correct in concluding that their clusters were very similar to the Goldstein and Linden clusters, one would expect two findings in this table: (1) First, the correlations of a cluster from one study should have a high value with one and only one cluster from the other study. Stated somewhat differently, one would expect that the correlation values in the diagonal of the lower left and upper right rectangular portions of the total matrix should contain large values. The location of these values are circled on Table 30. (2) The second expectation is that all other correlations would have small values. In other words, a cluster should correlate with only one cluster from the other study. In addition, no cluster would be highly correlated with another cluster found in the same study (if so, the clusters within a study would fail to be homogeneous).

Concerning the first expectation, of large diagonal values, most of these values were high. For instance, Goldstein and Linden's Cluster I correlated at $r = .77$ with the Whitelock *et al.* Cluster A. One exception was noted to the large diagonal values, and that was the correlation of Type IV from Goldstein and Linden with Type D from Whitelock *et al.*, which had a correlation of $r = .16$. However, regarding the second expectation, not all correlations in the rectangular "between-studies" portion of the matrix were low in the off-diagonal values. For instance, Cluster II from Goldstein and Linden correlated $r = .80$ with Cluster C from Whitelock *et al.* and $r = .88$ with Cluster D. Thus, Cluster II from Goldstein and Linden, was highly similar to three of the Whitelock *et al.* clusters. Another interesting feature to note from Table 30 is how large the correlations were among the clusters from the same study. These correlations are shown in the two triangular subsections of the matrix. For instance, Clusters B and C from Whitelock *et al.* correlated at $r = .80$. This result suggests that some of the clusters within each study were not very different; thus, the homogeneity of the clusters could be challenged.

The third cluster study using the MMPI was performed by Stein, Rozynko, and Pugh (1971). These investigators collected MMPIs on 302 consecutive admissions to an alcohol treatment program at a state hospital. Six variables were used as input for the clustering: three were factor-derived scales from the MMPI (i.e., no standard clinical scales from the MMPI were used); one was a scale from the California Psychological Inventory; and the last two were from the Motoric-Ideational Activity Preference Scale. An iterative partitioning method called OTYPE was used (Tryon & Bailey, 1970). The results indicated the presence of 11 clusters. Stein *et al.* then compared the clusters in terms of a number of variables not used to form the clusters, such as the standard MMPI clinical scales, demographic variables, and alcohol use data. They concluded, "Each of the types was found to be highly homogeneous and capable of differential predictions in relation to a host of psychological and social variables" (p. 253).

The conclusion of Stein *et al.* was unwarranted. Stein *et al.* made available an appendix that contains the mean MMPI profiles on the 11 standard clinical scales. A correlation matrix was formed among the 11 cluster profiles. The intercluster correlations ranged from $r = .98$ to $r = .64$, with a median intercluster correlation of .88. Thus, all 11 clusters had MMPI profiles with very similar patterns, too similar for these clusters to be considered either separate or reasonably homogeneous.

Another alcoholism MMPI study was done by Skinner *et al.* (1974). These researchers studied 282 male alcoholics in a state hospital. Two self-report tests were administered to all the subjects: the MMPI and the Differential Personality Inventory (DPI) (Jackson & Messick, 1971). On the basis of the DPI data, Skinner *et al.* performed inverse factor analysis to locate different types of alcoholics. Only factors that were replicated across two subsets of the data were kept in the final solution. Skinner *et al.* decided that eight factors were consistently noted among the DPI data. Using these eight factors, Skinner *et al.* then analyzed the MMPI profiles associated with each factor. These eight MMPI profiles were correlated with the cluster profiles found by Goldstein and Linden (1969b) and by Whitelock *et al.* (1971). In their discussion section, Skinner *et al.* concluded:

> The finding that eight typal dimensions . . . could be identified and replicated across three samples lends support to the hypothesis that unique personality subgroups exist within a population of alcoholic patients. These results are further corroborated by earlier studies of Goldstein and Linden (1969b) and Whitelock *et al.* (1971), and empirical commonalities demonstrated among the typologies. (p. 665)

The final clustering study to be analyzed is a recent article by researchers at the University of Washington. In this study, Donovan, Chaney, and O'Leary (1978) performed a cluster analysis on MMPI data gathered from 102 male veterans in an alcohol treatment program. The clustering method they used was an iterative partitioning method. Their clustering procedure located four clusters with the following patterns (all scales above 70 are listed in order):

Type a Pd-D
Type b D-Pt-Pd
Type c D-Sc-Pd-Pt-Hy-F
Type d Sc-F-Ma-Pt-Pd-D-Pa

Notice that these clusters appear to have different profiles from those found by either Goldstein and Linden or Whitelock *et al.*

Interestingly, Donovan *et al.* did not devote much of their discussion to the clusters that they had created. Instead, they focused on the clusters originally found by Goldstein and Linden (1969b). They assigned all of their patients to the Goldstein and Linden clusters containing patients with whom their patients were most similar. Donovan *et al.* had also gathered data on these male patients by administering the Alcohol Use Inventory (AUI). They found that the four Goldstein and Linden clusters were significantly different on the AUI scales. Most of these differences were associated with the heavy alcohol use found among patients assigned to Cluster II from Goldstein and Linden ("neurosis with anxiety reaction or depression"). In a subsequent follow-up study, O'Leary, Donovan, Chaney, and O'Leary (1980) reported that the Goldstein and Linden Cluster II alcoholics showed higher levels of psychopathology when studied 6 to 12 months after participating in an alcohol treatment program. Based on these data plus their review of the MMPI clustering literature on alcoholics, Donovan *et al.* concluded that the results clearly supported the validity of Cluster II from Goldstein and Linden (1969b). They also believed that there was sufficient evidence to support the viability of Goldstein and Linden's Cluster I ("emotionally unstable").

What conclusions can be drawn from the clustering literature on alcoholics? Frankly, there is a decided lack of any major substantive conclusions. The only conclusion that does seem to have sufficient support across studies is the apparent existence of two general subtypes of personality styles associated with alcoholism: (1) a neurotic style that involves numerous complaints of depression/anxiety and relatively heavy alcohol use and (2) a personality disorder style characterized by an antisocial stance and relatively moderate alcohol abuse. As Donovan *et al.* (1978) reported, these two clusters have appeared in some form in virtually all MMPI studies. The description of these clusters is disappointingly general and does little to stimulate the use of new concepts when thinking about the classification of alcoholism. It should be noted that most studies have also found additional clusters, but the constitution of these additional clusters has not been consistent.

Methodologically, conclusions from these studies are easier to make, but these conclusions do not yield a positive view of the clustering literature. These five MMPI studies demonstrate the four problems with cluster-analytic methodology that were outlined above:

1. The first problem is the large number of clustering methods. In

these five studies, five different methods of cluster analysis were used. The importance of noting that different methods were used is that different methods of cluster analysis can yield distinctly different solutions to a data set. The five studies showed clear differences in their solutions. Because these studies used different methods, it is possible that the variability in the solutions was simply a function of the choice of the clustering methods. Or it is possible that the different solutions were found because MMPI data on alcoholics are so "noisy" that finding consistent, homogeneous clusters is not possible. Distinctions between the two hypotheses cannot be made from the available results.

2. The second conclusion is that no method of clustering has gained clear ascendancy. Interestingly, the clustering method that was most frequently used in the 27 studies on alcoholics was inverse factor analysis. This fact is interesting because this method is quite controversial, and there is a sizable literature that discusses its pros and cons (Stephenson, 1953; Lorr, 1966; Fleiss, Lawlor, Platman, & Fieve, 1971; Morf, Miller, & Syrotuik, 1976; Skinner & Lei, 1980; Edelbrock & Reed, 1981). Less controversial, yet popular, methods such as average linkage or Ward's method were rarely utilized in the 27 studies.

3. The failure to follow the use of cluster analysis with an adequate validation procedure is the third problem. The five MMPI studies of alcoholics were not as bad as many cluster-analytic studies are in terms of this problem. Three of the five used a replication procedure (Goldstein & Linden, 1969b; Whitelock *et al.*, 1971; Skinner *et al.*, 1974), and three studies used external validation (Whitelock *et al.*, 1971; Stein *et al.*, 1971; Donovan *et al.*, 1978). None of the five failed to use a formal validation procedure. However, one-third of all 27 alcoholism studies did fail to use any type of validation—a serious deficiency, given the methodological problems associated with forming a cluster solution. Even though all five of the MMPI studies discussed in detail did use some validation procedure, the authors of these studies did not seem appropriately critical of their solutions. For instance, the correlations among the cluster centroids from the Stein *et al.* (1971) study were so great that the distinctiveness of these clusters must be seriously questioned.

4. The last problem that is apparent in the cluster-analytic literature is that this literature tends to be a fragmented collection of single studies. These studies have lacked the cumulative effect that is necessary if scientific research is to have an impact on other areas of science and on clinical practice. The lack of integration can be demonstrated by searching the references in each of the total 30 articles. Some articles cite Goldstein and Linden (1969b); others cite a study by Part-

ington and Johnson (1969) that used the 16PF as a measurement instrument. Very few studies cite both. Rarely do any studies cite any of the other remaining 25 articles that used clustering to form a classification of alcoholics. In other words, most of the researchers seemed unaware of the large number of articles that had been published on this topic, so they published another one.

Another interesting feature is that of the 57 authors involved in these 30 studies, only a single person (Nerviano) has been an author of more than three publications. This finding occurs more generally among the over 200 cluster-analytic articles listed in Appendix II. There are a total of 385 authors for these studies, of which 53 (13%) have been involved in more than two published cluster-analytic studies. Only 1 of the 385 authors (Lorr) has been involved in more than ten clustering studies of psychopathology.

CLUSTER ANALYSIS AND STUDIES OF DEPRESSION

Not all clustering studies of psychopathology are as beset by methodological problems as the analysis of the alcoholism literature would suggest. Probably the best positive example of the use of cluster analysis to study psychopathology is contained in a series of studies performed by E. S. Paykel. This British psychiatrist had been concerned with understanding the psychopathology of depression and became intrigued by the one-depression versus two-depression controversy that had been a long-standing issue in the British view of affective disorders (Kendell, 1968). This controversy was introduced earlier, and the factor-analytic literature relevant to the controversy was briefly discussed in Chapter 7. As noted in this discussion, methodological criticisms had suggested that factor analysis was not a good methodology for resolving the controversy, and that cluster analysis would be more appropriate.

Because of the complicated and often conflicting studies that had evolved from this controversy about depression, Paykel became interested in cluster analysis as a technique that could empirically form homogeneous groups of depressed patients. Because cluster analysis is a statistical procedure, it would not be biased by the various arguments concerning the classification of depression but would isolate whatever natural categories of depression might exist.

Paykel's initial study (1971) involved 165 depressed patients who were seen primarily as out-patients in one of four different mental-health settings in New Haven, Connecticut. All the patients were suffering at least a mild depressive syndrome, as rated by their clinicians, even though the primary diagnosis for these patients did not have to be depression. Patients were excluded if their primary diagnoses were organic brain syndrome, schizophrenia, alcoholism, drug addiction, or some independent form of neurosis. Three major types of information were gathered on these patients: (1) a clinical interview, which was scored by means of the Hamilton Rating Scale for Depression (Hamilton, 1960); (2) an interview about recent stressful life events; and (3) a self-report personality test, the Maudsley Personality Inventory (Eysenck, 1959).

The reason for providing some details about the sample and the measures used by Paykel is that this information has important implications for the type of cluster analysis solution that can be found. Specific types of psychopathology cannot be found by cluster analysis if the data submitted to the cluster-analytic method are not relevant to those types. For instance, no clusters were found in the MMPI studies that concerned the interpersonal aspects of alcohol abuse, despite the obvious importance of interpersonal issues (family, job, marriage, and so on) in alcoholism. No clusters could be found of this nature because the MMPI does not directly tap information about interpersonal relationships. However, a review of the studies concerning alcoholism that used other measures, such as the 16PF and the PRF, did find such clusters (Morey & Blashfield, 1981).

Paykel chose his subjects in such a way as to sample a reasonably broad spectrum of depressed patients, rather than choosing patients too narrowly and thereby perhaps unintentionally biasing the results in favor of the one-depression perspective. Also Paykel's sampling of measures was intended to represent the descriptive information that is usually considered relevant to the psychotic-neurotic distinction about depression: stressors, personality data, and clinician ratings of symptoms. Thus, Paykel designed his sampling so that the cluster-analytic results would be as relevant as possible to the problem he was addressing.

The method of cluster analysis that Paykel used was a version of iterative partitioning that had been designed by Friedman and Rubin (1967). This program is one of the most sophisticated iterative partitioning methods available. Paykel chose to optimize the homogeneity of the clusters using Wilks's lambda, a statistic that measures the multivariate homogeneity of the clusters. The results suggested that four clusters were present in the data:

Cluster I—Psychotic ($N = 24$)

Depressive delusions, retardation, anorexia, guilt, and delayed insomnia

Cluster II—Anxious ($N = 38$)

Moderate depression, high psychological and somatic anxiety, fatigue, depersonalization, obsessional symptoms

Cluster III—Hostile ($N = 46$)

Hostility and self-pity out of proportion to general severity

Cluster IV—Young depressives with a personality disorder ($N = 57$)

Relatively mild depression with marked fluctuations

The variables that best discriminated among the clusters were, in order, age, the patient's level of insight into his or her problem, hostility, and psychological/somatic anxiety. Of these four clusters, the two that were most similar were the hostile depressives and the young depressives with a personality disorder.

Paykel concluded that the cluster-analytic results favored neither the one-depression nor the two-depression position. What Paykel did find was a "psychotic" cluster, but he did not find that the remainder of the subjects were organized into one "neurotic" cluster. Instead, his results suggested that the appellation *neurotic depression* obscures the presence of three descriptive clusters.

In a second publication, Paykel (1972a) studied the differences among the four clusters across 48 variables that had not been used in forming the clusters. The subjects for this study were the same 165 depressed patients who had been used in forming the clusters. The 48 independent variables were organized into six groups: background and personality data, self-reports of symptoms, clinical diagnoses, treatment setting, medication, and outcome data.

Concerning the background and personality data, the results showed a significant association between cluster membership and race, with black patients tending to congregate in the hostile and young depressives clusters. Personality data also suggested that the hostile and young depressives had higher scores on a personality measure for hysteria. Most other background and personality variables did not show significant differences among the clusters.

The second collection of variables was the self-report of symptoms. Of the total 18 variables in this group, significant differences were found on 11. As might be expected, the anxious depressives rated themselves as being highest on a scale for anxiety, and the hostile and young depressives rated themselves as being highest on a hostility scale. The reader should note that these results did not have to occur from the self-report data, as the symptom data used to form the clusters were based on symptom ratings made by clinicians. If a clinician believes a patient to be high on hostility, it is not necessarily true that the patient will rate herself or himself in the same way.

Clinical diagnoses as made using the DSM-I showed a significant association with the cluster membership. Patients with a clinical diagnosis of psychotic depression were likely to be placed in the psychotic or anxious clusters. Patients with a clinical diagnosis of neurotic depression were placed in any of the three nonpsychotic clusters.

Concerning the clinical setting in which the patients were treated, the most striking result was that the young depressives were generally seen in outpatient settings, whereas most psychotic depressives were assigned to inpatient settings.

The medication data also showed differences among the clusters. The psychotic and anxious depressives were most likely to receive medication, either tricyclics or tricyclics plus phenothiazines. The hostile depressives frequently were given phenothiazines. Commonly, the young depressives were not given any medication at all.

The final set of variables was on outcome data: length of time to first remission, global severity of symptoms at follow-up, and social adjustment at follow-up. Disappointingly, none of the outcome variables showed any significant differences among the clusters.

To summarize, Paykel's second study tested for differences among the clusters of depressives by using 48 independent variables. He found significant differences among the clusters on 24 of the 48 variables. Finding significant differences among the clusters on half of the independent variables was certainly a better outcome than would be expected by chance. Nonetheless, half the variables showed no differences between the clusters. Thus, Paykel (1972a) concluded that this study provided "partial validation" of his classification of depression.

In the third study by Paykel (1972b), the four clusters were compared in terms of their responses to a commonly used antidepressant drug, amitriptyline (Elavil). The patients in this study were 85 depressed women who were consecutively seen at two different treatment settings. Most of the women were treated as outpatients. The outcome data that Paykel used were from a global rating of the severity of depression made by the clinicians treating the patients. Paykel found that all the depressed patients, regardless of their cluster membership, showed improvement while on amitriptyline. However, the four clusters were significantly different in terms of their improvement when an analysis of covariance was performed on the data. In particular, the anxious depressive cluster was notable because the patients in this group showed the least change. Later Paykel, Prusoff, Klerman, Haskell, and DiMascio (1973) published another study, using these same 85 female patients, that showed that the cluster-analytic solution was nearly as good at predicting the response to amitriptyline as a regression equation using 29 predictor variables. Given the well-demonstrated power of linear regression as a prediction system, the results of Paykel *et al.* with the four-cluster solution are impressive.

The final set of two studies by Paykel are important because they constituted attempts to replicate the initial findings, thereby testing the generality of these clusters. Prusoff and Paykel (1977) studied 143 depressed women who had been given amitriptyline as part of a controlled research study. Again, the clusters showed significant differences on an outcome measure of the global rating of severity. As found earlier by Paykel (1972b), the patients assigned to the anxious depressive cluster showed the least improvement while on amitriptyline. Hostile and young depressives showed the most improvement. In the second replication study, Paykel and Henderson (1977) used four different clustering methods on the data from the original 165 patients in order to see if different clustering methods would yield different solutions to these data. They concluded, "Three groups, psychotic depressives, anxious depressives, and young depressives with a personality disorder, replicated well, while hostile depressives failed to separate clearly" (p. 111). The Paykel and Henderson study was relevant because it showed that Paykel's original cluster solution was not just a function of a particular clustering method.

Before terminating this discussion of cluster-analytic studies of depression, the research by Overall, Hollister, and their colleagues deserves mention. Overall, Hollister, Johnson, and Pennington (1966) studied depressed patients using the BPRS as the measurement instrument. They found three clusters: (1) a retarded depression cluster; (2) an anxious depression cluster; and (3) a hostile depression cluster. These results are quite similar to the results of Paykel (1971), especially as the retarded depression cluster appeared descriptively to be like the psychotic depressive cluster of Paykel.

The literature examining the validity of the classification generated by Overall *et al.* (1966) has focused almost exclusively on the issue of drug-treatment response. Generally, these studies have found that patients in the anxious depression cluster respond best to the phenothiazines and the minor tranquilizers. In contrast, retarded depressives generally show the most positive change if administered tricyclics. Studies indicating success in the treatment of anxious depressives have looked at thioridazine (Mellaril), perphenazine (Trilafon), diazepam (Valium), and other tranquilizers (Overall *et al.*, 1966; Hollister, Overall, & Shelton, 1967; Henry, Market, & Emken, 1969; Rickles, Fesbacher, & Dowing, 1970; Klerman, DiMascio, Weissman, Prusoff, & Paykel, 1974). Retarded depressives have been treated most successfully with antidepressants, such as imipramine (Tofranil) and amitriptyline (Overall *et al.*, 1966; Overall, Hollister, & Pokorny, 1973; Raskin, 1974). These results on Overall's three cluster classification are consistent with the conclusions from a general review of tricyclic medication that found that psychomotor retardation and the absence of neurotic symptoms are positive indicators for the use of tricyclics (Bielski & Friedel, 1976). Also, these results agree quite well with the studies of Paykel, who found that anxious depressives generally did not respond as well when administered the tricyclic medication amitriptyline. However, unlike Paykel, who found that hostile depressives did respond well to amitriptyline, Overall and his colleagues found that there is no drug treatment that surpasses a placebo in the management of hostile depression.

The collection of studies by Paykel and by Overall and Hollister is impressive. There is a high degree of uniformity in the cluster solutions, as compared with the striking variance in solutions found in the MMPI alcoholism studies. The clustering studies concerned with depression have engaged in relatively extensive replications and tests of the external validity of their clusters. Finally, their persistence in this research has led to a body of knowledge about the descriptive patterns of depression that did not exist before the studies were performed. In particular, reviewers of this literature agree that there is strong evidence for considering psychotic/retarded depression as a qualitatively separate form (Kendell, 1976; Blashfield & Morey, 1979; Matussek, Soldner, & Nagel, 1981). In addition, it appears that what had been called *neurotic depression* can be subdivided into at least two or three descriptive subtypes (Roth & Barnes, 1981).

However, more needs to be accomplished concerning these cluster-analytic studies of depression if this new classification is to gain widespread acceptance. For instance, neither Paykel nor Overall has provided clear descriptions of the clusters using standard clinical terminology. Nor have these investigators presented any case histories of patients who fit the clusters. Also, it is possible that the parallel results of Paykel's cluster solution and Overall–Hollister's cluster solution are more apparent than real. In the only study to make a direct comparison of these two classifications, Paykel (1972b) found a nonsignificant association between his four-cluster solution and the three-cluster solution of Overall *et al.*

In conclusion, the research applying cluster analysis to the study of psychopathology is in its infancy and is still attempting to develop to a point where a meaningful body of knowledge will result. The methodological problems associated with cluster analysis are serious and can interfere with the discovery of a new and useful classification—this has been demonstrated in the application of cluster analysis to studies of alcoholism. On the other hand, the methodological problems associated with clustering are not insurmountable, and as the review of the studies on depression have indicated, the use of cluster analysis can generate meaningful solutions that have important implications for the classification of psychopathology (Chignell & Stacey, 1980; Paykel, 1981).

THE FUTURE OF CLUSTER ANALYSIS

At this point, the reader is probably somewhat confused and uncertain about what to make of the cluster-analytic literature. As Hartigan (1975) stated, it is in the area of psychiatric classification that cluster analysis has the greatest potential for leading to productive advances in classification. Yet, as Meehl (1979) stated, it is also true that, to date, cluster analysis has not led to the discovery of any new category of psychopathology.

Given all of this, what is the future of cluster analysis in regard to the classification of psychopathology? In the remainder of this chapter, three major areas that will need careful consideration in the future are noted. These are: (1) the need for systematic studies of the various methodologies associated with cluster analysis; (2) suggestions on how to perform and report future clustering studies; and (3) the importance of integrating theoretical views about psychopathology when using cluster analysis to create new classificatory systems.

Studies of Clustering Methodology

In his 1979 paper entitled "A Funny Thing Happend to Us on the Way to Latent Entities," Meehl mentioned the problems associated with quantitative methods and suggested that a valuable avenue for future research would be "a large scale study of the presently available taxonomic search methods" (p. 569). Cluster methods are statistical tools, and as with all tools, an artisan must first learn about their characteristics before she or he can reasonably expect to apply them to practical situations. Meehl suggested that there are three potentially profitable approaches to these studies of clustering methods: (1) Monte Carlo studies; (2) studies involving real data sets with known classificatory structures (which Cattell, 1966, called "plasmodes"); and (3) real data sets with unknown structures but for which meaningful classificatory issues exist. Because Monte Carlo studies have demonstrated a widespread utility in the study of a variety of statistical procedures, this section will focus on Monte Carlo studies of cluster-analytic methods.

Monte Carlo studies are studies in which computer programs are used to generate data from known statistical models. Various statistical techniques are tested on these artificial data. In effect, these studies are analogous to testing linear measurement devices, such as a yardstick or a ruler, against a standard that has a precisely known length. In Monte Carlo studies, the characteristics of the data are precisely known, and the statistical techniques are tested for accuracy. These studies are called *Monte Carlo studies* because they use random number generators (like the roulette wheel at Monte Carlo) to create the artificial data.

In Monte Carlo studies, a statistical model must be specified. In the studies of cluster analysis methods, the mixture model is the most frequently used statistical model. According to this model, cluster analysis is a procedure that attempts to find the underlying populations in any given data set. That is, the mixture model assumes that the data set to be clustered is comprised of subsets of data from a number of populations, where the exact number of populations and the distributional parameters of the populations are unknown. Thus, any data set that is being clustered is assumed to represent a "mixture" of samples from these underlying populations. According to this model, the purpose of cluster analysis is to estimate the characteristics of the underlying populations.

Typically, users of cluster analysis are interested in making inferences about the populations underlying the structures of their data sets. For example, the purpose of a cluster analysis of data from a group of schizophrenics is the discovery of what populations (disorders, diseases) comprise the more inclusive population of schizophrenia. In the same way, the purpose of the cluster analysis of fossilized bone fragments from animals related to the family of *Canis* (dogs, wolves, coyotes, and so on) is most likely to be the discovery of what species (populations) related to *Canis* existed in the past.

To show how Monte Carlo studies of cluster analysis are performed, an example is presented here. Because this chapter has focused heavily on clustering studies that have used MMPI data, this example of the Monte Carlo study mimics MMPI data.

Skinner and Jackson (1978) performed a study that suggested that the MMPI is sensitive to three broad categories of psychopathology: (1) psychotic patients whose typical profile has elevations on the Sc (schizophrenia) and Pa (paranoia) scales; (2) neurotic patients who exhibit elevations on the Hs-Hy-D (hypochondriasis, hysteria, depression) scales; and (3) patients with a personality disorder whose typical profile has elevations on Pd-Ma (psychopathic deviate and mania).

The structure of the MMPI as suggested by Skinner and Jackson was used in a Monte Carlo study to create mixture model data (Blashfield & Morey, 1980). The parent sample for each population in the mixture was created by means of a computer routine that generated multivariate normal deviates having a specified covariance matrix. The parameters needed for creating each parent sample were a vector of means (the centroid profile), a vector of standard deviations, and a matrix of MMPI scale intercorrelations. The mean vectors for the three parent samples were the values specified for the Gilberstadt and Duker (1965) Hs-Hy-D profile; the Marks and Seeman (1963) Sc-Pa-F-Pt-D-Pd profile; and the Gilberstadt and Duker Pd-Ma-Sc profile. The MMPI interpretation manuals by Gilberstadt and Duker and by Marks and Seeman were chosen as the sources of the mean profiles because these two "cookbooks" have become accepted among clinical psychologists who use the MMPI in their clinical practices. The standard deviations were estimated from a large data set studied by Goldberg (1965). The correlation matrix for each parent sample was the matrix of MMPI scale intercorrelations reported by Swenson, Pearson, and Osborne (1973) from a sample of over 25,000 medical patients.

To represent each parent sample, 30 MMPI profiles were created by the computer program; 11 of the MMPI clinical scales were used (Mf and Si were not included). The three groups of 30 profiles were then mixed together to form a data set that had 90 "patients" measured along 11 variables. In this data set, there was a known three-category structure underlying the mixture. The three-category structure of the data is shown in Figure 26 (p. 233), which contains a plot of the 90 artificial data points plotted in a space defined by the first two discriminant vectors. Notice that the mixture structure of the data is reasonably easy to visualize in this space, but that clear boundaries between the clusters do not exist.

The first method to be used on these data was single linkage with Euclidean distance for the similarity measure. The results of this method are shown in the dendrogram of Figure 27. The dendrogram illustrates a chronic difficulty in the application of single linkage to mixture model data. Single linkage tends to "chain." In other words, it forms only one large cluster and adds individual cases to that large cluster one by one. In other words, applied uses of single linkage often generate dendrograms similar to that shown in Figure 27, in which no clusters are obvious. The chaining that occurs with single linkage means that if a three-cluster solution to this mixture is examined, one cluster is found that contains 88 cases, a second cluster is noted that contains only 1 case, and a third cluster has 1 case. Other Monte Carlo studies of single linkage have suggested that this method has had little practical value for solving mixture-model data sets.

Another method that was applied to these data was Ward's method (with Euclidean distance). The dendrogram for this method is shown in Figure 28. Notice how clearly this dendrogram suggests the presence of three clusters.

The first of the three clusters found by Ward's method had a mean profile that was elevated on the Hs-Hy-D scales. There were 29 profiles in this cluster, 28 of which were from the 30 "neurotic patients" that were created by means of the Monte Carlo procedures. These results suggested that Ward's method adequately represented the neurotic category in the mixture.

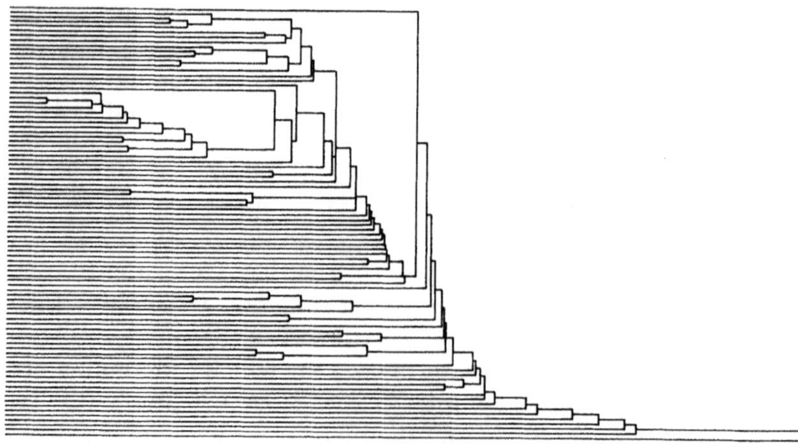

FIGURE 27. Dendrogram of data using single linkage with Euclidean distance.

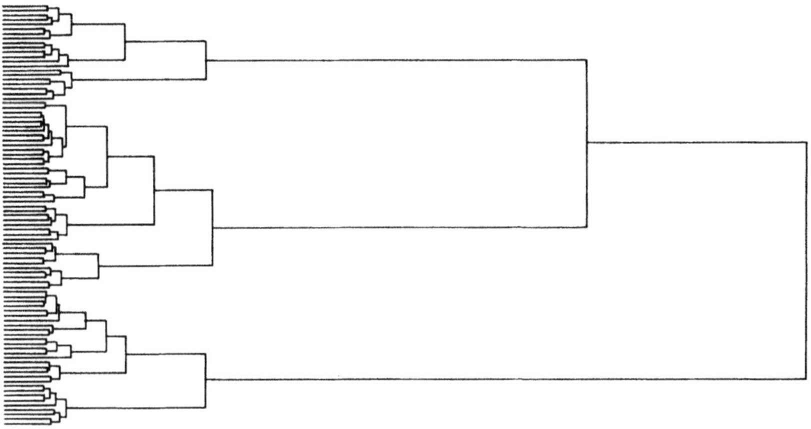

FIGURE 28. Dendrogram of data using Ward's method with Euclidean distance.

The second cluster found by Ward's method contained 21 cases and had a mean profile that was elevated on the following scales: Sc-Pa-F-Pt-D-Pd-Hs-Hy. This was the cluster that was most similar to the "psychotic" category in the parent Sc-Pa-F-Pt-D-Pd profile. However, the psychotic category had not been adequately captured by Ward's method, as 9 of the 30 patients who were created for this category were not classified in the second cluster. In addition, one patient in this cluster was from the personality disorder category.

The third cluster found by Ward's method had 40 cases and a mean profile of Pd-Ma-Sc. This cluster was most similar to the personality disorder patients who were created for the mixture, and it contained all of the cases from this category. However, there were an additional 10 cases in this cluster, most of which came from the psychotic category, that did not belong to the personality disorder category.

In summary, Ward's method formed a more accurate representation of the three-category structure of the MMPI mixture than did single linkage. However, Ward's method did have difficulties recovering the known structure of the data. In particular, it did not completely differentiate between the personality disorder cases and the psychotic cases.

Finally, a third clustering method was performed on these data. This method was average linkage; Pearson correlation was used as the similarity measure. The dendrogram resulting from this method is shown in Figure 29. This dendrogram does not clearly suggest how many clusters might exist in the data. Because there were known to be three categories in the mixture, a three cluster solution could be

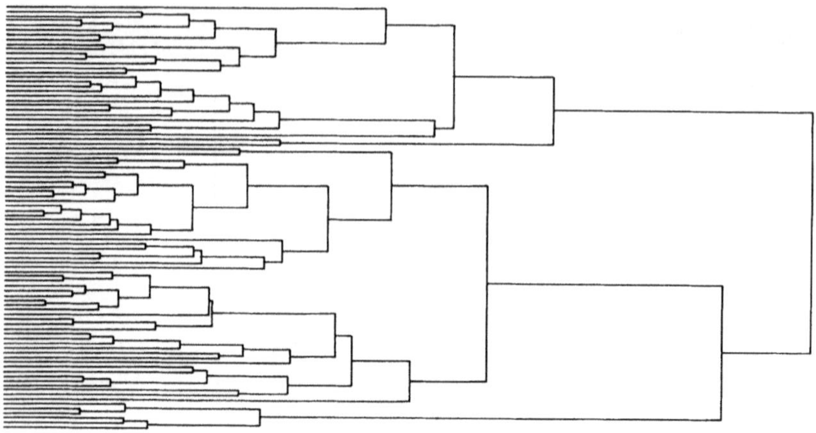

FIGURE 29. Dendrogram of data using average linkage with Pearson correlations.

examined. This solution was quite accurate. The sizes of these three clusters were almost equal. The mean profiles for the clusters were (1) Sc-Pa-F-Pt-D-Pd, (2) Hs-Hy-D, and (3) Pd-Ma-Sc, which match almost exactly the mean profiles of the parent categories in the mixture. Only six patients were misclassified (two per cluster). In other words, the average linkage method was close to making a perfect recovery of the known structure of this MMPI mixture.

In conclusion, the results of this MMPI Monte Carlo study again showed that different clustering methods can reach quite different solutions to the same data. Average linkage with Pearson correlation generated a very accurate solution to this data set. Ward's method found a somewhat less accurate solution. Finally, the solution generated by single linkage was practically useless because of its tendency to chain.

Up to now, about a dozen Monte Carlo studies of cluster analysis have been performed. Most of these studies have been concerned with comparing the performance of different methods of cluster analysis. For instance, one of the earliest Monte Carlo studies was done by Blashfield (1976), using 50 mixtures. He compared the four most popular hierarchical agglomerative methods: single linkage, complete linkage, average linkage, and Ward's method. The results of the study are shown in Table 31. To determine how well a cluster method was recovering the known cluster structure of the data, kappa values were computed that compared the cluster membership as found by the particular clustering method with the known category membership. Thus, the higher the value of kappa, the better was the cluster solution

TABLE 31. Median Kappa Values to Show Accuracy of Four Clustering Methods Across 50 Monte Carlo Data Sets (Blashfield, 1976)

	Median kappa
Single linkage	.06
Complete linkage	.42
Average linkage	.17
Ward's method	.77

in recovering the known structure of the mixture. The results in Table 31 present the medians of the kappa values across all 50 mixtures for each clustering method. Ward's method clearly obtained the highest kappa value, indicating that the clusters found by Ward's method were most similar to the known category structure of the mixture. In contrast, single linkage created solutions that had very little relationship to the actual structure of the mixtures.

Somewhat later, Edelbrock (1979) studied 20 of the mixtures created by Blashfield (1976). Edelbrock became aware of a problem inherent in the way in which Blashfield had resolved the number-of-clusters problem. He suggested a correction procedure and compared a set of cluster-analytic methods using both Euclidean distance and Pearson correlation. In contrast to Blashfield, Edelbrock found that average linkage using Pearson correlation produced better solutions to the same data than had Ward's method. In further studies using the Monte Carlo data, Edelbrock has demonstrated the usefulness of intraclass correlations as similarity measures and of inverse factor analysis as a clustering method (Edelbrock & McLaughlin, 1980; Edelbrock & Reed, 1981).

Also in the late 1970s, another psychologist, Milligan, became interested in the use of Monte Carlo procedures to compare clustering methods. Milligan (1980) published a comparison of 11 clustering methods in a Monte Carlo study that had the best design of any study to date. Like Edelbrock (1979), Milligan found that average linkage performed well. He also noted that the *k*-means iterative-partitioning method did well.

Later Milligan (1981) published a review of Monte Carlo studies that compared the performance of clustering methods. Milligan documented the variability in the methodologies of these papers as well as the variability in their conclusions. However, Ward's method and average linkage consistently appeared among the lists of the best methods across all Monte Carlo studies. Milligan ended his review with a series of five suggestions for any future Monte Carlo research:

> (1) Future Monte Carlo studies should be designed to provide information about the data characteristics which affect the recovery of the cluster structure.
> (2) Certain clustering algorithms, such as the group average and Ward's method, should always be included in a study in order to provide a basis of comparison for other methods.
> (3) Monte Carlo studies should be expanded to include other classes or types of clustering methods. To date, most of the Monte Carlo research has dealt with hierarchical procedures.
> (4) A potentially useful application of Monte Carlo procedures involves the analysis of real life data.
> (5) The final suggested direction of research is concerned with the problem of determining the actual number of clusters in a data set. (pp. 404–405)

How to Perform Cluster Analysis

Providing the reader with an explicit "cookbook" on how to perform cluster analysis is a risky venture. Given the deficits in the current knowledge about clustering methods, any researcher should always approach their use cautiously. The following suggestions are proposed simply as general guidelines for future applications of cluster analysis.

The first major issue in any cluster-analytic study is the selection of the variables and the sampling of the patients. Both are crucial because the decisions regarding these selection issues have a major effect on the types of clusters that can appear. To understand this point, consider the alcoholism literature discussed earlier. None of these cluster studies found a cluster of female alcoholics or a cluster of teenagers who were polydrug users as well as heavy alcohol users. Neither of these clusters could have appeared (assuming that such clusters even exist) because the six MMPI studies sampled patients from VA and state hospitals. None contained females in the sample, and the average age of the patients invariably was about 45. The other important issue in a clustering study is the selection of the variables. In the six alcoholism studies, the instrument was the MMPI. A reason for the variability in the cluster solutions across the six studies could be the poor psychometric properties of the MMPI (Gynther & Gynther, 1976). In other words, given the "noise" inherent in MMPI data plus the potential heterogeneity of the personality styles of persons who abuse alcohol, the finding of results that had little generality should not be surprising. Consistent results would have been more likely if a better measurement instrument had been chosen.

Second, before any clustering is performed on a data set, the researcher is strongly encouraged to perform a number of descriptive statistics on the data. Computing the univariate frequency distributions; analyzing the data for outliers (i.e., data points that are very dissimilar from the remainder of the data set); performing exploratory factor analysis to search for relationships among the variables; and plotting the data into a two-dimensional principal-components space—all of these are procedures that can help the researcher to understand the structure of the data. In addition, these procedures provide information for two preliminary aspects of the data analysis that should be performed before the data are clustered: (1) transformation of the variables into a common metric and (2) deletion of any outliers. Transformation into a common metric is necessary so that variables with larger variances do not necessarily receive more weight in the calculation of the similarity measures. The deletion of outliers is important because many clustering methods seem quite sensitive to these unusual data points. Other exploratory data-analytic procedures that can be of use prior to any clustering are discussed by Tversky (1977).

The third major step in performing cluster analysis is to select the clustering method and the similarity measure. No clustering method or similarity measure clearly has become preferred. The Monte Carlo research has suggested that Ward's method with Euclidean distance and average linkage with correlation work reasonably well. A researcher might find it beneficial to use both of these methods along with whatever other method(s) he or she prefers. If the various clustering methods do not yield the same solution, the researcher should not be surprised. With real data for which the clusters are not obvious, disagreement among cluster methods should be expected. However, if the different clustering methods do agree on the solution, this would suggest that the solution does have some generality. In addition to average linkage and Ward's method, the other technique to be recommended is the k-means iterative-partitioning method, in which the researcher starts with an *a priori* estimate of the classificatory structure of the data.

The fourth step concerns the choice of the computer program. Usually, this choice is limited by what programs are already available at the user's computer center. Unfortunately, the major statistical packages have a rather limited selection of clustering procedures. For instance, both SAS and BMDP contain a hierarchical agglomerative clustering method. However, SAS does not clearly define what method it embodies (the method is complete linkage); and to effectively use the BMDP program for hierarchical clustering, the researcher must transpose the data set so that the subjects are columns and the variables occur across the rows. The most versatile program for cluster analysis is CLUSTAN. This program contains the widest range of clustering methods and similarity measures. Its manual is reasonably clearly written and well documented. Nonetheless, even CLUSTAN does not include all of the methods one might want to use (for example, neither inverse factor analysis nor Lorr's method is contained in CLUSTAN).

The fifth significant step in the clustering process is to estimate how many clusters exist in a data set. Most computer programs that contain cluster-analytic methods provide no indication of how to solve this problem. For instance, if the researcher decides to use average linkage on 200 patients, the computer program can list solutions of anywhere from 200 clusters to 1 cluster. Or if the researcher decides to use an iterative partitioning method, most computer programs require the researcher to specify *a priori* the expected number of clusters. Making a decision about the number of clusters in a data set is frankly one of the most difficult problems in the research process. Very few studies have attempted to explore this issue in a systematic manner (Mojena, 1977; Edelbrock & Reed, 1981). Everitt (1979) discussed some of the approaches that have been suggested to resolve this problem.

Crucial to the performance of any satisfactory cluster-analytic study is the sixth step of validating a cluster solution. The particular procedures that are invoked to validate a cluster solution can vary, depending on the purpose for which the researcher wants a classification or on the theoretical assumption about classification that the researcher is willing to make. For instance, Golden and Meehl (1979) suggested a series of "consistency tests" that provide an indication of how well a cluster solution meets their theoretical assumptions. In applied uses of cluster analysis in the study of psychopathology, a number of researchers have subdivided their data sets and have checked to see if cluster solutions were actually replicated across subsets of the data. Another appropriate, but less often used, validation step is to have a set of independent variables that are not used in forming the clusters and then to test for significant differences among the clusters on these variables. Paykel (1972b) and Finney and Moos (1979) have provided particularly good examples of this validation step as applied to the classifications of depression and alcoholism. The importance of the validation step cannot be overemphasized. Cluster analysis, almost invariably, finds a cluster solution to a data set even if the data are simply random numbers. The researcher must be careful to discriminate between a cluster solution that has been "forced" on the data and a solution in which the natural groups within the data have been "discovered." Dubes and Jain (1979) provided a more extensive discussion of the issues associated with validation and reviewed statistics and procedures that have been recommended for testing cluster solutions.

Theory Formulation and Cluster Analysis

The availability and the current popularity of cluster-analytic methods has an inherent danger that should not be overlooked. Researchers find it relatively easy, if they have access to a large multivariate data set, to submit their data to a computer program for cluster analysis and then to publish the solution. Computer programmers have a remark that characterizes such behavior: "Garbage in, garbage out."

As the reader should be aware by this point, cluster analysis methods are not magical statistical tools that can discover the underlying structure of any data set. Nonetheless, in our contemporary scientific culture, which requires increasing knowledge about and dependency on computer systems, there exists a tendency to reify any solution that a computer program generates. After all, the edict is "The computer never makes a mistake." In addition, clinical researchers are apt to make a plausible interpretation of a cluster solution even though the solution makes no sense. Overinterpretation is a clinical skill. The apparent meaningfulness of a cluster solution to clinicians provides no insurance for the adequacy of a solution.

Not only does there exist a danger of reifying any cluster solution that a computer program generates, but given the fascination of computers and the intrigue of creating cluster solutions, there may be a tendency to cluster every data set one can find. Hence, the Law of the Hammer: "Give a small boy a hammer and suddenly he will discover everything needs pounding."

The dangers of reification, overinterpretation, and overuse seem obvious; nonetheless, there are authors who appear to advocate a position best described as *naive empiricism*. This position was clearly enunciated in the book by Sokol and Sneath (1963) that simulated the sudden explosion of interest in cluster analysis. Essentially, these authors advocated gathering all possible data on as many organisms as possible within the scope of a biological classification, and then using the unbiased, statistical procedures called *cluster analysis* to find the natural classes within these organisms. This particular approach to classification led to the development of a school of thought concerning biological classification called *numerical taxonomy.* Lorr (1982) and Corning and Steffy (1979) have presented a somewhat similar view regarding the classification of psychopathology.

Hill and Evans (1972) gave a strong counterargument to the naive empiricism position. They argued that classification always involves theoretical assumptions. For instance, the theoretical perspective influences the selection of variables for patient assessment. Researchers from one theoretical perspective may insist on using the overt behaviors of the patients that can be rated by an independent observer; other researchers may prefer to use self-report data and statements by patients about their symptoms; other researchers may assess patients using a number of laboratory measures of the metabolites of various neurotransmitters; and so on.

A good example of the effect of adopting an approach to cluster analysis that was guided by theoretical considerations, as opposed to one that was not can be seen in the contrast between the studies of Paykel and the study by Goldstein and Linden. Paykel was concerned with a theoretical issue, the one-depression versus two-depression controversy, and his methodological decisions were guided by this theoretical issue. He used variables in the clustering process that were related to the psychotic/neurotic distinction. A clustering method was chosen that was designed to optimally isolate descriptively homogeneous groups of depressed patients. He tested his cluster results using independent variables concerning the patients' perceptions of their own symptoms, their clinical diagnoses, the drugs they were given, and follow-up information. In contrast, Goldstein and Linden had no particular theoretical orientation to alcoholism; they chose a commonly used, easily administered self-report instrument whose psychometric properties are best described as poor; after forming the clusters, they did not perform any subsequent research to assess the utility of their classification or to study its generality. As noted in an earlier section of this chapter, Paykel's research led to a descriptive classification of depression that does appear to have generality as well as implications for therapy. In contrast, the Goldstein and Linden paper stimulated a series of additional cluster-analytic studies on alcoholism, but little cumulative knowledge has resulted from this series of studies.

Unfortunately, the topic of cluster analysis itself needs a great deal more theoretical development before it is likely to be of major assistance in improving our knowledge about classification. The currently popular methods of cluster analysis, such as average linkage, Ward's method, and *k*-means partitioning, were all developed as heuristic techniques; that is, they were "rules of thumb" that seemed plausible procedures for creating a classification. Very few cluster-analytic methods have been derived from theories about classification. Exceptions do exist. In psychology, examples of clustering methods that are relatively theoretical in their basis are the HITMAX method of Golden and Meehl (1979) and the MAPCLUS technique of Arabie and Carroll (1979). In biology, Jardine and Sibson (1971) developed an approach to clustering from their theoretical view of classification and information theory. Bezdek (1974a,b) has worked on clustering algorithms that were developed from the theoretical perspective of "fuzzy sets." Unfortunately, most of this theoretical development has been relatively recent, has not attracted much interest beyond the theoreticians themselves, and has rarely been used in practical research.

Despite the need for theoretical advancement, the development of interest in cluster analysis—indeed, the whole growth of factor analysis and other multivariate statistical techniques—has had a useful effect on the literature about classification. Before this growth of interest, most discussions about classification were analytical, were typically evaluated in terms of the prominence and the political power of the person proposing the classification, and often degenerated into little more than obsessive thinking and armchair philosophizing. In part, cluster analysis has helped to revitalize the topic of classification by demonstrating that empirical methods can be applied to the problems of classification and by helping to make clear the rather staggering number of problems associated with the classification of psychopathology that have yet to be solved.

SUGGESTED READINGS

Everitt, B. S. *Cluster analysis* (2nd ed.). London: Halstead, 1980.

Although there are many reviews available about cluster analysis, finding a readable introduction for novices is not easy. Everitt's book is the best available. The book was written by a statistician who has a great deal of practical familiarity with research on psychopathology, and it covers most of the standard topics. In its paperback version, it is a useful investment for anyone who is interested in a good introduction to the methods, problems, and literature of cluster analysis.

Another valuable introduction is by Bailey (1974). This chapter was written for social scientists and is particularly good in its discussion of hierarchical agglomerative methods and their variants. Bailey's discussion gives a better sense than do most other introductory reviews of the range of heuristics that is associated with the generic title *cluster analysis.*

Cormack, R. M. A review of classification. *Journal of the Royal Statistical Society* (series A), 1971, *134,* 321–367.

The discussion by Cormack is decidedly more technical and difficult to read than the introductions by Bailey (1974) or Everitt (1980). However, Cormack's comments and skeptical analysis of clustering are adroit and as applicable today as they were more than 10 years ago, despite the growing literature on this topic. Cormack particularly emphasized the heuristic nature of clustering and pointed out the statistical complexities that most methods (and users) implicitly ignore.

Two other reasonably technical but useful resources about cluster analysis are the books by Anderberg (1973) and by Sneath and Sokol (1973). Anderberg emphasized the computer-programming aspects of cluster analysis and included a number of computer programs in FORTRAN to implement the techniques he discussed. The strongest part of the Anderberg book is his very clear discussion of iterative partitioning methods. The book by Sneath and Sokol (1973) represents a completely rewritten version of their classic (Sokol & Sneath, 1963). It was the 1963 version that was largely responsible for stimulating the initial "explosion" of interest in cluster analysis (see Blashfield & Aldenderfer, 1978). The 1973 book contains two chapters exclusively devoted to the methods of cluster analysis. Particularly good is its discussion of the many similarity coefficients that have been proposed for clustering research. The main difficulty with this book was that it was written for persons interested in biological classification, and it uses much of the jargon associated with this area of classification.

Wishart, D. *CLUSTAN* (release 2 of edition CLUSTAN 1C). St. Andrews, Scotland: Program Library Unit of Edinburgh University, 1978.

To perform cluster analysis on a data set requires the use of a computer program. Although general computer statistical packages (e.g., BMDP and SAS) are beginning to implement some cluster-analytic procedures, the range and power of these procedures are generally poor. The most versatile of the available clustering programs is a package devoted exclusively to cluster-analytic methods called *CLUSTAN*. This package contains 38 similarity measures, 8 hierarchical agglomerative methods, 2 hierarchical divisive methods, a complex iterative partitioning procedure, about a dozen miscellaneous clustering procedures, and a set of graphic techniques for displaying the output of these methods.

Simpson, G. G. *Principles of animal taxonomy*. New York: Columbia University Press, 1961.

This chapter, with its review of the cluster-analytic literature and the application of clustering to psychopathology, ends with a lament about the lack of theoretical development concerning both the methods of clustering and the classification of mental disorders. Interest in the theory of classification has been distinctly more elaborate in biology. In fact, biologists have a specific term for the theoretical study of classification: *taxonomy*. In contrast, biologists reserve the term *classification* for the activity of forming categories or classes of entities; and they use the word *identification* to refer to the activity of assigning a new entity to an already-existing set of categories. Thus, cluster analysis is a classification procedure, and discriminant analysis is an identification procedure. The standard psychiatric and medical name for what biologists call *identification* is *diagnosis*.

Simpson's book is the classic primer on the taxonomy of biological classification. The book introduced the terminology necessary for understanding the structure of biological classification. Simpson tied the principles of classification to evolutionary theory and to the broader understanding of relations among living organisms. The discussion of classification in this book is lucid, even to someone who knows little about biology, and still serves as an excellent introduction to taxonomy.

Since the publication of Simpson's book, three schools of thought have developed about biological classification.

The first of these schools is called *numerical taxonomy*, and its chief publications are the two books discussed in this chapter by Sokol and Sneath (1963; Sneath & Sokol, 1973). This school of thought has emphasized the use of cluster analysis in order to create classificatory categories in which different entities in the same category are morphologically similar.

The second approach to biological classification is known as *cladistics*. It insists that the hierarchical organization among categories in biological classification should represent the evolutionary histories of the categories. Cladists lump species together that appear morphologically dissimilar if the evolutionary history of these species shows them to have branched from a common ancestor.

Finally, the third school of thought about biological classification is known as *evolutionary systematics*. This is the traditional approach to classification that was championed by Simpson. This approach emphasizes that a major goal of biological classification is the organization and storing of information about living organisms (Mayr, 1981). Thus, this approach permits subjective compromises between morphological and evolutionary principles in the organization of an official classification.

Most contemporary classifiers in biology identify themselves as associated with the evolutionary systematics school; however, the most exciting theoretical developments in biological classification during the last decade have been stimulated by the cladists. The numerical taxonomists represent a small school of thought whose influence is waning (Nelson & Platnick, 1981).

Epilogue

When I was in graduate school, I trained in a number of settings. On one particular inpatient unit, I became involved in the family therapy of a woman who was admitted after a suicide attempt. This woman had a history of delinquent behaviors, all minor, as an adolescent. She was bright, and despite her avoidance of school, she had finished high school and entered college. Being attractive, she met a number of men and eventually married. Her husband was an energetic young businessman who entered politics and was elected to a surprisingly important post for his youth. They had one child. When in her late 20s, the patient was brought to our inpatient unit after a suicide attempt in which she had ingested five Valium tablets (not a lethal dose). She admitted later that she had taken only five because she had not wanted to die but was curious about how her husband and family would react.

Among the staff, this woman became known as the "China doll" because, like a China doll, her "exterior" behavior seemed appropriate, but inside she seemed to be "hollow." More specifically, this patient often reported that she was depressed. But her nonverbal behavior was inconsistent with depression. Also, she would report the symptom of depression only when talking to the doctors. Among the patients and the nursing staff, she appeared to be cheerful. Many on the unit expressed difficulty in being able to emphathize with any of her self-reported emotional states. Other patients commented that she would ask them how she was supposed to feel in various situations.

She was in the hospital for three months, during which time a number of incidents occurred. In general, these incidents involved rather pathetic suicide attempts. Once, she decided to kill herself and held her breath while in the bathtub. However, she said that she had not drowned because she would not open her mouth. Another time, she ran from a group of patients who were attending a museum and rolled down a flight of stairs, hoping that would break her neck. Also, she ate three newspaper pages and drank a cup of apple juice because she had been told by a psychotic patient that these in combination would create a toxic substance. In no instance did any of the suicide attempts require medical treatment.

Therapy with this woman was difficult. Her individual therapist was a social worker, and much of the psychotherapy time was absorbed by the patient's complaint that she was being seen by an insignificant social worker rather than the chief psychiatrist. At one point, the patient was started on an antipsychotic medication. Because attaining affective contact was so difficult with this woman and because her reasons for her behavior seemed strange, the unit psychiatrist was entertaining the hypothesis that she was schizophrenic. Later, a new psychiatrist joined the unit and became the administrator for her case. He did not believe that she was schizophrenic, and he discontinued the medication. Initially, the patient had been upset about taking the drug because she was afraid that she was losing her mind, and that the use of medication only seemed to confirm those fears. On the other hand, she appeared to enjoy standing in line for medication with the other patients and often compared notes with them about the side effects of the drugs.

The family therapy was also complicated. The husband seemed concerned about his wife; was quite angry at his father-in-law, who had been telling the patient to divorce the husband; and resented the effect that his wife's hospitalization could have on his public career. The new psychiatrist used a confrontational style in family therapy, and the husband interpreted the psychiatrist as siding with his father-in-law. The husband demanded that the patient be released, as he felt the psychiatrist was trying to turn his wife against him. The psychiatrist refused. The husband used his political contacts to pressure the medical director of the hospital. The medical director told the new psychiatrist to release the patient unless a strong case could be made that the patient was of danger to herself. The staff discussed the issue and were split over the necessity of holding the patient. No one, however, wanted to release her under the administrative and political pressure. The psychiatrist decided that he must acquiesce to the dictates of the medical director, and the patient was given a choice of being released or staying. The patient was pleased about being given the choice and left immediately.

That night, after putting her son to bed, the patient picked up a revolver she had just purchased and killed herself.

We were devastated. The shock of this event left us angry, scared, depressed, and guilty. We had not predicted that this would occur. We had not prevented this woman's death. We had not understood her.

The misery associated with psychopathology is almost overwhelming. Most people prefer not to deal with psychiatric patients. These patients scare us. Often, they remind us of our own misery and the serious personal problems we prefer to ignore. Rationally, we do not generally understand the patients whom we describe as having a mental disorder. Emotionally, we often understand these people all too well.

To work as a clinician in the mental health field requires hope. As Lara Jefferson noted in her first account of being in a mental hospital, people with psychopathology sail "an unmarked sea" in which hope is a scarce commodity. The promise of being able to understand psychopathology gives us hope. For clinicians, the search through the mysterious phenomenological worlds of our patients is often a scary—but also sometimes a rewarding—experience. Like our patients, we do not have maps. We do not have a reliable compass. We rarely have more than our previous experiences from exploring the worlds of other patients. Yet, we daily enter into these worlds, trying to find the dragon, trying to create useful maps, trying to understand.

The task of classification involves a paradox. To classify people is inherently to sacrifice some of their individuality. Each patient is a human being, and the individuality of each person is essential to his or her humanity. On the other hand, classification is essential if there is to be a science of psychopathology. It is the science of psychopathology that will give us a map and a compass to use when we venture into the worlds of our patients. Classificatory concepts will help us to develop the ideas that we can use to understand patients. A classification is necessary for the development of a rational system, which can help our patients slay the irrational dragons in their worlds.

This book has discussed two contemporary schools of thought about classification. The book has not solved the issues of psychiatric classification. It has not explained how to understand psychopathology. It has not told mental health professionals how to treat patients more effectively. Nonetheless, classification is a fundamental topic, and improvements in the classification of mental disorders are necessary if we are to make many of the other gains that are needed in the field. My hope is that this book has suggested fruitful areas for research and study that can stimulate future improvements.

Currently, our knowledge about psychopathology is small. We do know more than was known when Kraepelin was a professor. But relative to the changes in other sciences, the amount of increased knowledge about mental disorders that has accumulated over the last century is meager. Our classificatory systems have shown relatively little change. The challenge is to search for the changes in classification that will provide a better foundation from which our knowledge about psychopathology can grow.

Appendix I

Classifications of Psychopathology

This appendix contains a listing of some categories from classification systems discussed in Chapter 1. The first is taken from the Table of Contents of Volume 2 of Kraepelin's sixth edition of his textbook. The second classification is the condensed form of the Standard Classified Nomenclature of Disease, which was the official American psychiatric classification before the DSM-I. Because of the historical importance of these classifications, all categories in both systems are listed.

The remaining four classifications presented in this appendix are the DSM-I, the DSM-II, the DSM-III, and the ICD-9. Because complete listings of these systems are readily available in the contemporary literature, complete listings of these systems are not provided here. Instead, only the categories subsumed under the general heading of "functional psychoses" are itemized.

At the appendix end is a brief comparison of the DSM-II, the DSM-III, the ICD-9, and the ICD-9-CM in an outline form.

"FORMS OF MENTAL DISEASES" FROM KRAEPELIN'S *TEXTBOOK OF PSYCHIATRY* (6TH ED., VOL. 2)*

I. Infection Psychoses
  A. Delirium
  B. Infection Delirium
  C. Post-infection Psychoses

* The terms in Kraepelin's classification were translated by A. Ross Diefendorf. Because Diefendorf did not adhere strictly to the structure presented by Kraepelin in his table of contents, the present author (RKB) has modified Diefendorf's organization.

II. Exhaustion Psychoses
  A. Collapse Delirium
  B. Acute Confusional Psychosis
  C. Chronic Nervous Exhaustion
III. Intoxication Psychoses
    1. Acute Intoxications
    2. Chronic Intoxications
  A. Alcoholism
  B. Morphinism
  C. Cocainism
IV. Thyroigenous Psychoses
  A. Myxedematous Psychosis
  B. Cretinism
V. Dementia Praecox
  Forms
    Hebephrenic
    Catatonic
    Paranoid
VI. Dementia Paralytica
  Forms
    Depressive
    Expansive
    Agitated
    Demented
VII. The Organic Psychoses
VIII. Involutional Psychoses
  A. Melancholia
  B. Presenile Delusional Psychosis
  C. Senile Dementia
IX. Manic-Depressive Psychosis
  A. Manic States
  B. Depressed States
  C. Mixed States
X. Paranoia
XI. The Psychogenic Neuroses
  A. Epileptic Psychosis
  B. Hysterical Psychosis
  C. War Neurosis (Dread Neurosis)
XII. The Psychopathic Conditions
  A. Constitutional Despondency
  B. Compulsive Psychosis
  C. Impulsive Psychosis
  D. Contrary Sexual Instincts
XIII. Defective Moral Development

A. Imbecility
B. Idiocy

STANDARD CLASSIFIED NOMENCLATURE OF DISEASES (1933)

1. Psychoses with syphilitic meningo-encephalitis (general paresis)
2. Psychoses with other forms of syphilis of the central nervous system
   Meningo-vascular type (cerebral syphilis)
   With intracranial gumma
   Other types (to be specified)
3. Psychoses with epidemic encephalitis
4. Psychoses with other infectious diseases
   With tuberculous meningitis
   With meningitis (unspecified)
   With acute chorea (Sydenham's)
   With other infectious disease (to be specified)
   Post-infectious psychoses (infection to be specified)
5. Alcoholic psychoses
   Pathological intoxication
   Delirium tremens
   Korsakow's psychosis
   Acute hallucinosis
   Other types (to be specified)
6. Psychoses due to drugs or other exogenous poisons
   Due to metals (to be specified)
   Due to gases (to be specified)
   Due to opium and derivatives
   Due to other drugs (to be specified)
7. Traumatic psychoses
   Traumatic delirum
   Post-traumatic personality disorders
   Post-traumatic mental deterioration
   Other types (to be specified)
8. Psychoses with cerebral arteriosclerosis
9. Psychoses with other disturbances of circulation
   With cerebral embolism
   With cardio-renal disease
   Other types (to be specified)
10. Psychoses with convulsive disorders (epilepsy)
    Epileptic deterioration
    Epileptic clouded states
    Other epileptic types
11. Senile psychoses
    Simple deterioration

Presbyophrenic type
Delirious and confused types
Depressed and agitated types
Paranoid types

12. Involutional psychoses
    Melancholia
    Paranoid types
    Other types (to be specified)
13. Psychoses due to other metabolic, etc. diseases
    With diseases of the endocrine glands (to be specified)
    Exhaustion delirium
    Alzheimer's disease
    With pellagra
    Other somatic diseases (to be specified)
14. Psychoses due to new growth
    With intracranial neoplasms
    With other neoplasms (to be specified)
15. Psychoses associated with organic changes of the nervous system
    With multiple sclerosis
    With paralysis agitans
    With Huntington's chorea
    With other brain or nervous diseases (to be specified)
16. Psychoneuroses
    Hysteria (anxiety hysteria, conversion hysteria and subgroups)
    Psychasthenia or compulsive states (and subgroups)
    Neurasthenia
    Hypochondriasis
    Reactive depression (simple situational reaction, others)
    Anxiety state
    Mixed psychoneurosis
17. Manic-depressive psychoses
    Manic-type
    Depressive type
    Circular type
    Mixed type
    Perplexed type
    Stuporous type
    Other types
18. Dementia praecox (schizophrenia)
    Simple type
    Hebephrenic type
    Catatonic type
    Paranoid type
    Other type
19. Paranoia and paranoid conditions

Paranoia
Paranoid conditions
20. Psychoses with psychopathic personality
21. Psychoses with mental deficiency
22. Undiagnosed psychoses
23. Without psychoses
Epilepsy
Alcoholism
Drug addiction
Mental deficiency
Disorders of personality due to epidemic encephalitis
Psychopathic personality
With pathological sexuality
With pathological emotionality
With asocial or amoral trends
Mixed types
24. Primary behavior disorders
Simple adult maladjustment
Primary behavior disorders in children
Habit disturbance
Conduct disturbance
Neurotic traits

CATEGORIES OF FUNCTIONAL PSYCHOSIS IN DSM-I

II. Disorders of psychogenic origin or without clearly defined physical cause or structural change in the brain.
A. Psychotic Disorders
1. Involutional psychotic reaction
2. Affective reactions
a. Manic depressive reaction
i. Manic depressive reaction, manic type
ii. Manic depressive reaction, depressive type
iii. Manic depressive reaction, other
b. Psychotic depressive reaction
3. Schizophrenic reactions
a. Schizophrenic reaction, simple type
b. Schizophrenic reaction, hebephenic type
c. Schizophrenic reaction, catatonic type
d. Schizophrenic reaction, paranoid type
e. Schizophrenic reaction, acute undifferentiated type
f. Schizophrenic reaction, chronic undifferentiated type
g. Schizophrenic reaction, schizo-affective type
h. Schizophrenic reaction, childhood type

i. Schizophrenic reaction, residual type
4. Paranoid reactions
   a. Paranoia
   b. Paranoid state
5. Psychotic reaction without clearly defined structural change, other than above

PSYCHOTIC CATEGORIES IN DSM-II

III. Psychoses not attributable to physical conditions listed previously
   A. Schizophrenia
      1. Schizophrenia, simple type (295.00)
      2. Schizophrenia, hebephrenic type (295.10)
      3. Schizophrenia, catatonic type (295.20)
         a. Schizophrenia, catatonic type, excited (295.23)
         b. Schizophrenia, catatonic type, withdrawn (295.24)
      4. Schizophrenia, paranoid type (295.30)
      5. Acute schizophrenic episode (295.40)
      6. Schizophrenia, latent type (295.50)
      7. Schizophrenia, residual type (295.60)
      8. Schizophrenia, schizo-affective type (295.70)
         a. Schizophrenia, schizo-affective type, excited (295.73)
         b. Schiozphrenia, schizo-affective type, depressed (295.74)
      9. Schizophrenia, childhood type (295.80)
      10. Schizophrenia, chronic undifferentiated type (295.90)
      11. Schizophrenia, other types (295.99)
   B. Major Affective Disorders
      1. Involutional melancholia (296.00)
      2. Manic-depressive illness, manic type (296.10)
      3. Manic-depressive illness, depressed type (296.20)
      4. Manic-depressive illness, circular type (296.30)
         a. Manic-depressive illness, circular type, manic (296.33)
         b. Manic-depressive illness, circular type, depressed (296.34)
      5. Other major affective disorder (296.80)
   C. Paranoid States
      1. Paranoia (297.00)
      2. Involutional paranoid state (297.10)
      3. Other paranoid state (297.90)
   D. Other Psychoses
      1. Psychotic depressive reaction (298.00)

CATEGORIES OF FUNCTIONAL PSYCHOSIS IN DSM-III

SCHIZOPHRENIC DISORDERS

Code in fifth digit: 1 = subchronic, 2 = chronic, 3 = subchronic with acute exacerbation, 4 = chronic with acute exacerbation, 5 = in remission, 0 = unspecified.

Schizophrenia
- 295.1x Disorganized
- 295.2x Catatonic
- 295.3x Paranoid
- 295.9x Undifferentiated
- 295.6x Residual

PARANOID DISORDERS
- 297.10 Paranoia
- 297.30 Shared paranoid disorder
- 298.30 Acute paranoid disorder
- 297.90 Atypical paranoid disorder

PSYCHOTIC DISORDERS NOT ELSEWHERE CLASSIFIED
- 295.40 Schizophreniform disorder
- 298.80 Brief reactive psychosis
- 295.70 Schizoaffective disorder
- 298.90 Atypical psychosis

AFFECTIVE DISORDERS

Major affective disorders

Code major depressive episode in fifth digit: 6 = in remission, 4 = with psychotic features, 3 = with melancholia, 2 = without melancholia, 0 = unspecified.

Code manic episode in fifth digit: 6 = in remission, 4 = with psychotic features, 2 = without psychotic features, 0 = unspecified.

Bipolar disorder
- 296.6x Mixed
- 296.4x Manic
- 296.5x Depressed

Major depression
- 296.2x Single episode
- 296.3x Recurrent

Other specific affective disorders
- 301.13 Cyclothymic disorder
- 300.40 Dysthymic disorder (or Depressive neurosis)

Atypical affective disorders
- 296.70 Atypical bipolar disorder
- 296.82 Atypical depression

CATEGORIES OF FUNCTIONAL PSYCHOSIS IN ICD-9

295 Schizophrenic psychoses
- 295.0 Simple type
- 295.1 Hebephrenic type
- 295.2 Catatonic type
- 295.3 Paranoid type
- 295.4 Acute schizophrenic episode

295.5 Latent schizophrenia
295.6 Residual schizophrenia
295.7 Schizoaffective type
295.8 Other
295.9 Unspecified
296 Affective psychoses
296.0 Manic-depressive psychosis, manic type
296.1 Manic-depressive psychosis, depressed type
296.2 Manic-depressive psychosis, circular type but currently manic
296.3 Manic-depressive psychosis, circular type but currently depressed
296.4 Manic-depressive psychosis, circular type, mixed
296.5 Manic-depressive psychosis, circular type, current condition not specified
296.6 Manic-depressive psychosis, other and unspecified
296.8 Other
296.9 Unspecified
297 Paranoid states
297.0 Paranoid state, simple
297.1 Paranoia
297.2 Paraphrenia
297.3 Induced psychosis
297.8 Other
297.9 Unspecified
298 Other nonorganic psychoses
298.0 Depressive type
298.1 Excitative type
298.2 Reactive confusion
298.3 Acute paranoid reaction
298.4 Psychogenic paranoid psychosis
298.8 Other and unspecified reactive psychosis
298.9 Unspecified psychosis

BRIEF OUTLINE COMPARISON OF DSM-II, DSM-III, ICD-9, AND ICD-9-CM

I. Description
A. The DSM-II (an abbreviated title for the *Diagnostic and Statistical Manual of Mental Disorders*, second edition) is the official classification system adopted by the American Psychiatric Association in 1968. Except for a few minor differences, the DSM-II is equivalent to the eighth edition of the *International Classification of Diseases* (ICD-8).
B. The DSM-III is the new classification system adopted by the

American Psychiatric Association in 1980. The DSM-III differs from the DSM-II in that the DSM-III uses (i) diagnostic criteria that are structured like a "Chinese menu"; (ii) a multiaxial system with five axes; (iii) a great deal of descriptive information about categories, including prevalence, sex ratio, associated features, course, and complications; and (iv) an expanded number of diagnostic categories.

C. The ICD-9 (an abbreviated title for the *International Classification of Diseases*, ninth edition) is the official classification system adopted by the World Health Organization in 1978.

D. The ICD-9-CM (the initials *CM* stand for "clinical modification") is the American version of the ICD-9. The American version has many more categories than the ICD-9. Most of the additional categories are recognized in the DSM-III but not in the ICD-9. For administrative and statistical purposes, the ICD-9-CM is an amalgamation of the DSM-III and ICD-9.

II. Categories

A. The DSM-II has a total of 182 categories, of which 158 are specific categories that have a unique code (e.g., 300.70 Hypochondriacal neurosis). The remaining categories are more general diagnostic terms (e.g., "Neuroses").

B. The DSM-III has 265 categories, of which 228 have a unique code and are specific categories.

C. The ICD-9 contains 209 categories, of which 180 are specific.

D. The ICD-9-CM has 344 categories, of which 315 are specific.

III. Organization

A. In terms of the hierarchical organization of categories, the DSM-III is distinctly different from the other systems. The ICD-9 and the ICD-9-CM have the same organization, with the ICD-9-CM simply being more specific. The DSM-II is reasonably similar to the ICD-9; the major difference is that the ICD-9 has a larger number of categories for childhood disorders.

B. *Affective Disorders.* This is an area of major difference between the DSM-III and the ICD-9. The ICD-9 maintains the distinction among manic-depressive psychosis, psychotic depression, and neurotic depression. In addition, the ICD-9 adds another category of depression: "Depressive disorders, not elsewhere classified." In contrast, the DSM-III organizes affective disorders into major affective disorders, affective personality disorders, and atypical affective disorders. Within major affective disorders, the DSM-III emphasizes the bipolar/unipolar distinction.

C. *Childhood Disorders.* All three recent classifications increased the numbers of categories concerned with childhood disorders. The DSM-II had only 11 specific categories concerned with childhood disorders. In contrast, the ICD-9 has 27, the ICD-9-CM has 39, and the DSM-III has 45. The DSM-III differs from the ICD version in that included under this general category are any specific disorders that usually start in childhood/adolescence. Thus, an adult with enuresis would be included under childhood disorders in the DSM-III, but not in the ICD-9.

D. Additional major areas of differences between the DSM-III and other systems concern the organic brain syndromes, the neuroses, and the psychosexual disorders.

IV. Coding

A. The ICD-9 uses a four-digit coding system. Each category in the ICD-9 has a unique code. The codes range from
   290.0 to 299.9—for the psychoses
   300.0 to 319.0—for nonpsychotic disorders

B. The ICD-9 code identifies each category uniquely, as well as the major categories of which it is a subset. For instance, the code 301.2 refers to "Schizoid personality disorder," which is a subset of the category "Personality disorders (301)."

C. The ICD-9-CM uses the ICD-9 four-digit codes. In addition, there are about 150 categories with five-digit codes. "Schizotypal personality" has the code 301.22; hence, it is a subset of "Schizoid personality disorder (301.2)."

D. The DSM-III uses the ICD-9-CM codes. The only exceptions are the "V codes," which refer to "nonpathological" reasons for seeing a mental health professional. For example, V61.10 refers to "Marital problems."

E. The DSM-III and the ICD-9 often assign different names to the same code. The code 313.0 is given the name "Disturbance of emotions specific to childhood and adolescence, with anxiety and fearfulness" by the ICD-9. In the DSM-III, the same code has the name "Overanxious Disorder."

V. Category definitions

A. The ICD-9-CM contains no definitions. Definitions of the ICD-9-CM diagnoses can be found in either the DSM-III or the ICD-9.

B. The ICD-9, like the DSM-II, contains prose definitions of categories.

C. The DSM-III definitions use diagnostic criteria that are structured like "Chinese menus." That is, the definition of "Overanxious Disorder" contains four major subdivisions. Under Subdivision 1, the child/adolescent must have four of seven symptoms. In addition, the patient must fit all the criteria in the remaining subdivisions.

D. The definitions in the DSM-III are more detailed than those in the ICD-9. For instance, the DSM criteria for "Overanxious Disorder" contain 11 sentences and 120 words. In the ICD-9 definition, there are 2 sentences and 27 words.

E. The DSM-III, unlike the other classifications, also contains sections that state with what other disorders a given diagnostic category is likely to be confused and how a differential diagnosis can be made.

Appendix II

Cluster-Analytic Studies of Psychopathology

Studies in which cluster analysis has been used to create new classification systems have shown a steady growth during the past two decades. This appendix contains a list of all known studies about psychopathology that have used clustering. The nearly 250 studies are organized according to the major classes of psychopathology reflected in their sampling of patients: alcoholism, childhood disorders, functional psychosis, general psychopathology, mental retardation, offenders, organic brain syndromes, personality disorders, schizophrenia, substance abuse, suicide, and other miscellaneous disorders.

The author has also worked on a comparable bibliography of factor-analytic studies of psychopathology. The latter bibliography is much longer (600 + references) and is less likely to be comprehensive. Interested readers can obtain a copy by writing to me.

ALCOHOLISM

Bean, K. L., & Karasievich, G. O. Psychological test results at three stages of inpatient alcoholism treatment. *Journal of Studies on Alcohol*, 1975, *36*, 838–852.

Costello, R. M., Lawlis, G. F., Manders, K. R., & Celistino, J. F. Empirical derivation of a partial personality typology of alcoholics. *Journal of Studies on Alcohol*, 1978, *39*, 1258–1266.

Donovan, D. M., & Marlatt, G. A. Personality subtypes among driving-while-intoxicated offenders: Relationship to drinking behavior and driving risk. *Journal of Consulting and Clinical Psychology*, 1982, *50*, 241–249.

Donovan, D. M., Chaney, E. F., & O'Leary, M. R. Alcoholic MMPI subtypes:

Relationship to drinking styles. *Journal of Nervous and Mental Disease,* 1978, *166,* 533–561.

Esbaugh, D. M., Hoyt, C., & Tosi, D. J. Some personality patterns and dimensions of male alcoholics: A multivariate description. *Journal of Personality Assessment,* 1978, *42,* 409–417.

Evenson, R. C., Altman, H., Sletten, I. W., & Knowles, R. R. Factors in the description and grouping of alcoholics. *American Journal of Psychiatry,* 1973, *130,* 49–54.

Finney, J. W., & Moos, R. H. Treatment and outcome for empirical subtypes of alcoholic patients. *Journal of Consulting and Clinical Psychology,* 1979, *47,* 25–38.

Glen, A., Royer, F. L., & Custer, R. A study of MMPI profile types of hospitalized alcoholic veterans. *Newsletter for Research on Mental Health and Behavioral Sciences,* 1973, *25,* 53–56.

Goldstein, S. G., & Linden, J. D. Multivariate classification of alcoholics by MMPI. *Journal of Abnormal Psychology,* 1969, *74,* 661–669.

Hart, L. S. Differentiation of rehabilitation needs and outcomes in alcoholics. *Dissertation Abstracts International,* 1977, *38(1-B),* 361.

Hart, L.S., & Stueland, D. An application of the multidimensional model of alcoholism: Differentiation of alcoholics by mode analysis. *Journal of Studies on Alcohol,* 1979, *40,* 283–290.

Hart, L. S., & Stueland, D. Relationship of sociodemographic and drinking variables to differentiated subgroups of alcoholics. *Community Mental Health Journal,* 1979, *15,* 47–57.

Hetler, J. H. A critical examination of the adequacy of a typological analysis provided by several clustering techniques. *Dissertation Abstracts International,* 1977, *37(12-B),* 6327.

Hill, H. E., Haertzen, C. A., & Davis, H. An MMPI factor analytic study of alcoholics, narcotic addicts and criminals. *Quarterly Journal of Studies on Alcohol,* 1962, *23,* 411–431.

Lawlis, G. F., & Rubin, S. E. 16-PF study of personality patterns in alcoholics. *Quarterly Journal of Studies of Alcohol,* 1971, *32,* 318–327.

Mulligan, M. J., Steer, R. A., & Fine, E. W. Psychiatric disturbances in drunk driving offenders referred for treatment of alcoholism. *Alcoholism: Clinical and experimental research,* Vol. 2, No. 2., 1978.

Nerviano, V. J. Common personality patterns among alcoholic males: A multivariate study. *Journal of Consulting and Clinical Psychology,* 1976, *44,* 104–110.

Nerviano, V. J. Personality patterns of alcoholics revisited: Delineation against the MMPI and clinical implications. *International Journal of the Addictions,* 1981, *16,* 723–730.

Nerviano, V. J., & Gross, W. F. A multivariate delineation of two alcoholic profile types on the 16PF. *Journal of Clinical Psychology,* 1973, *29,* 371–374.

Nerviano, V. J., McCarty, D., & McCarty, S. M. MMPI profile patterns of men alcoholics in two contrasting settings: Results and implications. *Journal of Studies on Alcohol,* 1980, *41,* 1143–1152.

Partington, J. T., & Johnson, F. G. Personality types among alcoholics. *Quarterly Journal of Studies on Alcohol,* 1969, *30,* 21–34.

Repogle, W. H. Multivariate approach to profiling alcoholic typologies. *Multivariate Experimental Clinical Research,* 1977, *3,* 157–164.

Skinner, H. A., Jackson, D. N., & Hoffman, H. Alcoholic personality types: Identification and correlates. *Journal of Abnormal Psychology,* 1974, *83,* 658–666.

Skinner, H. A., Reed, P. L., & Jackson, D. N. Toward the objective diagnosis of psychopathology: Generalizability of modal personality profiles. *Journal of Consulting and Clinical Psychology,* 1976, *44,* 111–117.

Steer, R. A., Fine, E. W., & Scoles, P. E. Classification of men arrested for driving while intoxicated and treatment implications: A cluster analytic study. *Journal of Studies on Alcohol,* 1979, *40,* 222–229.

Stein, K. B., Rozynko, V., & Pugh, L. A. The heterogeneity of personality among alcoholics. *British Journal of Social and Clinical Psychology,* 1971, *118,* 253–259.

Svanum, S., & Dallas, C. L. Alcoholic MMPI types and their relationship to patient characteristics, polydrug abuse, and abstinence following treatment. *Journal of Personality Assessment,* 1981, *45,* 278–287.

Warner, J. S. Personality clusters in MMPI protocols of alcoholics. *Dissertation Abstracts International,* 1977, *38*(*5-B*), 2389.

Whitelock, P. R., Overall, J. E., & Patrick, J. H. Personality patterns and alcohol abuse in a state hospital population. *Journal of Abnormal Psychology,* 1971, *78,* 9–16.

Zivich, J. M. Alcoholic subtypes and treatment effectiveness. *Journal of Consulting and Clinical Psychology,* 1981, *49,* 72–80.

CHILDHOOD DISORDERS

Aram, D. M., & Nation, J. E. Patterns of language behavior in children with developmental language disorders. *Journal of Clinical Psychology,* 1975, *12,* 263–269.

Ayres, A. J. Patterns of perceptual-motor dysfunction in children: A factor analytic study. *Perceptual Motor Skills,* 1965, *20,* 335–368.

Ayres, A. J. Deficits in sensory integration in educationally handicapped children. *Journal of Learning Disabilities,* 1969, *2,* 160–168.

Conners, C. K. Stimulant drugs and evoked cortical responses in learning and behavior disorders in children. In W. L. Smith (Ed.), *Drugs, development and cerebral function.* Ft. Lauderdale, Fla.: Charles C Thomas, 1972.

Conners, C. K. Psychological assessment of children with minimal brain dysfunction. *Annals of the New York Academy of Science,* 1973, *205,* 283–300.

Crinella, F. M. Identification of brain dysfunction syndromes in children through profile analysis: Patterns associated with so-called minimal brain dysfunction. *Journal of Abnormal Psychology,* 1973, *82,* 33–45.

Crinella, F. M., & Dreger, R. M. Tentative identification of brain dysfunction syndromes in children through profile analysis. *Journal of Consulting and Clinical Psychology,* 1972, *38,* 251–260.

Cunningham, W. G., & Pooley, R. C. Development of diagnostic/prescriptive system reveals childhood disabling behaviors vary with mismatches in families. *Journal of Instructional Psychology,* 1979, *6,* 29–44.

Curry, J. F., & Thompson, R. J. Patterns of behavioral disturbance in developmentally disabled children: A replicated cluster analysis. *Journal of Pediatric Psychology,* 1982, *7,* 61–73.

Darby, R. O. *Learning disabilities: A multivariate search for subtypes.* Doctoral dissertation, University of Florida, 1978.

DeHorn, A. B., Lachar, D., & Gdowski, C. L. Profile classification strategies for the Personality Inventory for Children. *Journal of Consulting and Clinical Psychology,* 1979, *47,* 874–881.

Doehring, D. G., & Hoshko, I. M. Classification of reading problems by the Q-technique of factor analysis. *Cortex,* 1977, *13,* 281–294.

Doehring, D. G., Hoshko, I. M., & Bryans, A. Statistical classification of children with reading problems. *Journal of Clinical Neuropsychology,* 1979, *1,* 5–16.

Dreger, R. M., Reid, M. P., Lewis, P. M., Overlade, D. C., Rich, T. A., Taffel, C., Miller, K. S., & Flemming, E. L. Behavioral classification project. *Journal of Consulting Psychology,* 1964, *28,* 1–13.

Edelbrock, C., & Achenbach, T. M. A typology of child behavior profile patterns: Distribution and correlates for disturbed children aged 6–16. *Journal of Abnormal Child Psychology,* 1980, *8,* 441–470.

Eisenberg, J. G., Gersten, J. C., Langer, T. S., McCarthy, E. D., & Simcha-Fagan, O. A behavioral classification of welfare children from survey data. *American Journal of Orthopsychiatry,* 1976, *46,* 447–463.

Forness, S. R., Guthrie, D., & Nihira, K. Clusters of observable behavior in high-risk kindergarten children. *Journal of Clinical Psychology,* 1975, *12,* 263–269.

Gay, J. R. *A linear typal analysis of the Children's Behavioral Classification Project.* Unpublished paper, Louisiana State University, 1974.

Gdowski, C. L. Typologies of childhood psychopathology and their relationships with the Personality Inventory for Children. *Dissertation Abstracts International,* 1978, *38(11-B),* 5566.

Jenkins, R. L. Psychiatric syndromes in children and their relation to family background. *American Journal of Orthopsychiatry,* 1966, *26,* 45–457.

Koller, K. M., & Williams, W. T. Early parental deprivation and later behavioral outcome: A cluster analysis study of normal and abnormal groups. *Australian and New Zealand Journal of Psychology,* 1974, *8,* 89–96.

Langer, T. S., Gersten, J. C., Eisenberg, J. G., Greene, E. L., & Herson, J. H.

Children under stress: Family and social factors in the behavior of urban children and adolescents. New York: Columbia University Press, 1979.

Lessing, E. E., Williams, V., & Gil, E. A cluster-analytically derived typology: Feasible alternative to clinical diagnostic classification of children. *Journal of Abnormal Child Psychology,* 1982, *10,* 451–482.

Luick, A., Agranowitz, A., Kirk, S. A., & Busby, R. Profiles of children with severe oral language disorders. *Journal of Speech and Hearing Disorders.* 1982, *47,* 88–92.

Lyon, R., & Watson, B. Empirically derived subgroups of learning disabled readers: Diagnostic characteristics. *Journal of Learning Disabilities,* 1981, *14,* 256–261.

Moore, J. S. An empirical study of four methods of classifying L/LD students in a metropolitan district. *Dissertation Abstracts International,* 1978, *38(10-A),* 6019.

Morris, R., Blashfield, R., & Satz, P. Neuropsychology and cluster analysis: Potentials and problems. *Journal of Clinical Neuropsychology,* 1981, *3,* 79–99.

Naidoo, S. *Specific dyslexia.* New York: Putnam, 1972.

Overall, J. E., & Pfefferbaum, B. An investigation of the consistency of diagnostic concepts among five child and adolescent psychiatrists. *Multivariate Behavioral Research,* 1982, *17,* 435–445.

Petrauskas, R., & Rourke, B. P. Identification of subgroups of retarded readers: A neuropsychological approach. *Journal of Clinical Neuropsychology,* 1979, *1,* 17–37.

Prior, M., Boulton, D., Gajzago, C., & Perry, D. The classification of childhood psychoses by numerical taxonomy. *Journal of Child Psychology and Psychiatry,* 1975, *16,* 321–330.

Prior, M., Perry, D., & Gajzago, C. Kanner's syndrome or early onset psychosis: A taxonomic analysis of 142 cases. *Journal of Autism and Child Schizophrenia,* 1975, *5,* 71–80.

Satz, P., & Morris, R. Learning disability subtypes: A review. In F. J. Pirozzolo & M. C. Wittrock (Eds.), *Neuropsychological and cognitive processes in reading.* New York: Academic Press, 1981.

Smith, D. E. P., & Carrigan, P. M. *The nature of reading disability.* New York: Harcourt, Brace & Company, 1959.

Smith, G. J., & Danielsson, A. An attempt at grouping non-organic mental disorders in childhood and adolescence using process-oriented instruments of description. *Psychological Research Bulletin,* 1977, Monograph #1, 1–28.

Soli, S. D., Nuechterlein, K. H., Garmezy, N., Devine, V. T., & Schaefer, S. M. A classification system for research in childhood psychopathology: I. An empirical approach using factor and cluster analyses and conjunctive decision rules. In *Progress in experimental personality research,* Vol. 10. New York: Academic Press, 1981.

Spivak, G., Swift, M., & Prewitt, J. Syndromes of disturbed classroom behavior: A behavioral diagnostic system for elementary schools. *Journal of Special Education*, 1971, *5*, 269–292.

Watson, B. U., & Goldgar, D. E. *Subtypes of specific reading disability.* Paper presented at the annual meeting of the American Psychological Association, Washington, D.C., Aug. 23–27, 1982.

Wolkind, S. N., & Everitt, B. A cluster analysis of the behavioral items in the pre-school child. *Psychological Medicine*, 1974, *4*, 422–427.

Wren, C. T. Identifying patterns of syntactic disorder in six-year-old children. *British Journal of Psychiatry*, 1982, *140*, 257–262.

DELINQUENCY/DISORDERS OF ADOLESCENCE

Butler, E. W., & Adams, S. W. Typologies of delinquent girls: Some alternative approaches. *Social Forces*, 1966, *44*, 401–407.

Cooper, B., & Gath, D. Psychiatric illness, maladjustment and juvenile delinquency: An ecological study in a London borough. *Psychological Medicine*, 1977, *7*, 465–474.

Kreuchauf, G. A. A typological study on institutionalized juvenile delinquents using hierarchical classification techniques. *Dissertation Abstracts International*, 1978, *39*(*2-B*), 956–957.

Kulik, J. A., Stein, K. B., & Sarbin, T. R. Dimensions and patterns of adolescent antisocial behavior. *Journal of Consulting and Clinical Psychology*, 1968, *32*, 375–382.

Pickett, L. K. MMPI modal profiles in a juvenile delinquent population. *Criminal Justice and Behavior*, 1981, *8*, 39–47.

Stein, K. B., Gough, H. G., & Sarbin, T. R. The dimensionality of the CPI socialization scale and an empirically derived typology among delinquent and non-delinquent boys. *Multivariate Behavioral Research*, 1966, *1*, 192–208.

DEPRESSION

Amodei, C., Bollea, E., & Vella, G. Statistical research as applied to the problem of classifying various states of depression. *Revista Di Psichiatria*, (Italian), 1973, *8*, 348–412.

Andreasen, N. C., & Grove, W. M. The classification of depression: Traditional versus mathematical approaches. *American Journal of Psychiatry*, 1982, *139*, 45–52.

Andreasen, N. C., Grove, W., & Maurer, R. Cluster analysis and the classification of depression. *British Journal of Psychiatry*, 1980, *137*, 256–265.

Ballinger, B. R., Reid, A. H., & Heather, B. B. Cluster analysis of symptoms in elderly demented patients. *British Journal of Psychiatry*, 1982, *140*, 257–262.

Byrne, D. G. Cluster analysis applied to self-reported depressive symptomatology. *Acta Psychiatrica Scandinavica,* 1978, *55,* 1–10.

Byrne, D. G. A note on cluster analysis and depression: Disparities in results produced by the application of different clustering methods. *Australian and New Zealand Journal of Psychology,* 1978, *8,* 89–96.

Demel, I., Gordesch, J., Katschnig, H., Poldinger, W., Scheiber, V., & Sint, P. Multivariate procedures in the statistical analysis of psychic and somatic symptoms in depression. In J. Ruesch & A. S. Schmale (Eds.), *Psychotherapy and psychosomatics.* Basel: Karger, 1973.

Fleiss, J. L. Classification of the depressive disorders by numerical typology. *Journal of Psychiatric Research,* 1972, *9,* 141–153.

Grinker, R. R., & Nunnally, J. C. The phenomena of depression. In M. M. Katz, J. O. Cole, & W. E. Barton (Eds.), *The role and methodology of classification in psychiatry and psychopathology.* Washington: HEW, 1968.

Grinker, R. R., Miller, J., Sabshin, M., Nunn, R., & Nunnally, J. C. *The phenomena of depression.* New York: Paul B. Hoeber, 1961.

Lorr, M., Pokorny, A. D., & Klett, C. J. Three depressive types. *Journal of Clinical Psychology,* 1973, *29,* 290–292.

Matussek, P., Soldner, M., & Nagel, D. Identification of the endogenous depressive syndrome based on the symptoms and the characteristics of the course. *British Journal of Psychiatry,* 1981, *138,* 361–372.

Matussek, P., Luks, O., & Nagel, D. Depression symptom patterns. *Psychological Medicine,* 1982, *12,* 765–773.

Matussek, P., Soldner, M. L., & Nagel, D. Neurotic depression. Results of cluster analysis. *Journal of Nervous and Mental Disease,* 1982, *170,* 588–597.

McLean, P. D., & Hakstian, A. R. Clinical depression: Comparative efficacy of outpatient treatments. *Journal of Consulting and Clinical Psychology,* 1979, *47,* 818–836.

Mezzich, A. C., & Mezzich, J. E. A data-based typology of depressed adolescents. *Journal of Personality Assessment,* 1979, *43,* 238–246.

Overall, J. E., Hollister, L. E., Johnson, M., & Pennington, V. Nosology of depression and differential response to drugs. *Journal of the American Medical Association,* 1966, *195,* 946–948.

Paykel, E. S. Classification of depressed patients: A cluster analysis derived grouping. *British Journal of Psychiatry,* 1971, *118,* 275–288.

Paykel, E. S., & Henderson, A. J. Application of cluster analysis in the classification of depression: A replication study. *Neuropsychobiology,* 1977, *3,* 111–119.

Pilowsky, I., Levine, S., & Boulton, D. M. The classification of depression by numerical taxonomy. *British Journal of Psychiatry,* 1969, *115,* 937–945.

Raskin, A., & Crook, T. H. The endogenous-neurotic distinction as a predictor of response to antidepressant drugs. *Psychological Medicine,* 1976, *6,* 59–70.

Raskin, A., Schulterbrandt, J. G., Boothe, H., Rentig, N., & McKeon, J. J. Some suggestions for selecting appropriate depression subgroups for biochemical studies. In T. A. Williams, M. M. Katz, & J. A. Shield (Eds.), *Recent advances in the psychobiology of depressive illness.* Washington: HEW, 1972.

FUNCTIONAL PSYCHOSES

Doherty, E. G. Labeling effects in psychiatric hospitalization: Study of diverging patterns of inpatient self-labeling processes. *Archives of General Psychiatry,* 1975, *32,* 562–568.

Fleiss, J. L., Lawlor, W., Platman, S. R., & Fieve, R. R. On the use of inverted factor analysis for generating typologies. *Journal of Abnormal Psychology,* 1971, *77,* 127–132.

Harmon, C. E., & Raymond, C. S. Computer prediction of chronic psychiatric patients. *Journal of Nervous and Mental Disease,* 1970, *150,* 490–503.

Lorr, M. M. (Ed.). *Explorations in typing psychotics.* New York: Pergamon Press, 1966.

Lorr, M. M. A typology for functional psychotics. In M. M. Katz, J. O. Cole, & W. E. Barton (Eds.), *The role and methodology of classification in psychiatry and psychopathology.* Washington: HEW, 1968.

Lorr, M. M. A typological conception of the behavior disorders. In A. R. Mahrer (Ed.), *New approaches to personality classification.* New York: Columbia University Press, 1970.

Lorr, M. M., & Klett, C. J. Psychotic behavioral types: A cross-cultural comparison. *Archives of General Psychiatry,* 1969, *20,* 592–597.

Lorr, M. M., Klett, C. J., & McNair, D. M. *Syndromes of psychosis.* New York: Pergamon Press, 1963.

Lorr, M., Bishop, P. F., & McNair, D. M. Interpersonal types among psychotic patients. *Journal of Abnormal Psychology,* 1965, *70,* 468–472.

Lorr, M. M., Caffey, E. M., & Gessner, T. L. Seven symptom profiles. *Journal of Nervous and Mental Disease,* 1968, *147,* 134–140.

McNair, D. M., Lorr, M., & Hemingway, P. Further evidence for syndrome-based psychotic types. *Archives of General Psychiatry,* 1964, *11,* 368–376.

Wittenborn, J., & Bailey, C. The symptoms of involutional psychosis. *Journal of Consulting Psychology,* 1952, *16,* 13–17.

Wittenborn, J. R., & Weiss, W. Patients diagnosed manic-depressive psychosis, manic state. *Journal of Consulting Psychology,* 1952, *16,* 193–198.

GENERAL PSYCHOPATHOLOGY

Bendig, A. W., & Hamlin, R. M. The psychiatric validity of an inverted factor analysis of Rorschach scoring categories. *Journal of Consulting Psychology,* 1955, *19,* 183–188.

Burger, G. K., & Cross, D. T. Personality types as measured by the California Personality Inventory. *Journal of Consulting and Clinical Psychology*, 1979, *47*, 65–71.

Coche, E., & Meehan, J. Factor and cluster analyses with the Rosenweig Picture Frustration Study. *Journal of Personality Assessment*, 1979, *43*, 39–44.

Everitt, B. S., Gourlay, A. J., & Kendell, R. E. An attempt at validation of traditional psychiatric syndromes by cluster analysis. *British Journal of Psychiatry*, 1971, *119*, 399–406.

Frank, G. H. The Wechsler-Bellevue and psychiatric diagnosis: A factor analytic study. *Journal of Consulting Psychology*, 1956, *20*, 67–69.

Freudenthal, K., Gebhardt, R., & Pietzcher, A. AMP (PAS) and BPRS: A comparison of two assessment methods. *Pharmakopsychiatrie und Neuropsychopharmakologie*, 1977, *10*, 57–66.

Gold, E. M., & Hoffman, P. J. Flange detection cluster analysis. *Multivariate Behavioral Research*, 1976, *11*, 217–235.

Horowitz, L. M. On the cognitive structure of interpersonal problems treated in psychotherapy. *Journal of Consulting and Clinical Psychology*, 1979, *47*, 5–15.

Lorei, T. W. Hierarchical grouping of opinions about release/retention of psychiatric patients. *Archives of General Psychiatry*, 1971, *25*, 69–73.

Lorr, M., & Suziedelis, A. A cluster analytic approach to MMPI profile types. *Multivariate Behavioral Research*, 1982, *17*, 287–299.

McNair, D. M., & Lorr, M. Differential typing of psychiatric outpatients. *Psychological Record*, 1965, *15*, 33–41.

Monro, A. B. Psychiatric types: A Q-technique study of 200 patients. *Journal of Mental Science*, 1955, *101*, 330–343.

Morf, M. E., Miller, C. M., & Syrotuik, J. M. A comparison of cluster analysis and Q-factor analysis. *Journal of Clinical Psychology*, 1976, *32*, 59–64.

Morf, M., Syrotuik, J., & Krznaric, S. Real data simulation of a two-stage statistical diagnostic system. *Journal of Consulting and Clinical Psychology*, 1977, *45*, 822–828.

Overall, J. E. Major phenomenological sub-types in a general psychiatric population. *Diseases of the Nervous System*, 1971, *32*, 383–387.

Overall, J. E., & Hollister, L. E. Studies of quantitative approaches to psychiatric classification. In M. M. Katz, J. O. Cole, & W. E. Barton (Eds.), *The role and methodology of classification in psychiatry and psychopathology*. Washington: HEW, 1968.

Rubinstein, E. I., & Lorr, M. Patient types in outpatient psychotherapy. *Journal of Clinical Psychology*, 1957, *13*, 356–361.

Skinner, H. A. A model of psychopathology based on the MMPI. In C. Newmark (Ed.), *MMPI: Clinical and research trends*. New York: Praeger, 1979.

Skinner, H. A., & Jackson, D. N. A model of psychopathology based on an integration of MMPI actuarial systems. *Journal of Consulting and Clinical Psychology*, 1978, *46*, 231–238.

Smith, P. M., & Barclay, A. G. Q analysis of the Holtzman Inkblot Technique. *Journal of Clinical Psychology*, 1975, *31*, 131–134.

Steinmeyer, E. M. Investigation of the automatic taxonomy (cluster analysis) of FPI test values with psychiatric practice. (German) *Zeitschrift fur Experimentelle und Angewandte Psychologie*, 1976, *23*, 140–150.

Stone, L. A. Groupings of psychiatric nomenclature classifications: Cluster analysis of psychiatric subjective ratio scales. *Psychological Reports*, 1969, *25*, 351–355.

Strauss, J. S., Bartko, J. J., & Carpenter, W. T. The use of clustering for the classification of psychiatric patients. *British Journal of Psychiatry*, 1973, *122*, 531–540.

Tatom, M. H. A factorial isolation of psychiatric outpatient syndromes. *Journal of Consulting Psychology*, 1958, *22*, 73–81.

Toenjes, C. M. Investigation of cluster analysis as a grouping procedure for the MMPI. *Dissertation Abstracts International*, 1978, *38*, (*11-B*), 5598.

Tryon, R. C. Person-clusters on intellectual abilities and on MMPI attributes. *Multivariate Behavioral Research*, 1967, *2*, 5–34.

Tryon, R. C., & Bailey, D. E. *Cluster analysis*. New York: McGraw-Hill, 1970.

Van Atta, R. E., & Ruppel, R. W. H-group method of profile analysis. *Journal of Clinical Psychology*, 1971, *27*, 327–333.

Wiener-Erlich, W. K. Hierarchical vs. generally overlapping models in psychiatric classification. *Multivariate Behavioral Research*, 1981, *16*, 455–482.

Williams, G. W., Barton, G. M., White, A. A., & Won, H. Cluster analysis applied to symptom ratings of psychiatric patients: An evaluation of its predictive ability. *British Journal of Psychiatry*, 1976, *129*, 178–185.

Zubin, J. Socio-biological types and methods for their isolation. *Psychiatry*, 1938, *1*, 237–247.

Zubin, J. A technique for measuring like-mindedness. *Journal of Abnormal and Social Psychology*, 1938, *33*, 508–516.

MENTAL RETARDATION

Clausen, J. *Ability structure and subgroups in mental retardation*. Washington: Spartan Books, 1966.

Culbertson, E., Guthrie, G. M., Butler, A. J., & Gorlow, L. Patterns of hostility among the retarded. *American Journal of Mental Deficiency*, 1962, *66*, 421–427.

Del Polito, G. A. A study of subgroups with the mentally retarded. *Dissertation Abstracts International*, 1973, *33* (*9-B*), 4567–4568.

Dingman, H. F. Some uses of descriptive statistics in population analysis. *Journal of Mental Deficiency*, 1959, *64*, 291–295.

Reid, A. H., Bullinger, B. R., & Heather, B. B. Behavioral syndromes identified by cluster analysis in a sample of 100 severely and profoundly retarded adults. *Psychological Medicine*, 1978, *8*, 399–412.

Stark, C. R., & White, N. B. Cluster analysis and racial differences in risk of Down's syndrome. In E. B. Hook & I. H. Porter (Eds.), *Population cytogenetics.* New York: Academic Press, 1977.

Stephens, W. B., Peck, J. R., & Veldman, D. J. Personality and success profiles characteristic of young adult male retardates. *American Journal of Mental Deficiency,* 1968, *73,* 405–413.

OFFENDERS

Anderson, W. P., Kunce, J. T., & Rich, B. Sex offenders: Three personality types. *Journal of Clinical Psychology,* 1979, *35,* 671–676.

Blackburn, R. Personality types among abnormal homicides. *British Journal of Criminology,* 1971, *11,* 14–31.

Carlson, K. A. Classes of adult offenders: A multivariate study. *Journal of Abnormal Psychology,* 1972, *79,* 84–93.

Kunce, J. T., Ryan, J. J., & Eckelman, C. C. Violent behavior and differential WAIS characteristics. *Journal of Consulting and Clinical Psychology,* 1976, *44,* 42–45.

Megargee, E. I. Development and validation of an MMPI-based system for classifying criminal offenders. In J. Butcher (Ed.), *New developments in the use of the MMPI.* Minneapolis: University of Minnesota Press, 1979.

Megargee, E. I., & Bohn, M. J. *Classifying criminal offenders: A new system based on the MMPI.* Beverly Hills, Calif.: Sage, 1979.

Meyer, J., & Megargee, E. I. A new classification system for criminal offenders: II. Initial development of the system. *Criminal Justice and Behavior,* 1977, *4,* 115–124.

Verberne, T. J. Blackburn's typology of abnormal homicides: Additional data and a critique. *British Journal of Criminology,* 1972, *12,* 88–89.

Widom, C. S. An empirical classification of female offenders. *Criminal Justice and Behavior,* 1978, *5,* 35–52.

ORGANIC BRAIN SYNDROME

Bear, D. M. The temporal lobes: An approach to the study of organic behavioral changes. In M. S. Gazzaniga (Ed.), *Handbook of behavioral neurobiology,* Vol. *2: Neuropsychology.* New York: Plenum Press, 1979.

Bear, D. M., & Fedio, P. Quantitative analysis of inter-ictal behavior in temporal lobe epilepsy. *Archives of Neurology,* 1977, *34,* 454–467.

Clark, C., Crockett, D. J., & Klonoff, H. Empirically derived groups in the assessment of recovery time from aphasia. *Brain and Language,* 1979, *7,* 240–251.

Crockett, D. J. A comparison of empirically derived groups of aphasic patients on the Neurosensory Center Comprehensive Examination for Aphasia. *Journal of Clinical Psychology,* 1977, *33,* 194–198.

Crockett, D., Clark, C., Spreen, O., & Klonoff, H. Severity of impairment or specific types of aphasia: An empirical investigation. *Cortex,* 1981, *17,* 83–96.

Goebel, R. A., & Satz, P. Profile analysis and the abbreviated Wechsler Adult Intelligence Scale: A multivariate approach. *Journal of Consulting and Clinical Psychology,* 1975, *43,* 780–785.

Guertin, W. H. A transposed factor analysis of the Bender Gestalts of brain disease cases. *Journal of Clinical Psychology,* 1954, *13,* 356–361.

Hanson, W. R., Riege, W. H., & Metter, J. H. Factor derived categories of chronic aphasia. *Brain and Language,* 1982, *15,* 369–380.

Hecaen, H., & Kremin, H. Neurolinguistic research on reading disorders resulting from left hemisphere lesions: Aphasia and "pure" alexia. In H. Whitaker & H. Whitaker (Eds.), *Studies in neurolinguistics* (Vol. 2). New York: Academic Press, 1976.

John, E. R. Neurometrics. *Science,* 1977, *196,* 1393–1410.

Kertesz, A. *Aphasia and associated disorders: Taxonomy, localization and recovery.* New York: Grune & Stratton, 1979.

Kertesz, A., & Phipps, J. B. Numerical taxonomy of aphasia. *Brain and Language,* 1977, *4,* 1–10.

Peyser, J. M., Edwards, K. R., & Poser, C. M. Psychological profiles in patients with multiple sclerosis: A preliminary investigation. *Archives of Neurology,* 1980, *37,* 437–440.

Powell, G. E., Clark, E., & Bailey, S. Categories of aphasia: Cluster analysis of Schull test profiles. *British Journal of Disabilities,* 1979, *14,* 111–122.

Wexler, N. S. Perceptual-motor, cognitive, and emotional characteristics of persons at risk for Huntington's Disease. In T. N. Chase, N. S. Wexler, & A. Barbeau (Eds.), *Advances in neurology* (Vol. 23). New York: Raven Press, 1979.

Wittenborn, J. R. The behavioral symptoms for certain organic psychoses. *Journal of Consulting Psychology,* 1952, *16,* 104–106.

PERSONALITY DISORDERS

Blackburn, R. An empirical classification of psychopathic personality. *British Journal of Psychiatry,* 1975, *127,* 456–460.

Grinker, R. R., Webble, B., & Drye, R. C. *The borderline syndrome.* New York: Basic Books, 1968.

Holland, T. R., Levi, M., & Watson, C. G. Personality patterns among hospitalized vs. incarcerated psychopaths. *Journal of Clinical Psychology,* 1980, *36,* 826–832.

Moore, T. V., Stafford, J. W., & Hsu, E. H. Obverse analysis of personality. *Journal of Personality,* 1947, *16,* 11–48.

Mullins, B. N. A cognitive and perceptual taxonomy of character disorders. *Journal of Clinical Psychology,* 1972, *28,* 9–13.

SCHIZOPHRENIA

Bartko, J. J., Carpenter, W. T., & Strauss, J. S. Statistical basis for exploring schizophrenia. *American Journal of Psychiatry,* 1981, *138,* 941–947.

Beck, S. J. The six schizophrenias. *American Orthopsychiatric Association Research Monograph,* #6, 1954.

Beck, S. J., & Nunnally, J. C. Two researches in schizophrenia. *American Journal of Orthopsychiatry,* 1953, *23,* 223–225.

Conger, J. J., Sawrey, W. L., & Krause, L. F. A reanalysis of Beck's "Six Schizophrenias." *Journal of Consulting Psychology,* 1956, *20,* 83–87.

Gerard, R. W. The nosology of schizophrenia: A co-operative study. *Behavioral Science,* 1964, *9,* 311–333.

Guertin, W. H. An inverted factor-analytic study of schizophrenics. *Journal of Consulting Psychology,* 1952, *16,* 371–375.

Guertin, W. H. A transposed factor analysis of schizophrenic performance on the Bender-Gestalt. *Journal of Clinical Psychology,* 1954, *10,* 225–228.

Guertin, W. H. A factor analytic study of the adjustment of chronic schizophrenics. *Journal of Clinical Psychology,* 1955, *11,* 174–177.

Guertin, W. H. A transposed analysis of the Bender-Gestalts of paranoid schizophrenics. *Journal of Clinical Psychology,* 1955, *11,* 73–76.

Guertin, W. H. A factor analysis of schizophrenics rated on the Activity Rating Scale. *Journal of Clinical Psychology,* 1956, *12,* 163–166.

Guertin, W. H. A transposed analysis of paranoid schizophrenics. *Psychological Reports,* 1958, *4,* 591–594.

Guertin, W. H. Empirical syndrome groupings of schizophrenic hospital admissions. *Journal of Clinical Psychology,* 1961, *17,* 268–275.

Guertin, W. H., & Jenkins, R. L. A transposed factor analysis of a group of schizophrenic patients. *Journal of Clinical Psychology,* 1956, *12,* 64–68.

Guertin, W. H., & Zilaitis, V. A transposed factor analysis of paranoid schizophrenics. *Journal of Consulting Psychology,* 1953, *17,* 455–458.

Hays, P. Taxonomic map of the schizophrenias with special reference to puerperal psychosis. *British Medical Journal,* 1978, *6139,* 755–757.

Jonsson, H. A factorial study of schizophrenia. *Scandinavian Journal of Psychology,* 1975, *16,* 125–130.

Jonsson, H., & Franzen, G. Evaluation of some factor-analytically derived subclasses of schizophrenia. *Psychological Research Bulletin,* 1976, *16,* 1–17.

Jonsson, H., & Franzen, G. Evaluation of two factor-analytically derived subclasses of schizophrenia. *Scandinavian Journal of Psychology,* 1978, *19,* 309–315.

Katz, M. M. A phenomenological typology of schizophrenia. In M. M. Katz, J. O. Cole, & W. E. Barton (Eds.), *The role and methodology of classification in psychiatry and psychopathology.* Washington: HEW, 1968.

Katz, M. M., & Cole, J. D. A phenomenological approach to the classification of schizophrenic disorders. *Diseases of the Nervous System,* 1963, *24,* 147–154.

Koh, S. D., Kayton, I., & Schwarz, C. The structure of work storage in the permanent memory of nonpsychotic schizophrenics. *Journal of Consulting and Clinical Psychology,* 1974, *42,* 879–887.

Labreche, G. P. A comparative analysis of three clustering algorithms applied to four similarity coefficients for the purpose of classifying schizophrenics in epidemiological research. *Dissertation Abstracts International,* 1974, *34* (10-B), 5198–5199.

Mattsson, N. B., & Gerard, R. W. Typology of schizophrenia based on multidisciplinary observational vectors. In M. M. Katz, J. O. Cole, & W. E. Barton (Eds.), *The role and methodology of classification in psychiatry and psychopathology.* Washington: HEW, 1968.

Meyer, L., & Kline, P. On the use of delegate scores in the G analysis of clinical data: A Q factor method for diagnostic classification. *Multivariate Behavioral Research,* 1977, *12,* 479–486.

SUBSTANCE ABUSE

Berzins, J. I., Ross, W. F., English, G. E., & Haley, J. V. Subgroups among opiate addicts: A typological investigation. *Journal of Abnormal Psychology,* 1974, *83,* 65–73.

Collins, H. A., Burger, G. K., & Taylor, G. A. An empirical typology of heroin abusers. *Journal of Clinical Psychology,* 1976, *32,* 473–476.

Joe, G. W., & Simpson, D. D. Research on patient classification for drug users in the DARP. *American Journal of Drug and Alcohol Abuse,* 1975, *2,* 29–35.

Lorr, M. M. Cluster and typological analysis. In P. M. Bentler, D. J. Lettieri, & G. A. Austin (Eds.), *Data analysis strategies and designs for substance abuse research.* Washington: National Institute for Drug Abuse, 1976.

Stimson, G. V. *Heroin and behaviour: Diversity among addicts attending London clinics.* London: Halstead Press, 1973.

SUICIDE

Bagley, C., Jacobsen, S., & Rehin, A. Completed suicide: A taxonomic analysis of clinical and social data. *Psychological Medicine,* 1976, *6,* 429–438.

Clopton, J. R., Pallis, D. J., & Birtchnell, J. Minnesota Multiphasic Personality Inventory profile patterns of suicide attempters. *Journal of Consulting and Clinical Psychology,* 1979, *47,* 135–139.

Colson, C. An objective-analytic approach to the classification of suicidal motivation. *Acta Psychiatrica Scandinavica,* 1973, *49,* 105–113.

Henderson, A. S., Hartigan, J., Davidson, J., Lance, G. N., Duncan-Jones, P., Koller, K. M., Ritchie, K., McAuley, H., Williams, C. L., & Slaghuis, W. A typology of parasuicide. *British Journal of Psychiatry,* 1977, *131,* 45–52.

Katschnig, H., & Sint, P. Are there different types of attempted suicide? A cluster analytic approach. In N. Speyer, R. Diekstra, & K. van de Loo (Eds.), *Proceedings 7th International Conference for Suicide Prevention.* Amsterdam: Swets and Zeitlinger, 1974.

Kiev, A. Cluster analysis profiles of suicide attempters. *American Journal of Psychiatry,* 1976, *133,* 150–153.

Paykel, E. S., & Rassaby, E. Classification of suicide attempters by cluster analysis. *British Journal of Psychiatry,* 1978, *135,* 45–52.

OTHER

Bradley, L. A., Prokop, C. K., Margolis, R., & Gentry, W. D. Multivariate analysis of the MMPI profiles of low back pain patients. *Journal of Behavioral Medicine,* 1978, *1,* 253–272.

Burger, G. K., & Kabakoff, R. I. Personality types as measured by the 16PF. *Journal of Personality Assessment,* 1982, *46,* 175–180.

Horowitz, M., Wilner, N., & Alvarez, W. Impact of Event Scale: A measure of subjective stress. *Psychosomatic Stress,* 1979, *41,* 209–218.

Moos, R. H., & Moos, B. A typology of family environments. *Family Process,* 1976, *15,* 357–371.

Murray, R. M., Cooper, J. E., & Smith, A. The Leyton Obsessional Inventory: An analysis of the responses of 73 obsessional patients. *Psychological Medicine,* 1979, *9,* 305–311.

Prokop, C. K., Bradley, L. A., Margolis, R., & Gentry, W. D. Multivariate analysis of the MMPI profiles of patients with multiple pain complaints. *Journal of Personality Assessment,* 1980, *44,* 246–252.

Schulling, D., Cronholm, B., Asberg, M., & Espmark, S. Ratings of psychic and somatic anxiety indicants: Interrater reliability and relations to personality variables. *Acta Psychologica Scandinavica,* 1973, *49,* 353–361.

Snyder, D. K., & Power, D. G. Empirical descriptors of unelevated MMPI profiles among chronic pain patients: A typological approach. *Journal of Clinical Psychology,* 1981, *37,* 602–607.

Stephenson, W. Introduction to inverted factor analysis, with some applications to studies of orexsis. *Journal of Educational Psychology,* 1936, *27,* 353–367.

Woodruff, R. A., Robins, L. N., Taibleson, M., Reich, T., Schwin, R., & Frost, N. A computer assisted deviation of a screening interview for hysteria. *Archives of General Psychiatry,* 1973, *29,* 450–454.

References

Achenbach, T. M. The classification of children's psychiatric symptoms: A factor analytic study. *Psychological Monographs*, 1966, *80*, (Whole No. 615).

Achenbach, T. M. The Child Behavior Profile: I. Boys aged 6–11. *Journal of Consulting and Clinical Psychology*, 1978, *46*, 478–488.

Achenbach, T. M. DSM-III in light of empirical research on the classification of child psychopathology. *Journal of the American Academy of Child Psychiatry*, 1980, *19*, 395–412.

Achenbach, T. M., & Edelbrock, C. S. The classification of child psychopathology: A review and analysis of empirical effects. *Psychological Bulletin*, 1978, *85*, 1275–1301.

Achenbach, T. M., & Edelbrock, C. S. The Child Behavior Profile: II. Boys aged 12 through 16 and girls aged 6–11 and 12–16. *Journal of Consulting and Clinical Psychology*, 1979, *47*, 223–233.

Ackerson, L. *Children's behavior problems, Vol. 1: Incidence, genetic and intellectual factors.* Chicago: University of Chicago Press, 1931.

Ackerson, L. *Children's behavior problems, Vol. 2: Relative importance and intercorrelation among traits.* Chicago: University of Chicago Press, 1942.

Akiskal, H. S., & McKinney, W. T. Overview of recent research into depression: Integration of ten conceptual models into a comprehensive clinical frame. *Archives of General Psychiatry*, 1975, *32*, 285–306.

Albert, J. M. *Psychiatry, psychology and social work: The clinical professions' twentieth-century search for identity.* Doctoral dissertation, School of Social Work, University of Southern California, 1978.

American Psychiatric Association. Notes and Comment: Revised classified nomenclature of mental disorders. *American Journal of Psychiatry*, 1933, *90*, 1369–1376.

American Psychiatric Association. *Diagnostic and statistical manual of mental disorders* (1st ed.). Washington, D.C.: Author, 1952.

American Psychiatric Association. *Diagnostic and statistical manual of mental disorders* (2nd ed). Washington, D.C.: Author, 1968.

American Psychiatric Association. *Diagnostic and statistical manual of mental disorders* (3rd ed.). Washington, D.C.: Author, 1980.

Anderberg, M. R. *Cluster analysis for applications.* New York: Academic Press, 1973.

Anderson, T. W. *An introduction to multivariate statistical analysis.* New York: Wiley, 1958.

Andreasen, N. C. Thought, language and communication disorders: I. Clinical assessment, definition of terms and evaluation of their reliability. *Archives of General Psychiatry,* 1979, *36,* 1315–1321. (a)

Andreasen, N. C. Thought, language and communication disorders: II. Diagnostic significance. *Archives of General Psychiatry,* 1979, *36,* 1325–1330. (b)

Arabie, P., & Carroll, J. D. MAPCLUS: A mathematical programming approach to fitting the ADCLUS model. *Harvard-Yale Preprints in Mathematical Sociology,* #16, 1979.

Armstrong, J. S., & Soelberg, P. On the interpretation of factor analysis. *Psychological Bulletin,* 1968, *70,* 361–364.

Ash, P. The reliability of psychiatric diagnosis. *Journal of Abnormal and Social Psychology,* 1949, *44,* 272–276.

Bagshaw, V. E., & McPherson, F. M. The applicability of the Foulds and Bedford hierarchy model to mania and hypomania. *British Journal of Psychiatry,* 1978, *132,* 293–295.

Bailey, K. D. Cluster analysis. In D. Heise (Ed.), *Sociological methodology.* San Francisco: Jossey-Bass, 1974.

Barrios, B. A., Hartmann, D. P., & Shigetomi, C. Fears and anxieties in children. In E. J. Mash & L. G. Terdal (Eds.), *Behavioral assessment of childhood disorders.* New York: Guilford Press, 1981.

Bartko, J. J., & Carpenter, W. T. On the methods and theory of reliability. *Journal of Nervous and Mental Disease,* 1976, *163,* 307–317.

Bartko, J. J., Strauss, J. S., & Carpenter, W. T. An evaluation of taxometric techniques for psychiatric data. *Classification Society Bulletin,* 1971, *2,* 2–28.

Bayer, R. *Homosexuality and American psychiatry: The politics of diagnosis.* New York: Basic Books, 1981.

Beck, A. T. Reliability of psychiatric diagnosis: A critique of systematic studies. *American Journal of Psychiatry,* 1962, *119,* 210–216.

Beck, A. T. *Depression: Clinical, theoretical and experimental aspects.* New York: Harper & Row, 1967.

Beck, A. T., Ward, C. H., Mendelson, M., Mock, J. E., & Erbaugh, J. K. Reliability of psychiatric diagnosis: II. A study of consistency of clinical judgments and ratings. *American Journal of Psychiatry,* 1962, *119,* 351–357.

Bedford, A., & Presly, A. S. Symptom patterns among chronic schizophrenic in-patients. *British Journal of Psychiatry,* 1978, *133,* 176–178.

Benjamin, L. S. Structural analysis of differential failure. *Psychiatry,* 1979, *42,* 1–23.

Berlin, B. Ethnobiological classification. In E. Rosch & B. B. Lloyd (Eds.), *Cognition and categorization.* Hillsdale, N.J.: Erlbaum Associates, 1978.

Bezdek, J. C. Cluster validity with fuzzy sets. *Journal of Cybernetics,* 1974, *3,* 58–73. (a)

Bezdek, J. C. Numerical taxonomy with fuzzy sets. *Journal of Mathematical Biology,* 1974, *1,* 57–71. (b)

Bielski, R. J., & Friedel, R. D. Prediction of antidepressant response: A critical review. *Archives of General Psychiatry,* 1976, *33,* 1479–1489.

Blashfield, R. An evaluation of the DSM-II classification of schizophrenia as a nomenclature. *Journal of Abnormal Psychology,* 1973, *82,* 382–389.

Blashfield, R. K. Mixture model tests of cluster analysis: Accuracy of four hierarchical agglomerative methods. *Psychological Bulletin,* 1976, *83,* 377–388.

Blashfield, R. K., & Aldenderfer, M. S. The literature on cluster analysis. *Multivariate Behavioral Research*, 1978, *13*, 271–295.

Blashfield, R. K., & Draguns, J. G. Evaluative criteria for psychiatric classification. *Journal of Abnormal Psychology*, 1976, *85*, 140–150. (a)

Blashfield, R. K., & Draguns, J. G. Towards a taxonomy for psychopathology. *British Journal of Psychiatry*, 1976, *129*, 574–583. (b)

Blashfield, R. K., & Morey, L. C. The classification of depression through cluster analysis. *Comprehensive Psychiatry*, 1979, *20*, 516–527.

Blashfield, R. K., & Morey, L. C. A comparison of four clustering methods using MMPI Monte Carlo data. *Applied Psychological Measurement*, 1980, *4*, 57–64.

Bleuler, E. *Dementia praecox; or, The group of schizophrenias.* New York: International Universities Press, 1950.

Boisen, A. Types of dementia praecox: A study in psychiatric classification. *Psychiatry*, 1938, *1*, 233–236

Bowman, K. M., & Jellinek, E. M. Alcohol addiction and its treatment. *Quarterly Journal of Studies on Alcohol*, 1941, *2*, 98–176.

Broad, W. J. Priority war: Discord in pursuit of glory. *Science*, 1981, *211*, 465–467.

Brockington, I. F., Kendell, R. E., & Leff, J. P. Definitions of schizophrenia: Concordance and prediction of outcome. *Psychological Medicine*, 1978, *8*, 387–398.

Buck, R., & Hull, D. The logical structure of the Linnean hierarchy. *Systematic Zoology*, 1966, *15*, 97–111.

Byrne, D. G. A note on cluster analysis and depression: Disparities in results produced by the application of different clustering methods. *Australian and New Zealand Journal of Psychiatry*, 1978, *12*, 99–102.

Cantor, N., Smith, E. E., French, R., & Mezzich, J. Psychiatric diagnosis as prototype categorization. *Journal of Abnormal Psychology*, 1980, *89*, 181–193.

Cantwell, D. P., Russell, A. T., Mattison, R., & Will, L. A comparison of DSM-II and DSM-III in the diagnosis of childhood psychiatric disorders. *Archives of General Psychiatry*, 1979, *36*, 1208–1213.

Caplan, A. L., Englehardt, H. T., & McCartney, J. J. (Eds.). *Concepts of health and disease: Interdisciplinary perspectives.* Reading, Mass.: Addison-Wesley, 1981.

Carey, G., & Gottesman, I. I. Reliability and validity in binary ratings. *Archives of General Psychiatry*, 1978, *35*, 1454–1459.

Carmines, E. G., & Zeller, R. A. *Reliability and validity assessment.* Beverly Hills, Calif.: Sage, 1980.

Carney, M. W. P., Roth, M., & Garside, R. T. The diagnosis of depressive symptoms and the prediction of E.C.T. response. *British Journal of Psychiatry*, 1965, *111*, 659–674.

Carpenter, W. T., & Stephens, J. H. An attempted integration of information relevant to schizophrenic subtypes. *Schizophrenia Bulletin*, 1979, *5*, 490–506.

Cattell, R. B. Factor analysis: An introduction to the essentials. *Biometrics*, 1965, *21*, 190–215.

Cattell, R. B. The scree test for the number of factors. *Multivariate Behavioral Research*, 1966, *1*, 140–161.

Chapman, L. J., & Chapman, J. P. *Disordered thought in schizophrenia.* Englewood Cliffs, N.J.: Prentice-Hall, 1973.

Chignell, M., & Stacey, B. G. Practical problems associated with the use of cluster analysis. *Psychological Reports*, 1980, *46*, 131–134.

Clare, A. *Psychiatry in dissent: Controversial issues in thought and practice.* London: Tavistock, 1976.

Clifford, H. T., & Stephenson, W. *An introduction to numerical classification.* New York: Academic Press, 1975.

Cobb, S. *Borderland of psychiatry.* Cambridge, Mass.: Harvard University Press, 1943.

Cohen, J. A coefficient of agreement for nominal scales. *Educational and Psychological Measurement,* 1960, *20,* 37–46.

Cole, J. R., & Zuckerman, H. The emergence of a scientific speciality: The self-exemplifying case of the sociology of science. In L. A. Coser (Ed.), *The idea of social structure: Papers in honor of Robert K. Merton.* New York: Harcourt-Brace-Jovanovich, 1975.

Comrey, A. L. Common methodological problems in factor analysis. *Journal of Consulting and Clinical Psychology,* 1978, *46,* 648–659.

Cooper, J. E., Kendell, R. E., Gurland, B. J., Sharpe, L., Copeland, J. R. M., & Simon, R. Psychiatric diagnosis in New York and London. *Maudsley Monographs, #20,* 1972.

Cormack, R. M. A review of classification. *Journal of the Royal Statistical Society (Series A),* 1971, *134,* 321–367.

Corning, W. C., & Steffy, R. A. Taximetric strategies applied to psychiatric classification. *Schizophrenia Bulletin,* 1979, *5,* 294–305.

Coryell, W., & Tsuang, M. T. DSM-III schizophreniform disorder: Comparisons with schizophrenia and affective disorder. *Archives of General Psychiatry,* 1982, *39,* 66–69.

Costello, C. G. Classification and psychopathology. In C. G. Costello (Ed.), *Symptoms of psychopathology: A handbook.* New York: Wiley, 1970.

Costello, C. G. Childhood depression. In E. J. Mash & L. G. Terdal (Eds.), *Behavioral assessment in children.* New York: Guilford Press, 1981.

Cronbach, L. J., Rajaratnam, N., & Gleser, G. C. *The dependability of behavioral measurements.* New York: Wiley, 1973.

Crown, S. "On being sane in insane places": A comment from England. *Journal of Abnormal Psychology,* 1975, *84,* 453–455.

Cunningham, K. M., & Ogilive, J. C. Evaluation of hierarchical grouping techniques—A preliminary study. *Computer Journal,* 1972, *15,* 209–215.

Degan, J. W. Dimensions of functional psychosis. *Psychometric Monographs, #6,* 1952.

Degerman, R. L. The geometric representation of some simple structures. In R. N. Shepard, A. K. Romeny, & S. B. Nerlove (Eds.), *Multidimensional scaling* (Vol. I). New York: Seminar Press, 1971.

DeLeon, P. H. Psychology and the Carter administration. *American Psychologist,* 1977, *32,* 750–751.

Derogatis, L. R., Bonato, R. R., & Yang, K. C. The power of the IMPS in psychiatric drug research as a function of sample size, number of raters, and choice of treatment comparison. *Archives of General Psychiatry,* 1968, *19,* 689–699.

Deutsch, K. W. On theories, taxonomies and models as communication codes for organizing information. *Behavioral Science,* 1966, *11,* 1–17.

Doering, C. R., & Raymond, A. F. Reliability of observation in psychiatric and related characteristics. *American Journal of Orthopsychiatry,* 1934, *4,* 249–257.

Donovan, D. M., Chaney, E. F., & O'Leary, M. R. Alcoholic MMPI subtypes: Relationships to drinking styles, benefits and consequences. *Journal of Nervous and Mental Disease,* 1978, *166,* 553–561.

Dorken, H. CHAMPUS ten-state claim experience for mental disorders: Fiscal year 1975. *American Psychologist,* 1977, *32,* 697–710.

Dorken, H., & Webb, J. T. 1976 third-party reimbursement experience: An interstate comparison by insurance carrier. *American Psychologist*, 1980, *35*, 355–363.

Drapes, T. On the maniacal-depressive insanity of Kraepelin. *Journal of Mental Science*, 1909, *55*, 58–64.

Dubes, R., & Jain, A. K. Validity studies in clustering methodologies. *Pattern Recognition*, 1979, *11*, 235–254.

Eagle, P. F., & Marcos, L. R. Factors in medical students' choice of psychiatry. *American Journal of Psychiatry*, 1980, *137*, 423–427.

Edelbrock, C. Mixture model tests of hierarchical clustering algorithms. *Multivariate Behavioral Research*, 1979, *14*, 367–384.

Edelbrock, C., & McLaughlin, B. Hierarchical cluster analysis using intraclass correlations: A mixture model study. *Multivariate Behavioral Research*, 1980, *15*, 299–318.

Edelbrock, C., & Reed, M. *Inverse factor analysis: An evaluation using benchmark data sets*. Unpublished paper, 1981.

Endicott, J., Nee, J., Fleiss, J., Cohen, J., Williams, J. B. W., & Simon, R. Diagnostic criteria for schizophrenia: Reliabilities and agreement between systems. *Archives of General Psychiatry*, 1982, *39*, 884–889.

Essen-Moller, E. On the classification of mental disorders. *Acta Psychiatrica Scandinavica*, 1961, *37*, 119–126.

Essen-Moller, E. Suggestions for further improvement of the international classification of mental disorders. *Psychological Medicine*, 1971, *1*, 308–311.

Everitt, B. S. Unresolved problems in cluster analysis. *Biometrics*, 1979, *35*, 169–181.

Everitt, B. S. *Cluster analysis* (2nd ed.). London: Halstead Press, 1980.

Everitt, B. S., Gourlay, A. J., & Kendell, R. E. An attempt at validation of traditional syndromes by cluster analysis. *British Journal of Psychiatry*, 1971, *119*, 399–412.

Eysenck, H. J. *The scientific study of personality*. London: Routledge & Kegan Paul, 1952.

Eysenck, H. J. *The structure of human personality*. London: Mithsun, 1953.

Eysenck, H. J. *Manual of the Maudsley Personality Inventory*. London, 1959.

Eysenck. H. J. Classification and the problem of diagnosis. In H. J. Eysenck (Ed.), *Handbook of abnormal psychology*. London: Pitman, 1960.

Farber, I. E. Sane and insane: Constructions and misconstructions. *Journal of Abnormal Psychology*, 1975, *84*, 589–620.

Farrell, B. A. Mental illness: A conceptual analysis. *Psychological Medicine*, 1979, *9*, 21–35.

Feighner, J. P., Robins, E., Guze, S. B., Woodruff, R. A., Winokur, G., & Munoz, R. Diagnostic criteria for use in psychiatric research. *Archives of General Psychiatry*, 1972, *26*, 57–63.

Fenton, W. S., Mosher, L. R., & Matthews, S. M. Diagnosis of schizophrenia: A critical review of current diagnostic system. *Schizophrenia Bulletin*, 1981, *7*, 452–476.

Finn, S. E. Base rates, utilities, and DSM-III: Shortcomings of fixed-rule systems of psychodiagnosis. *Journal of Abnormal Psychology*, 1982, *91*, 294–302.

Finney, J. W., & Moos, R. H. Treatment and outcome for empirical subtypes of alcohol patients. *Journal of Consulting and Clinical Psychology*, 1979, *47*, 25–38.

Fisher, L., & Van Ness, J. W. Admissible clustering procedures. *Biometrika*, 1971, *58*, 91–104.

Fisher, R. A. The use of multiple measurements in taxonomic problems. *Annals of Eugenics,* 1936, *7,* 179–188.

Fleiss, J. L. *Statistical methods for rates and proportions.* New York: Wiley, 1973.

Fleiss, J. L., & Zubin, J. On the methods and theory of clustering. *Multivariate Behavioral Research,* 1969, *4,* 235–250.

Fleiss, J. L., Cohen, J., & Everitt, B. S. Large sample standard errors of kappa and weighted kappa. *Psychological Bulletin,* 1969, *72,* 323–327.

Fleiss, J. L., Lawlor, W., Platman, S. R., & Fieve, R. R. On the use of inverted factor analysis for generating typologies. *Journal of Abnormal Psychology,* 1971, *77,* 127–132.

Fleiss, J. L., Spitzer, R. L., Endicott, J., & Cohen, J. Quantification of agreement in multiple psychiatric diagnosis. *Archives of General Psychiatry,* 1972, *26,* 168–171.

Foulds, G. The reliability of psychiatric and validity of psychological diagnosis. *Journal of Mental Science,* 1955, *101,* 851–862.

Foulds, G. A. Personal continuity and psychopathological disruption. *British Journal of Psychology,* 1964, *55,* 269–276.

Foulds, G. A. *Personality and personal illness.* London: Tavistock, 1965.

Foulds, G. A. Personality deviance and personal symptomatology. *Psychological Medicine,* 1971, *1,* 222–233.

Foulds, G. A., & Bedford, A. *Hierarchy of classes of personal illness.* Unpublished manuscript, 1974.

Foulds, G. A., & Bedford, A. Hierarchy of classes of personal illness. *Psychological Medicine,* 1975, *5,* 181–192.

Fowler, R. C., Liskow, B. I., Tanna, V. L., Lytle, L., & Mezzich, J. Schizophrenia-primary affective disorder-discrimination: I. Development of data-based diagnostic index. *Archives of General Psychiatry,* 1980, *37,* 811–814.

Fox, H. A. The DSM-III concept of schizophrenia. *British Journal of Psychiatry,* 1981, *138,* 60–63.

Freedman, A. M., Kaplan, H. I., & Sadock, B. J. (Eds.). *Comprehensive textbook of psychiatry* (2nd ed.). Baltimore: Williams & Wilkins, 1975.

Friedman, H. P., & Rubin, J. On some invariant criteria for grouping data. *Journal of the American Statistical Association,* 1967, *62,* 1159–1178.

Garmezy, N. DSM-III. Never mind psychologists: Is it good for children? *Clinical Psychologist,* 1978, *31,* #3.

Gilberstadt, H., & Duker, J. *A handbook for clinical and actuarial MMPI interpretation.* Philadelphia: Saunders, 1965.

Goffman, E. The moral career of the mental patient. *Psychiatry,* 1959, *22,* 123–142.

Goldberg, L. R. Diagnosticians vs. diagnostic signs: The diagnoses of psychosis vs. neurosis from the MMPI. *Psychological Monographs,* 1965, *79* (Whole No. 602).

Goldberg, S. C., Mattison, N., Cole, J. O., & Klerman, G. L. Prediction of improvement in schizophrenia under 4 phenothiazines. *Archives of General Psychiatry,* 1967, *16,* 107–110.

Golden, R. R., & Meehl, P. E. Detection of the schizoid taxon with MMPI indicators. *Journal of Abnormal Psychology,* 1979, *88,* 217–253.

Goldfarb, A. Reliability of diagnostic judgments by psychologists. *Journal of Clinical Psychology,* 1959, *15,* 392–396.

Goldstein, S. G., & Linden, J. D. A comparison of multivariate grouping techniques commonly used with profile data. *Multivariate Behavioral Research,* 1969, *4,* 103–114. (a)

Goldstein, S. G., & Linden, J. D. Multivariate classification of alcoholics by the MMPI. *Journal of Abnormal Psychology,* 1969, *74,* 661–669. (b)

Gorham, D. R., & Overall, J. E. Drug action profiles based on an abbreviated psychiatric rating scale. *Journal of Nervous and Mental Disease,* 1960, *132,* 528–535.

Gorsuch, R. L. *Factor analysis.* Philadelphia: Saunders, 1974.

Gottfredson, D. W., & Dyer, S. E. Health service providers in psychology. *American Psychologist,* 1978, *33,* 314–338.

Gregg, J. R. *The language of taxonomy.* New York: Columbia University Press, 1954.

Guertin, W. H. A factor-analytic study of schizophrenic symptoms. *Journal of Consulting Psychology,* 1952, *16,* 308–312.

Gulliksen, H. Looking back and ahead in psychometrics. *American Psychologist,* 1974, *29,* 251–261.

Gurney, C., Roth, M., Garside, R. F., Kerr, T. A., & Schapira, K. Studies in the classification of affective disorders: The relationship between anxiety states and depressive illness—II. *British Journal of Psychiatry,* 1972, *121,* 162–166.

Guy, W. *ECDEU assessment manual for psychopharmacology.* Rockville, Md.: National Institute of Mental Health, 1976.

Guze, S. B. The need for toughmindedness in psychiatric thinking. *Southern Medical Journal,* 1970, *63,* 662–671.

Guze, S. B. *Criminality and psychiatric disorders.* London: Oxford University Press, 1976.

Guze, S. B. Nature of psychiatric illness: Why psychiatry is a branch of medicine. *Comprehensive Psychiatry,* 1978, *19,* 295–307.

Gynther, M. D., & Gynther, R. A. Personality inventories. In I. B. Weiner (Ed.), *Clinical methods in psychology.* New York: Wiley, 1976.

Hamilton, M. A rating scale for depression. *Journal of Neurology, Neurosurgery and Psychiatry,* 1960, *23,* 56–62.

Hamilton, M., & White, J. Clinical syndromes in depressive states. *Journal of Mental Science,* 1959, *105,* 985–998.

Harman, H. H. *Modern Factor Analysis* (2nd ed.). Chicago: University of Chicago Press, 1967.

Harman, H. H. *Modern Factor Analysis* (3rd ed.). Chicago: University of Chicago Press, 1976.

Harms, E. Introduction. In E. Kraepelin *Dementia praecox and paraphrenia,* transl. by Mary Barclay from the 8th ed. of Kraepelin's *Textbook.* New York: Robert E. Krieger Publishing Company, 1971.

Harris, R. J. *A primer of multivariate statistics.* New York: Academic Press, 1975.

Hartigan, J. *Clustering algorithms.* New York: Wiley, 1975.

Hartigan, J. A. Representation of similarity matrices by trees. *Journal of the American Statistical Association,* 1967, *62,* 1140–1158.

Hay, A. J., & Forrest, A. D. The diagnosis of schizophrenia and paranoid psychosis: An attempt at clarification. *British Journal of Medical Psychology,* 1972, *45,* 233–241.

Hedlund, J. L., & Vieweg, B. W. The Brief Psychiatric Rating Scale (BPRS): A comprehensive review. *Journal of Operational Psychiatry,* 1980, *11,* 48–65.

Helmchen, H. Schizophrenia: Diagnostic concepts in the ICD 8. *British Journal of Psychiatry,* 1975, special publication #10, 10–18.

Helmchen, H. Multiaxial systems of classification. *Acta Psychiatrica Scandinavica,* 1980, *61,* 43–55.

Helzer, J. E., Clayton, P. J., Pambakian, R., Reich, T., Woodruff, R. A., & Reveley, M. A. Reliability of psychiatric diagnosis: II. The test/retest reliability of diagnostic classification. *Archives of General Psychiatry,* 1977, *34,* 136–141.

Helzer, J. E., Robins, L. N., Taibleson, M., Woodruff, R. A., Reich, T., & Wish, E. D. Reliability of psychiatric diagnosis: I. Methodological review. *Archives of General Psychiatry,* 1977, *34,* 129–133.

Henry, B. W., Market, J. R., & Emken, R. I. Drug treatment of anxious depression in psychiatric outpatients. *Diseases of the Nervous System,* 1969, *30,* 678–679.

Hill, K. M., & Evans, R. K. A model for classification and typology. In D. C. Clark (Ed.), *Models in archaelogy.* London: Methuen, 1972.

Hollingshead, A. B., & Redlich, F. C. *Social class and mental illness.* New York: Wiley, 1958.

Hollister, L. E., Overall, J. E., & Shelton, J. Amitriptyline, perphenazine and amitriptyline-perphenazine in different depressive syndromes. *Archives of General Psychiatry,* 1967, *17,* 489–493.

Hollister, L. E., Overall, J. E., Shelton, J., Pennington, V., Kimbell, I., & Johnson, M. Drug therapy of depression—Amitriptyline, perphenazine and their combination in different syndromes. *Archives of General Psychiatry,* 1967, *17,* 486–488.

Hubert, L. J. Kappa revisited. *Psychological Bulletin,* 1977, *84,* 289–297.

Hubert, L. J. A general formula for the variance of Cohen's weighted kappa. *Psychological Bulletin,* 1978, *85,* 183–184.

Hull, D. L., Tessner, P. D., & Diamond, A. M. Plank's principle. *Science,* 1978, *202,* 717–722.

Hunt, W. A. A rationale for psychiatric selection. *American Psychologist,* 1955, *10,* 199–204.

Hunt, W. A., Wittson, C. L., & Hunt, E. B. A theoretical and practical analysis of the diagnostic process. In P. H. Hoch & J. Zubin (Eds.), *Current problems in psychiatric diagnosis.* New York: Grune & Stratton, 1953.

Hurd, H. M. A plea for systematic therapeutical, clinical and statistical study. *Journal of Insanity,* 1881, *38,* 16–31.

Jackson, D. N. The dynamics of structured personality tests: 1971. *Psychological Review,* 1971, *78,* 229–248.

Jackson, D. N. Interpreter's guide to the Jackson Personality Inventory. In P. McReynolds (Ed.), *Advances in psychological assessment* (Vol. 4). San Francisco: Jossey-Bass, 1978.

Jackson, D. N., & Messick, S. *Differential Personality Inventory.* London. Canada: Authors, 1971.

Jackson, L. Personal communication from the American Nursing Association, 1980.

Janes, C. L. Agreement measurement and the judgement process. *Journal of Nervous and Mental Disease,* 1979, *167,* 343–347.

Jardine, N., & Sibson, R. *Mathematical taxonomy.* New York: Wiley, 1971.

Jefferson, L. *These are my sisters.* Tulsa: Vickers Publishing, 1948.

Jenkins, R. L., & Glickman, S. Common syndromes in child psychiatry I. Deviant behavior traits. *American Journal of Ortho-Psychiatry,* 1946, *16,* 244–261.

Johnstone, T. The case for dementia praecox. *Journal of Mental Science,* 1909, *55,* 64–74.

Joreskog, K. G. A general approach to confirmatory maximum likelihood factor analysis. *Psychometrika,* 1969, *34,* 183–202.

Joreskog, K. G. A general method for analysis of covariance structures. *Biometrika,* 1970, *57,* 239–251.

Joreskog, K. G. Structural analysis of covariance and correlation structures. *Psychometrika,* 1978, *43,* 443–477.

Joreskog, K. G., & Sorbon, D. *LISREL IV.* Chicago: National Educational Resources, 1978.

Kahn, E. The Emil Kraepelin memorial lecture. In D. Pasamanick (Ed.). *Epidemiology of mental disorders.* Washington, D.C.: American Association for the Advancement of Science, 1959.

Kaplan, A. *The conduct of inquiry.* San Francisco: Chandler, 1964.

Kaplan, H. I., Freedman, A. M., & Sadock, B. J. (Eds.). *Comprehensive textbook of psychiatry* (3rd ed.). Baltimore: Williams & Wilkins, 1980.

Katz, M. M., Cole, J. O., & Lowery, H. A. Non-specificity of diagnosis of paranoid schizophrenia. *Archives of General Psychiatry,* 1964, *11,* 197–202.

Katz, M., Cole, J. O., & Lowery, H. A. Studies of the diagnostic process: The influence of symptom perception, past experience and ethnic background on diagnostic decisions. *American Journal of Psychiatry,* 1969, *125,* 937–947.

Kelly, J. T., Koch, M., & Buegel, D. Lithium-carbonate in juvenile manic-depressive illness. *Diseases of the Nervous System,* 1976, *37,* 90–92.

Kendell, R. E. *The classification of depressive illness.* London: Oxford University Press, 1968.

Kendell, R. E. The continuum model of depressive illness. *Proceedings of the Royal Society of Medicine,* 1969, *62,* 335–339.

Kendell, R. E. *The role of diagnosis in psychiatry.* London: Blackwell Scientific Publications, 1975.

Kendell, R. E. The classifications of depressions: A review of contemporary confusion. *British Journal of Psychiatry,* 1976, *129,* 15–28.

Kendell, R. E., & Gourlay, A. J. The clinical distinction between psychotic and neurotic depression. *British Journal of Psychiatry,* 1970, *117,* 257–260.

Kendell, R. E., Cooper, J. E., Gourlay, A. J., Copeland, J. R. M., Sharpe, L., & Gurland, B. J. Diagnostic criteria of American and British psychiatrists. *Archives of General Psychiatry,* 1971, *25,* 123–130.

Kim, J., & Mueller, C. W. *Introduction to factor analysis: What is it and how to do it.* Beverly Hills, Calif.: Sage Publications, 1978.

Klecka, W. R. *Discriminant analysis.* Beverly Hills, Calif.: Sage Publications, 1980.

Klein, D. F., & Davis, J. M. *Diagnosis and drug treatment of psychiatric disorders.* Baltimore: Williams & Wilkins, 1969.

Klein, D. F., Gittelman, R., Quitkin, F., & Rifkin, A. *Diagnosis and drug treatment of psychiatric disorders: Adults and children* (2nd ed.). Baltimore: Williams & Wilkins, 1980.

Klerman, G. L. Mental illness, the medical model and psychiatry. *Journal of Medicine and Philosophy,* 1977, *2,* 220–243.

Klerman, G. L. The evolution of a scientific nosology. In J. C. Shershow (Ed.), *Schizophrenia: Science and practice.* Cambridge, Mass.: Harvard University Press, 1978.

Klerman, G. L., DiMascio, A., Weissman, M., Prusoff, B., & Paykel, E. S. Treatment of depression by drugs and psychotherapy. *American Journal of Psychiatry,* 1974, *131,* 186–191.

Klett, C. J., & McNair, D. M. Reliability of acute psychotic types. In M. Lorr (Ed.), *Explorations in typing psychotics.* New York: Pergamon Press, 1966.

Klett, C. J., & Moseley, E. C. The right drug for the right patient. *Journal of Consulting Psychology*, 1965, *29*, 546–551.

Kondo, Y. History of calculus (letter). *Science*, 1981, *211*, 1258.

Koran, L. M. The reliability of clinical methods, data and judgments: Part I. *New England Journal of Medicine*, 1975, *293*, 642–646. (a)

Koran, L. M. The reliability of clinical methods, data and judgments: Part II. *New England Journal of Medicine*, 1975, *293*, 695–701. (b)

Kraepelin, E. *Compendium der Psychiatrie: Zum Gebrauche fur Studirende und Aerzte* (1st ed). Leipzig: Verlag von Ambr. Abel, 1883.

Kraepelin, E. *Psychiatrie: Ein Lehrbuch fur Studirende und Aerzte* (6th ed, in 2 vols.). Leipzig: Verlag von Johann Ambrosius Barth, 1899.

Kraepelin, E. *Clinical psychiatry: A textbook for students and physicians*, trans. by A. Ross Diefendorf from the 6th ed. of Kraepelin's *Textbook*. London: MacMillan, 1902.

Kraepelin, E. *Lectures on clinical psychiatry*, trans. and ed. by Thomas Johnstone. New York: William Wood, 1904.

Kraepelin, E. *Clinical psychiatry: A textbook for students and physicians*, trans. by A. Ross Diefendorf from the 7th ed. of Kraepelin's *Textbook*. London: MacMillan, 1907.

Kraepelin, E. *Dementia praecox and paraphrenia*. Melbourne, Fla.: Krieger, 1971. (Originally published, 1919.)

Kramer, M. Introduction: The historical background of ICD-8. In American Psychiatric Association, *Diagnostic and statistical manual of mental disorders* (2nd ed.). Washington, D.C.: American Psychiatric Association, 1968.

Kraupl-Taylor, F. *Psychopathology: Its causes and symptoms*. Baltimore: John Hopkins University Press, 1979.

Kreitman, N. The reliability of psychiatric diagnosis. *Journal of Mental Science*, 1961, *107*, 878–886.

Krippendorf, K. Bivariate agreement coefficients for reliability of data. In E. F. Borgatta (Ed.), *Sociological methodology*. San Francisco: Jossey-Bass, 1970.

Krippendorf, K. Reliability of recording instructions: Multivariate agreement for nominal data. *Behavioral Science*, 1971, *16*, 222–235.

Kruskal, J. B., & Wish, M. *Multidimensional scaling*. Beverly Hills, Calif.: Sage Publications, 1978.

Kuiper, F. K., & Fisher, L. A Monte Carlo comparison of six clustering procedures. *Biometrics*, 1975, *31*, 777–783.

Lanyon, R. I. *A handbook of MMPI group data*. Minneapolis: University of Minnesota Press, 1968.

Lasky, J. J., Klett, C. J., Caffey, E. M., Bennett, J. L., Rosenblum, M. P., & Hollister, L. E. Drug treatment of schizophrenic patients: A comparative evaluation of chlorpromazine, chlorprothixene, fluphenazine, reserpine, thioridazine, and triflupromazine. *Diseases of the Nervous System*, 1962, *23*, 698–706.

Leary, T. *Interpersonal diagnosis of personality*. New York: Ronald Press, 1957.

Lefkowitz, M. M., & Burton, N. Childhood depression: A critique of the concept. *Psychological Bulletin*, 1978, *85*, 716–726.

Levy, L. *Conceptions of personality: Theory and research*. New York: Random House, 1970.

Lief, A. *The commonsense psychiatry of Dr. Adolph Meyer*. New York: McGraw-Hill, 1948.

Lipkowitz, M. H., & Idupuganti, S. Diagnosing schizophrenia: Criteria sets and reality. *Hospital and Community Psychiatry*, 1981, *32*, 344–345.

Lipowski, Z. J. Holistic-medical foundations of American psychiatry: A bicentennial. *American Journal of Psychiatry*, 1981, *138*, 88–895.

Liss, J. L., Welner, A., & Robins, E. Psychiatric symptoms in black and white patients: I. Record study. *Comprehensive Psychiatry*, 1973, *4*, 475–481.

Loevinger, J. The attenuation paradox in test theory. *Psychological Bulletin*, 1954, *51*, 493–504.

Lorr, M. Multidimensional Scale for Rating Psychotic Patients. *Veterans Administration Technical Bulletin*, 1953, *10*, 507.

Lorr, M. Rating scales and checklists for the evaluation of psychopathology. *Psychological Bulletin*, 1954, *51*, 119–127.

Lorr, M. The Wittenborn psychiatric syndromes: An oblique rotation. *Journal of Consulting Psychology*, 1957, *21*, 439–444.

Lorr, M. (Ed.). *Explorations in typing psychotics*. New York: Pergamon Press, 1966.

Lorr, M. On the use of cluster analytic techniques. *Journal of Clinical Psychology*, 1982, *38*, 461–462.

Lorr, M., & Cave, R. The equivalence of psychotic syndromes across two media. *Multivariate Behavioral Research*, 1966, *1*, 189–195.

Lorr, M., & Hamlin, R. M. A multimethod factor analysis of behavioral and objective measures of psychopathology. *Journal of Consulting and Clinical Psychology*, 1971, *36*, 136–141.

Lorr, M., & Jenkins, R. Patterns of maladjustment in children. *Journal of Clinical Psychology*, 1953, *9*, 16–19.

Lorr, M., & Klett, C. Constancy of psychotic syndromes in men and women. *Journal of Consulting Psychology*, 1965, *29*, 309–313.

Lorr, M. & McNair, D. M. Expansion of the interpersonal behavior circle. *Journal of Personality and Social Psychology*, 1965, *2*, 823–830.

Lorr, M., & O'Connor, J. P. Psychotic symptom patterns in a behavior inventory. *Educational and Psychological Measurement*, 1962, *22*, 139–146.

Lorr, M., & Radhakrishnan, B. K. A comparison of two methods of cluster analysis. *Educational and Psychological Measurement*, 1967, *27*, 47–53.

Lorr, M., Jenkins, R., & O'Conner, J. Factors descriptive of psychopathology and behavior of hospitalized psychotics. *Journal of Abnormal and Social Psychology*, 1955, *50*, 78–86.

Lorr, M., O'Connor, J. P., & Stafford, J. W. Confirmation of nine psychotic symptom patterns. *Journal of Clinical Psychology*, 1957, *13*, 252–257.

Lorr, M., Klett, C. J., & McNair, D. M. *Syndromes of psychosis*. New York: Pergamon Press, 1963.

MacQueen, J. B. Some methods for classification and analysis of multivariate observations. *Proceedings of the Fifth Berkeley Symposium on Mathematical Statistics and Probability*, 1967, *1*, 281–297.

Maher, B. A. *Principles of psychopathology*. New York: McGraw-Hill, 1966.

Mariotto, M. J., & Paul, G. L. A multimethod validation of the Inpatient Multidimensional Psychiatric Scale with chronically institutionalized patients. *Journal of Consulting and Clinical Psychology*, 1974, *42*, 497–508.

Marks, P. A., & Seeman, W. *Actuarial description of abnormal personality*. Baltimore: Williams & Wilkins, 1963.

Masserman, J. H., & Carmichael, H. T. Diagnosis and prognosis in psychiatry: With a follow-up study of the results of short term general hospital therapy in psychiatric cases. *Journal of Mental Science*, 1938, *84*, 893–946.

Matarazzo, J. D. The interview: Its reliability and validity in psychiatric diagnosis. In B. B. Wolman (Ed.), *Clinical diagnosis of mental disorders: A handbook*. New York: Plenum Press, 1978.

Mattera, M. D. Why psychiatrists are behind the economic eight ball. *Medical Economics*, 1979, *56*, 158–167.

Mattison, R., Cantwell, D. P., Russell, A. T., & Will, L. A comparison of DSM-II and DSM-III in the diagnosis of childhood psychiatric disorders: II. Interrater agreement. *Archives of General Psychiatry*, 1979, *36*, 1217–1222.

Matussek, P., Soldner, N., & Nagel, D. Identification of the endogneous depressive syndrome based on the symptoms and the characteristics of the course. *British Journal of Psychiatry*, 1981, *138*, 361–372.

Matza, D. *Becoming deviant*. Englewood Cliffs, N.J.: Prentice-Hall, 1968.

Maxwell, A. E. Coefficients of agreement between observers and their interpretation. *British Journal of Psychiatry*, 1977, *130*, 79–83.

Mayer-Gross, W. Die Entwicklung der klinishen Anschauung KRAEPELINS. *Archiv fur Psychiatrie und Nervenkrankheiten*, 1929, *67*, 30–42.

Mayr, E. Biological classification: Toward a synthesis of opposing methodologies. *Science*, 1981, *214*, 510–516.

McLemore, C. W., & Benjamin, L. S. Whatever happened to interpersonal diagnosis? A psychosocial alternative to DSM-III. *American Psychologist*, 1979, *34*, 17–34.

McPherson, F. M., Antram, M. C., Bagshaw, V. E., & Carmichael, S. K. A test of the hierarchical model of personal illness. *British Journal of Psychiatry*, 1977, *131*, 56–58.

Meehl, P. E. A funny thing happened to us on the way to the latent entities. *Journal of Personality Assessment*, 1979, *43*, 564–580.

Meehl, P. E., & Rosen, A. Antecedent probability and the efficiency of psychometric signs, patterns or cutting scores. *Psychological Bulletin*, 1955, *52*, 194–216.

Mefferd, R. B. Technique for minimizing the instrument factor. *Multivariate Behavioral Research*, 1968, *3*, 339–354.

Mellon, T. Reflections on Rosenhan's "On being sane in insane places." *Journal of Abnormal Psychology*, 1975, *84*, 456–461.

Menard, H. W. *Science: Growth and change*. Cambridge, Mass.: Harvard University Press, 1973.

Mendels, J., & Cochrane, C. The nosology of depression: The endogenous-reactive concept. *American Journal of Psychiatry*, 1968, *124*, (May supplement), 1–11.

Menninger, K. *The vital balance*. New York: Viking Press, 1963.

Menninger, K. Sheer verbal Mickey Mouse. *International Journal of Psychiatry*, 1969, *7*, 415.

Meredith, W. M., Frederiksen, C. H., & McLaughlin, D. H. Statistics and data analysis. *Annual Review of Psychology*, 1974, *25*, 453–505.

Merton, R. K. The Matthew effect in science. *Science*, 1968, *199*, 55–63.

Meyer, A. *Psychobiology*. Springfield, Ill.: Charles C Thomas, 1957.

Mezzich, J. E. Evaluating clustering methods for psychiatric diagnosis. *Biological Psychiatry*, 1978, *13*, 265–281.

Mezzich, J. E. Patterns and issues in multiaxial psychiatric diagnosis. *Psychological Medicine*, 1979, *9*, 125–137.

Mezzich, J. E., & Solomon, H. *Taxonomy and behavioral science.* New York: Academic Press, 1980.

Mickle, W. J. Catatonia: In relation to dementia praecox. *Journal of Mental Science,* 1909, *55,* 22–36.

Milligan, G. W. An examination of the effect of six types of error perturbation on fifteen clustering algorithms. *Psychometrika,* 1980, *45,* 325–342.

Milligan, G. W. A review of Monte Carlo tests of cluster analysis. *Multivariate Behavioral Research,* 1981, *16,* 379–407.

Millon, T. Reflections on Rosenhan's "On being sane in insane places." *Journal of Abnormal Psychology,* 1975, *84,* 456–461.

Millon, T. *Disorders of personality: DSM-III, Axis II.* New York: Wiley, 1982.

Mojena, R. Hierarchical grouping methods and stopping rules: An evaluation. *Computer Journal,* 1977, *20,* 359–363.

Moore, T. V. The empirical determination of certain syndromes underlying praecox and manic-depressive psychoses. *American Journal of Psychiatry,* 1930, *86,* 719–738.

Moore, T. V. The essential psychoses and their fundamental syndromes. *Catholic University Studies in Psychology and Psychiatry,* 1933, *3,* 1–28.

Morales, A., & Sheafor, B. W. *Social work: A profession of many faces* (2nd ed.). Boston: Allyn & Bacon, 1980.

Morey, L. C. *Diagnostic information and DSM-III.* Master's thesis, University of Florida, 1979.

Morey, L. C., & Blashfield, R. K. The empirical classification of alcoholism: A review. *Journal of Studies on Alcohol,* 1981, *42,* 925–936.

Morf, M. E., Miller, C. E., & Syrotuik, J. M. A comparison of cluster analysis and Q-factor analysis. *Journal of Clinical Psychology,* 1976, *32,* 59–64.

Morrison, D. F. *Multivariate statistical methods.* New York: McGraw-Hill, 1967.

Morrison, J., Clancy, J., Crowe, R., & Winokur, G. The Iowa 500: I. Diagnostic validity in mania, depression and schizophrenia. *Archives of General Psychiatry,* 1972, *27,* 457–461.

Morse, S. J. Crazy behavior, morals, and science: An analysis of mental health law. *Southern California Law Review,* 1978, *51,* 527–654.

Muliak, S. A. *The foundations of factor analysis.* New York: McGraw-Hill, 1972.

Nathan, P. E. DSM-III and schizophrenia: Diagnostic delight or nosological nightmare? *Journal of Clinical Psychology,* 1979, *35,* 477–481.

National Institute of Mental Health. *Staffing of mental health facilities in the United States, 1974.* Washington, D.C.: U.S. Government Printing Office, 1976.

Nelson, G. & Platnick, N. I. *Systematics and biogeography: Cladistics and vicariance.* New York: Columbia University Press, 1981.

Nichols, C. H. Statistics of suicides. *Journal of Insanity,* 1848, *4,* 247–253.

Nielsen, A. C. Choosing psychiatry, importance of psychiatric education in medical school. *American Journal of Psychiatry,* 1980, *137,* 428–431.

Nolan, M. J. Some aspects of "maniacal-depressive insanity." *Journal of Mental Science,* 1909, *55,* 45–51.

Norris, V. Mental illness in London. *Maudsley Monograph,* #6, 1959.

O'Leary, M. R., Donovan, D. M., Chaney, E. F., & O'Leary, D. E. Relationship of alcoholic personality subtypes to treatment follow-up measures. *Journal of Nervous and Mental Disease,* 1980, *168,* 475–480.

Ottosson, J. D., & Perris, C. Multidimensional classification of mental disorders. *Psychological Medicine,* 1973, *3,* 238–243.

Overall, J. E. Cluster oriented factor analysis: Oblique powered vector factor analysis. *Multivariate Behavioral Research,* 1968, *3,* 479–488.

Overall, J. E., & Gorham, D. R. The Brief Psychiatric Rating Scale. *Psychological Reports,* 1962, *10,* 799–812.

Overall, J. E., & Hollister, L. E. Comparative evaluation of research diagnostic criteria for schizophrenia. *Archives of General Psychiatry,* 1979, *36,* 1198–1205.

Overall, J. E., & Klett, C. J. *Applied multivariate analysis.* New York: McGraw-Hill, 1972.

Overall, J. E., & Porterfield, J. L. Powered vector method of factor analysis. *Psychometrika,* 1963, *28,* 415–422.

Overall, J. E., Hollister, L. E., Johnson, M., & Pennington, V. Nosology of depression and differential response to drugs. *Journal of the American Medical Association,* 1966, *195,* 946–948.

Overall, J. E., Hollister, L. E., & Pichot, P. Major psychiatric disorders: A four-dimensional model. *Archives of General Psychiatry,* 1967, *16,* 146–151.

Overall, J. E., Hollister, L. E., & Pokorny, A. D. Alcohol history in drug treatment. *Diseases of the Nervous System,* 1973, *34,* 175–180.

Overall, J. E., Pull, C., Carranza, J., & Cassano, G. Phenomenological classification of psychiatric patients: Consistency of syndrome interpretation by psychologists in Italy, France, Mexico and the United States. *Journal of Psychiatric Research,* 1977, *13,* 225–236.

Panzetta, A. F. Towards a scientific psychiatric nosology: Conceptual and pragmatic issues. *Archives of General Psychiatry,* 1974, *30,* 154–161.

Partington, J. T., & Johnson, F. G. Personality types among alcoholics. *Quarterly Journal of Studies on Alcohol,* 1969, *30,* 21–34.

Pauker, J. D. Inpatient Multidimensional Psychiatric Scale. In O. K. Buros (Ed.), *Mental measurements yearbook* (7th ed.). Highland Park, N.J.: Gryphon Press, 1972.

Paykel, E. S. Classification of depressed patients: A cluster analysis derived grouping. *British Journal of Psychiatry,* 1971, *118,* 275–288.

Paykel, E. S. Correlates of a depressive typology. *Archives of General Psychiatry,* 1972, *20,* 203–210. (a)

Paykel, E. S. Depressive typologies and response to amitriptyline. *British Journal of Psychiatry,* 1972, *120,* 147–156. (b)

Paykel, E. S. Have multivariate statistics contributed to classification? *British Journal of Psychiatry,* 1981, *139,* 357–362.

Paykel, E. S., & Henderson, A. J. Application of cluster analysis in the classification of depression: A replication study. *Neuropsychobiology,* 1977, *3,* 111–119.

Paykel, E. S., Prusoff, B. A., Klerman, G. L., Haskell, D., & DiMascio, A. Clinical response to amitriptyline among depressed women. *Journal of the Nervous and Mental Disease,* 1973, *156,* 149–165.

Pfohl, B., & Andreasen, N. C. Development of classification systems in psychiatry. *Comprehensive Psychiatry,* 1978, *19,* 197–207.

Price, D. S. *Little science, big science.* New York: Columbia University Press, 1963.

Price, D. S. Networks of scientific papers. *Science,* 1965, *149,* 510–515.

Price, D. S. R & D and productivity (letter). *Science,* 1981, *211,* 1116.

Price, R. H. *Abnormal behavior: Perspectives in conflict.* New York: Holt, Rinehart and Winston, 1971.

Procci, W. R. Schizo-affective psychosis: Fact or fiction. *Archives of General Psychiatry,* 1976, *33,* 1167–1178.

Prusoff, B. A., & Paykel, E. S. Typological prediction of response to amitriptyline: A replication. *International Journal of Pharmacopsychiatry,* 1977, *12,* 153–159.

Raines, C. N. Forward, in American Psychiatric Association, *Diagnostic and statistical manual of mental disorders* (1st ed.). Washington, D.C.: Author, 1952.

Raskin, A. A guide for drug use in depressive disorders. *American Journal of Psychiatry,* 1974, *131,* 181–185.

Raskin, A., & Crook, T. H. The endogenous-neurotic distinction as a predictor of response to antidepressant drugs. *Psychological Medicine,* 1976, *6,* 59–70.

Redlich, F., & Kellert, S. R. Trends in American mental health. *American Journal of Psychiatry,* 1978, *135,* 22–28.

Reid, J. R., & Finesinger, J. E. The role of definitions in psychiatry. *American Journal of Psychiatry,* 1952, *109,* 413–420.

Rice, C. E., & Lorr, M. *An empirical comparison of typological methods.* Contract #0014-67-A-0214 Task 005. Office of Naval Research, 1969.

Rickels, K., Eesbacher, P., & Dowing, R. W. Differential drug effects in neurotic depression. *Diseases of the Nervous System,* 1970, *31,* 468–475.

Rie, H. Psychology, mental health and the public interest. *American Psychologist,* 1977, *32,* 1–4.

Rivers, W. H. R. Experimental psychology in relation to insanity. *Journal of Mental Science,* 1895, *41,* 591–599.

Robins, E., & Guze, S. B. Establishment of diagnostic validity in psychiatric illness: Its application to schizophrenia. *American Journal of Psychiatry,* 1970, *126,* 983–987.

Rogers, G., & Linden, J. D. Use of multiple discriminant function analysis in the evaluation of three multivariate grouping techniques. *Educational and Psychological Measurement,* 1973, *33,* 787–802.

Rosch, E. Principles of categorization. In E. Rosch & B. B. Lloyd (Eds.), *Cognition and categorization.* Hillsdale, N.J.: Lawrence Erlbaum, 1978.

Rosenhan, D. L. On being sane in insane places. *Science,* 1973, *179,* 250–258.

Rosenhan, D. L. The contextual nature of psychiatric diagnosis. *Journal of Abnormal Psychology,* 1975, *84,* 462–472.

Roth, M., & Barnes, T. R. E. The classification of affective disorders: A synthesis of old and new concepts. *Comprehensive Psychiatry,* 1981, *22,* 54–77.

Rummel, R. J. *Applied factor analysis.* Evanston, Ill.: Northwestern University Press, 1970.

Rutter, M., & Shaffer, D. DSM-III: A step forward or back in terms of the classification of child psychiatric disorders? *Journal of the American Academy of Child Psychiatry,* 1980, *19,* 371–394.

Rutter, M., & Shopler, E. (Eds.). *Autism: A reappraisal of concepts and treatment.* New York: Plenum Press, 1978.

Rutter, M., Lebovici, S., Eisenberg, L., Sneznevskij, A. V., Sadoun, R., Brooke, E., & Lin, T. Y. A tri-axial classification of mental disorders in childhood. *Journal of Child Psychology, Psychiatry and Related Disciplines,* 1969, *10,* 41–61.

Rutter, M., Shaffer, D., & Shepherd, M. *A multi-axial classification of child psychiatric disorders.* Geneva: World Health Organization, 1975.

Saghir, T., & Robins, E. *Male and female homosexuality.* Baltimore: Williams & Wilkins, 1973.

Samelson, F. J. B. Watson's Little Albert, Cyril Burt's twins, and the need for a critical science. *American Psychologist,* 1980, *35,* 619–625.

Sandifer, M. G., Pettus, C., & Quade, D. A study of psychiatric diagnosis. *Journal of Nervous and Mental Disease,* 1964, *139,* 350–356.

Sandifer, M. G., Hordern, A., Timbury, G. C., & Green, L. M. Psychiatric diagnosis: A comparative study in North Carolina, London and Glasgow. *British Journal of Psychiatry,* 1968, *114,* 1–9.

Sanson-Fisher, R. W., & Martin, C. J. Standardized interviews in psychiatry: Issues of reliability. *British Journal of Psychiatry,* 1981, *139,* 138–143.

Sarason, I. G., & Ganzer, V. T. Concerning the medical model. *American Psychologist,* 1968, *23,* 507–510.

Schact, T., & Nathan, P. E. But is it good for psychology? Appraisal and status of the DSM-III. *American Psychologist,* 1977, *32,* 1017–1025.

Scheff, T. J. *Being mentally ill: A sociological theory.* Chicago: Aldine, 1966.

Schmidt, H., & Fonda, C. The reliability of psychiatric diagnosis: A new look. *Journal of Abnormal and Social Psychology,* 1956, *52,* 262–267.

Schulterbrandt, J. G., & Raskin, A. *Depression in childhood: Diagnosis, treatment and conceptual models.* New York: Raven Press, 1977.

Schwartz, R. A. Psychiatry's drift away from medicine. *American Journal of Psychiatry,* 1974, *131,* 129–134.

Scoltock, J. Survey of the literature of cluster analysis. *Computer Journal,* 1982, *25,* 130–134.

Scott, W. A. Reliability of content analysis: The case of nominal scale coding. *Public Opinion Quarterly,* 1955, *19,* 321–325.

Simon, R. J., Fleiss, J. L., Gurland, B. J., Stiller, P. R., & Sharpe, L. Depression and schizophrenia in hospitalized black and white mental patients. *Archives of General Psychiatry,* 1973, *28,* 509–512.

Simpson, G. G. *Principles of animal taxonomy.* New York: Columbia University Press, 1961.

Simpson, G. M., Angus, J. W. S., & Edwards, J. G. A controlled study of haloperidol in chronic schizophrenia. *Current Therapeutic Research, Clinical and Experimental,* 1967, *9,* 407–412. (a)

Simpson, G. M., Angus, J. W. S., & Edwards, J. G. A preliminary study of CI-601 in chronic schizophrenia. *Current Therapeutic Research, Clinical and Experimental,* 1967, *9,* 486–491. (b)

Singerman, B., Stoltzman, R. K., Robins, L. N., Helzer, J. E., & Croughan, J. L. Diagnostic concordance between DSM-III, Feighner, and RDC. *Journal of Clinical Psychology,* 1981, *42,* 422–426.

Skinner, H. A. "The eyes that fix you": A model for classification research. *Canadian Psychological Review,* 1977, *18,* 142–151.

Skinner, H. A. Dimensions and clusters: A hybrid approach to classification. *Applied Psychological Measurement,* 1979, *3,* 327–341. (a)

Skinner, H. A. A model of psychopathology based on the MMPI. In C. S. Newmark (Ed.), *MMPI: Clinical and research trends.* New York: Praeger, 1979. (b)

Skinner, H. A. Towards the integration of classification theory and methods. *Journal of Abnormal Psychology,* 1981, *90,* 68–87.

Skinner, H. A., & Jackson, D. N. A model of psychopathology based on an interpretation of MMPI actuarial systems. *Journal of Consulting and Clinical Psychology,* 1978, *46,* 231–238.

Skinner, H. A., & Lei, H. Modal Profile Analysis: A computer program for classification research. *Educational and Psychological Measurement*, 1980, *40*, 769–772.
Skinner, H. A., Jackson, D. N., & Hoffman, H. Alcoholic personality types: Identification and correlates. *Journal of Abnormal Psychology*, 1974, *83*, 658–666.
Smith, K., Surphlis, W. R., Gynther, M. D., & Shimkunas, A. M. ECT-chlorpromazine and chlorpromazine compared in treatment of schizophrenia. *Journal of Nervous and Mental Disease*, 1967, *144*, 284–289.
Sneath, P. H. A. A vector model of disease for teaching and diagnosis. *Medical Hypotheses*, 1975, *1*, 12–22.
Sneath, P. H. E., & Sokol, R. R. *Numerical taxonomy*. San Francisco: Freeman, 1973.
Sokol, R. R., & Sneath, P. H. E. *Principles of numerical taxonomy*. San Francisco: Freeman, 1963.
Sorenson, T. A method of establishing groups of equal amplitude in plant sociology based on similarity of species content and its application to analysis of the vegetation of Danish commons. *Biologiske Skrifter*, 1948, *5*, 1–34.
Speigel, D. E., & Keith-Speigel, P. Effects of carphenazine, trifluoperazine and chlorpromazine on ward behavior, physiological functioning and psychological test scores in chronic schizophrenia. *Journal of Nervous and Mental Disease*, 1967, *144*, 111–116.
Spitzer, R. L. A proposal about homosexuality and the APA nomenclature: Homosexuality as an irregular form of sexual behavior and sexual orientation disturbance as a psychiatric disorder. *American Journal of Psychiatry*, 1973, *130*, 1214–1216.
Spitzer, R. L. On pseudoscience in science, logic in remission and psychiatric diagnosis. *Journal of Abnormal Psychology*, 1975, *84*, 442–452.
Spitzer, R. L. Letter to the editor. *Schizophrenia Bulletin*, 1982, *8*, 592.
Spitzer, R. L., & Endicott, J. DIAGNO: A computer program for psychiatric diagnosis utilizing the differential diagnostic procedure. *Archives of General Psychiatry*, 1968, *18*, 746–756.
Spitzer, R. L., & Endicott, J. DIAGNO II: Further developments in a computer program for psychiatric diagnosis. *American Journal of Psychiatry*, 1969, *125*, 12–21.
Spitzer, R. L., & Fleiss, J. L. A re-analysis of the reliability of psychiatric diagnosis. *British Journal of Psychiatry*, 1974, *125*, 341–347.
Spitzer, R. L., & Forman, J. B. W. DSM-III field trials: II. Initial experience with the multiaxial system. *American Journal of Psychiatry*, 1979, *136*, 818–820.
Spitzer, R. L., & Sheehy, M. DSM-III: A classification system in development. *Psychiatric Annals*, 1976, *6*, 102–109.
Spitzer, R. L., & Wilson, P. T. A guide to the American Psychiatric Association's new diagnostic nomenclature. *American Journal of Psychiatry*, 1968, *124*, 1619–1629.
Spitzer, R. L., & Wilson, P. T. Nosology and the official psychiatric nomenclature. In A. M. Freedman, H. I. Kaplan, & B. J. Sadock (Eds.), *Comprehensive textbook of psychiatry, Vol. 2* (2nd ed.). Baltimore: Williams & Wilkins, 1975.
Spitzer, R. L., Kramer, Y., & Rabkin, R. The neck-righting reflex in children. *Journal of Pediatrics*, 1958, *52*, 149–151.
Spitzer, R. L., Cohen, J., Fleiss, J. L., & Endicott, J. Quantification of agreement in psychiatric diagnosis: A new approach. *Archives of General Psychiatry*, 1967, *17*, 83–87.
Spitzer, R. L., Fleiss, J. L., Endicott, J., & Cohen, J. Mental Status Schedule: Properties of a factor-analytically derived scales. *Archives of General Psychiatry*, 1967, *16*, 479–493.

Spitzer, R. L., Endicott, J., Fleiss, J. L., & Cohen, J. The Psychiatric Status Schedule: A technique for evaluating psychopathology and impairment of role functioning. *Archives of General Psychiatry*, 1970, *23*, 41–55.

Spitzer, R. L., Endicott, J., Cohen, J., & Fleiss, J. L. Constraints on the validity of computer diagnosis. *Archives of General Psychiatry*, 1974, *31*, 197–203.

Spitzer, R. L., Endicott, J., & Robins, E. *Research Diagnostic Criteria (RDC) for a selected group of functional disorders.* New York: New York State Psychiatric Institute, 1975.

Spitzer, R. L., Sheehy, M., & Endicott, J. DSM-III: Guiding principles. In V. M. Rakoff, H. C. Stancer, & H. B. Kedward (Eds.), *Psychiatric diagnosis.* New York: Brunner/Mazel, 1977.

Spitzer, R. L., Forman, J. B. W., & Nee, J. DSM-III field trials: I. Initial interrater diagnostic reliability. *American Journal of Psychiatry*, 1979, *136*, 815–817.

Spitzer, R. L., Williams, J. B. W., & Skodol, A. E. DSM-III: A major achievement and an overview. *American Journal of Psychiatry*, 1979, *137*, 151–164.

Stein, K. B., Rozynko, V., & Pugh, L. A. The heterogeneity of personality among alcoholics. *British Journal of Social and Clinical Psychology*, 1971, *10*, 253–259.

Stengel, E. Classification of mental disorders. *Bulletin of the World Health Organization*, 1959, *21*, 601–663.

Stephenson, W. *The study of behavior.* Chicago: University of Chicago Press, 1953.

Stone, A. A., Hopkins, R., Mahnke, M. W., Shapiro, D. W., & Silvergate, H. A. Simple schizophrenia—Syndrome or shibboleth. *American Journal of Psychiatry*, 1968, *125*, 305–312.

Strauss, J. S. A comprehensive approach to psychiatric diagnosis. *American Journal of Psychiatry*, 1975, *132*, 1193–1197.

Strauss, J. S., Gabriel, K. R., Kokes, R. F., Ritzler, B. A., VanOrd, A., & Tarana, E. Do psychiatric patients fit their diagnoses? Patterns of symptomatology as described with the biplot. *Journal of Nervous and Mental Disease*, 1979, *167*, 105–113.

Surtees, P. G., & Kendell, R. E. The hierarchy model of psychiatric symptomatology: An investigation based on Present State Examination ratings. *British Journal of Psychiatry*, 1979, *135*, 438–443.

Swenson, W. M., Pearson, J. S., & Osborne, D. *An MMPI source book: Basic item, scale, and pattern data on 50,000 medical patients.* Minneapolis: University of Minnesota Press, 1973.

Szasz, T. S. *The myth of mental illness.* New York: Hoeber-Harper, 1961.

Tatsuoka, M. M. *Multivariate analysis: Techniques for educational and psychological research.* New York: Wiley, 1971.

Thorndike, R. M. *Correlational procedures for research.* New York: Gardner Press, 1978.

Thurstone, L. L. The vectors of the mind. *Psychological Review*, 1934, *41*, 1–32.

Thurstone, L. L. *Multiple-factor analysis.* Chicago: University of Chicago Press, 1947.

Tryon, R. C., & Bailey, D. E. *Cluster analysis.* New York: McGraw-Hill, 1970.

Tversky, A. Features of similarity. *Psychological Review*, 1977, *84*, 327–352.

Verma, R. M., & Eysenck, H. J. Severity and type of psychotic illness as a function of personality. *British Journal of Psychiatry*, 1973, *122*, 573–585.

Ward, C. H., Beck, A. T., Mendelson, M., Mock, J. E., & Erbaugh, J. K. The psychiatric nomenclature. *Archives of General Psychiatry*, 1962, *7*, 198–205.

Ward, J. H. Hierarchical grouping to optimize an objective function. *Journal of the American Statistical Association*, 1963, *58*, 236–244.

Watson, J. B., & Rayner, R. Conditioned emotional reactions. *Journal of Experimental Psychology*, 1920, *3*, 1–14.

Watson, R. I. *The clinical method in psychology*. New York: Harper, 1951.

Weiner, B. "On being sane in insane places": A process (attributional) analysis and critique. *Journal of Abnormal Psychology*, 1975, *84*, 433–441.

White, H. L., McLeod, M. N., & Davidson, J. R. Platelet monamine-oxidase activity in schizophrenia. *American Journal of Psychiatry*, 1976, *133*, 1191–1193.

White, W. A. *Twentieth century psychiatry: Its contribution to man's knowledge of himself*. New York: Arno, 1973.

Whitelock, P. R., Overall, J. E., & Patrick, J. H. Personality patterns and alcohol abuse in a state hospital population. *Journal of Abnormal Psychology*, 1971, *78*, 9–16.

Wiggins, J. S. *Personality and prediction: Principles of personality assessment*. Reading, Mass.: Addison-Wesley, 1973.

Wiggins, J. S. A psychological taxonomy of trait-descriptive terms: The interpersonal domain. *Journal of Personality and Social Psychology*, 1979, *37*, 395–412.

Wiggins, J. S. Circumplex models of interpersonal behavior in clinical psychology. In P. C. Kendell & J. N. Butcher (Eds.), *Handbook of research methods in clinical psychology*. New York: Wiley, 1982.

Williams, W. T., Lance, G. N., Dale, M. B., & Clifford, H. T. Controversy concerning the criteria for taxonomic strategies. *Computer Journal*, 1971, *14*, 162–165.

Wing, L. Observations on the psychiatric section of the International Classification of Diseases and the British Glossary of Mental Disorders. *Psychological Medicine*, 1970, *1*, 79–85.

Wing, J. K. *Reasoning about madness*. London: Oxford University Press, 1978.

Wing, J. K., & Brown, G. W. *Institutionalism and schizophrenia*. London: Cambridge University Press, 1970.

Wing, J. K., Cooper, J. E., & Sartorius, N. *The description and classification of psychiatric symptoms: An instruction manual for the PSE and Catego system*. London: Cambridge University Press, 1974.

Winokur, G., Clayton, P. J., & Reich, T. *Manic-depressive insanity*. St. Louis: Mosby, 1969.

Winslow, W. W. The changing role of psychiatrists in community mental health centers. *American Journal of Psychiatry*, 1979, *136*, 24–27.

Wishart, D. *CLUSTAN user manual*. Edinburgh, Scotland: Edinburgh University, 1978.

Wittenborn, J. R. Symptom patterns in a group of mental hospital patients. *Journal of Consulting Psychology*, 1951, *15*, 290–302.

Wittenborn, J. R. The behavioral symptoms for certain organic psychoses. *Journal of Consulting Psychology*, 1952, *16*, 104–106.

Wittenborn, J. R. *Psychiatric Rating Scales*. New York: Psychological Corporation, 1955.

Wittenborn, J. R. The dimensions of psychosis. *Journal of Nervous and Mental Disease*, 1962, *134*, 117–128.

Wittenborn, J. R., & Bailey, C. The symptoms of involutional psychosis. *Journal of Consulting Psychology*, 1952, *16*, 13–17.

Wittenborn, J. R., & Holzberg, J. D. The generality of psychiatric syndromes. *Journal of Consulting Psychology*, 1951, *15*, 372–380.

Wittenborn, J. R., & Weiss, W. Patients' diagnosed manic-depressive psychosis-manic state. *Journal of Consulting Psychology*, 1952, *16*, 193–198.

Wittenborn, J. R., Herz, M. I., Kurtz, K. H., Mandell, W., & Tatz, A. The effect of rater differences on symptom rating scale clusters. *Journal of Consulting Psychology,* 1952, *16,* 107–109.

Woodruff, R. A., Goodwin, D. W., & Guze, S. B. *Psychiatric diagnosis.* London: Oxford University Press, 1974.

Woodruff, R. A., Clayton, P. J. & Guze, S. B. Is everyone depressed? *American Journal of Psychiatry,* 1975, *132,* 627–628.

World Health Organization. *International Classification of Diseases* (6th ed.). Geneva: Author, 1948.

World Health Organization. *International Classification of Diseases* (8th ed.). Geneva: Author, 1967.

Zubin, J. Classification of behavior disorders. *Annual Review of Psychology,* 1967, *18,* 373–406.

Author Index

Subject Index

GPSR Compliance

The European Union's (EU) General Product Safety Regulation (GPSR) is a set of rules that requires consumer products to be safe and our obligations to ensure this.

If you have any concerns about our products, you can contact us on ProductSafety@springernature.com

In case Publisher is established outside the EU, the EU authorized representative is:

Springer Nature Customer Service Center GmbH
Europaplatz 3
69115 Heidelberg, Germany

Batch number: 10844655

Printed by Printforce, the Netherlands